Diagnostic Ultraso

Physics and Equipment

Third Edition

Edited by

Peter R Hoskins
Professor of Medical Physics and Biomechanics
Edinburgh University
Edinburgh, United Kingdom

Kevin Martin
Retired Consultant Medical Physicist
Leicester, United Kingdom

Abigail Thrush
Clinical Scientist (Medical Physics)
Associate Lecturer at University of Salford and University of Derby
Imaging Department, Chesterfield Royal Hospital NHS Trust
Chesterfield, United Kingdom

CRC Press
Taylor & Francis Group
Boca Raton London New York

CRC Press is an imprint of the
Taylor & Francis Group, an **informa** business

T0230801

CRC Press
Taylor & Francis Group
6000 Broken Sound Parkway NW, Suite 300
Boca Raton, FL 33487-2742

© 2019 by Taylor & Francis Group, LLC
CRC Press is an imprint of Taylor & Francis Group, an Informa business

International Standard Book Number-13: 978-1-138-89293-4 (Paperback)
978-0-367-19041-5 (Hardback)

Library of Congress Cataloging-in-Publication Data

Names: Hoskins, P. R. (Peter R.), editor. | Martin, Kevin, editor. | Thrush, Abigail, editor.
Title: Diagnostic ultrasound: Physics and equipment / edited by Peter R. Hoskins, Kevin Martin and Abigail Thrush.
Description: Third edition. | Boca Raton, FL: CRC Press/Taylor & Francis Group, [2019] | Includes bibliographical references and index.
Identifiers: LCCN 2018053785 | ISBN 9781138892934 (pbk. : alk. paper) | ISBN 9780367190415 (hardback : alk. paper) | ISBN 9781138893603 (ebook : alk. paper)
Subjects: | MESH: Ultrasonography--methods | Ultrasonography--instrumentation | Physical Phenomena
Classification: LCC RC78.7.U4 | NLM WN 208 | DDC 616.07/543--dc23
LC record available at https://lccn.loc.gov/2018053785

Visit the Taylor & Francis Web site at
http://www.taylorandfrancis.com

and the CRC Press Web site at
http://www.crcpress.com

Contents

Preface to the third edition

The aim of this book is to provide underpinning knowledge of physics and instrumentation needed in order to practise ultrasound in a clinical setting. The book is primarily aimed at sonographers and clinical users in general, and will also serve as a first textbook for physicists and engineers.

The text concentrates on explanations of principles which underpin the clinical use of ultrasound systems. The book contains relatively few equations and even fewer derivations. The book has been updated throughout to include recent developments in technology, quality assurance and safety. As with previous editions the emphasis is on technology available commercially rather than in research labs. Examples of new material in this edition include high frame rate imaging, small vessel imaging, arterial wall motion measurement, vector Doppler and automated measurement packages. We hope that this third edition of *Diagnostic Ultrasound: Physics and Equipment* will meet the needs of sonographers, physicists and engineers in their training and practice.

Peter R Hoskins
Kevin Martin
Abigail Thrush
Autumn 2018

Preface to the second edition

The aims and intended audience of this second edition remain unchanged from the first edition. The aim is to provide the underpinning knowledge of physics and instrumentation needed in order to practise ultrasound in a clinical setting. The book is primarily aimed at sonographers and clinical users in general, and will also serve as a first textbook for physicists and engineers. The text concentrates on explanations of principles which underpin the clinical use of ultrasound systems. The book contains relatively few equations and even fewer derivations. In the last 7 years a number of techniques which existed in embryo form in 2002 have become available on commercial ultrasound systems, and are used in a sufficient number of hospitals to justify inclusion in this book. There are additional chapters dedicated to 3D ultrasound, contrast agents and elastography. The other chapters have been updated to include developments in technology, quality assurance and safety. We hope that this second edition of *Diagnostic Ultrasound Physics and Equipment* will meet the needs of sonographers, physicists and engineers in their training and practice.

Peter R Hoskins
Kevin Martin
Abigail Thrush
Autumn 2009

Preface to the first edition

This book is an introductory text in the physics and instrumentation of medical ultrasound imaging. The level is appropriate for sonographers and clinical users in general. This will also serve as a first textbook for physicists and engineers. The text concentrates on explanations of principles which underpin the clinical use of ultrasound systems, with explanations following a 'need to know' philosophy. Consequently, complex techniques, such as Doppler frequency estimation using FFT and 2D autocorrelation, are described in terms of their function, but not in terms of their detailed signal processing. The book contains relatively few equations and even fewer derivations. The scope of the book reflects ultrasound instrumentation as it is used at the time of submission to the publishers. Techniques which are still emerging, such as tissue Doppler imaging (TDI) and contrast agents, are covered in a single chapter at the end of the book. Techniques which are even further from commercial implementation, such as vector Doppler, are not covered. We hope this book fills the gap in the market that we perceive from discussions with our clinical colleagues, that of a text which is up to date and at an appropriate level.

Peter R Hoskins
Abigail Thrush
Kevin Martin
Tony Whittingham
Summer 2002

Contributors

Mairéad Butler
Research Associate in Medical Physics
Heriot Watt University
Edinburgh, United Kingdom

Aline Criton
Clinical and Regulatory Affairs Director
Mauna Kea
Paris, France

Francis Duck
Retired Consultant Medical Physicist
Bath, United Kingdom

Nick Dudley
Principal Medical Physicist
United Lincolnshire Hospitals
Lincoln, United Kingdom

Tony Evans
Retired Senior Lecturer in Medical Physics
Leeds, United Kingdom

Peter R Hoskins
Professor of Medical Physics and Biomechanics
University of Edinburgh
Edinburgh, United Kingdom

Tom MacGillivray
Senior Research Fellow
University of Edinburgh
Edinburgh, United Kingdom

Kevin Martin
Retired Consultant Medical Physicist
Leicester, United Kingdom

Carmel Moran
Professor of Translational Ultrasound
University of Edinburgh
Edinburgh, United Kingdom

Kumar V Ramnarine
Deputy Head
Non-Ionising Radiation
Guy's and St Thomas' NHS Foundation Trust
London, United Kingdom

Abigail Thrush
Clinical Scientist (Medical Physics)
Associate Lecturer at University of Salford and
University of Derby
Imaging Department, Chesterfield Royal Hospital
NHS Trust
Chesterfield, United Kingdom

Tony Whittingham
Retired Consultant Medical Physicist
Newcastle-upon-Tyne, United Kingdom

Introduction to B-mode imaging

KEVIN MARTIN

INTRODUCTION

Ultrasound is one of the most widely used non-invasive imaging techniques in medical diagnosis. In the year from July 2016 to July 2017, over 9.2 million ultrasound scans were carried out on National Health Service (NHS) patients in England (NHS England 2017). This is almost twice the number of computed tomography (CT) scans and almost three times the number of magnetic resonance imaging (MRI) scans carried out in the same period. Ultrasound is an interactive imaging technique in which the operator holds the ultrasound probe in contact with the patient and observes images of internal anatomy in real time. In most instances, interpretation of the images is carried out simultaneously, rather than being reported by someone else at a later time. This, and the ultrasound examination's lower cost (£40–£49) compared to CT (£71–£199) and MRI (£116–£225) (NHS Improvement 2016) makes it attractive as a first line of investigation in many circumstances. As it is radiation free, ultrasound is the preferred imaging modality in obstetric and paediatric investigations. Ultrasound is most useful for imaging soft tissue anatomy and is widely used in the abdomen, pelvis, heart and neck. It has become a widely used imaging tool in musculo-skeletal investigations of muscles, tendons and joints. In addition to its well-established use in imaging soft tissue anatomy, ultrasound is a powerful tool in the study of blood flow in arteries and veins, where use of the Doppler effect enables blood velocity to be measured and imaged.

Commercial ultrasound systems incorporate many complex technologies which provide improved image quality and enable the system to image subtle changes in tissue and finer detail at greater depths in tissue. Additional techniques include elastography, in which the stiffness of the tissue can be evaluated. The use of contrast agents with ultrasound imaging can increase its diagnostic accuracy in differentiating benign from malignant lesions. Standard ultrasound systems produce two-dimensional cross-sectional images of anatomy, but these may be extended to acquire data in three dimensions. The range of technologies available to the user to enhance the imaging performance is ever growing. However, these additional tools bring with them a need to understand when and where they should be used and how they affect what is seen on the screen. All such technologies are still ruled by the laws of physics, some of which can lead to misleading artefacts in the image. The safe use of ultrasound in medical imaging has two aspects. First, ultrasound imaging involves depositing ultrasound energy within the patient, which can lead to some warming of exposed tissue. Clearly, it is important to minimise such effects in obtaining a diagnosis. Second, perhaps a greater hazard is misdiagnosis. The risk of misdiagnosis must be controlled by ensuring that the ultrasound system is performing as intended and is operated competently.

The intention of this text is to provide the reader with an understanding of the physics and technology necessary to use ultrasound imaging systems safely and efficiently. It might also serve as a first-level text for those training as scientists. The text may be divided into four sections: Chapters 2 through 6 describe the physics and technology of ultrasound imaging systems in addition to their limitations and use in measurement. Chapters 7 through 11 describe the principles and applications of Doppler ultrasound, including a description of flow in blood vessels. This is followed by descriptions of three-dimensional ultrasound, contrast agents and elastography in Chapters 12 through 14. Ultrasound quality assurance and safety are covered in Chapters 15 and 16. This introductory chapter explains some of the basic concepts involved in forming an ultrasound image.

BASIC PRINCIPLES OF ULTRASOUND IMAGE FORMATION

We begin with a description of the ultrasound image and an explanation of the basic principles of its formation. In essence, these simple principles are still used in commercial ultrasound imaging systems, but they are likely to be enhanced by numerous additional image-forming and processing techniques designed to improve image quality and performance. These are described in later chapters.

A B-mode image is a cross-sectional image representing tissues and organ boundaries within the body (Figure 1.1). It is constructed from echoes, which are generated by reflection of ultrasound waves at tissue boundaries, and scattering from small irregularities within tissues. Each echo is displayed at a point in the image, which corresponds to the relative position of its origin within the body cross section, resulting in a scaled map

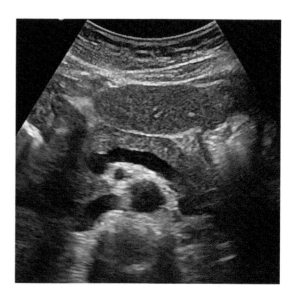

Figure 1.1 An example of a B-mode image showing reflections from organ and blood vessel boundaries and scattering from tissues.

of echo-producing features. The brightness of the image at each point is related to the strength or amplitude of the echo, giving rise to the term *B-mode* (brightness mode). Usually, the B-mode image bears a close resemblance to the anatomy which might be seen by eye if the body could be cut through in the same plane. Abnormal anatomical boundaries and alterations in the scattering behaviour of tissues can indicate pathology.

To form a B-mode image, a source of ultrasound, the transducer, is placed in contact with the skin and short bursts or pulses of ultrasound are sent into the patient. These are directed along narrow beam-shaped paths. As the pulses travel into the tissues of the body, they are reflected and scattered, generating echoes, some of which travel back to the transducer along the same path, where they are received. These echoes are used to form the image. To display each echo in a position corresponding to that of the interface or feature (known as a target) that produced it, the B-mode system needs two pieces of information. These are as follows:

1. The range (distance) of the target from the transducer.
2. The direction of the target from the active part of the transducer, i.e. the position and orientation of the ultrasound beam.

The pulse-echo principle

The range of the target from the transducer is measured using the pulse-echo principle. The same principle is used in echo-sounding equipment in boats to measure the depth of water. Figure 1.2 illustrates the measurement of water depth using the pulse-echo principle. Here, the transducer transmits a short burst or pulse of ultrasound, which travels through the water to the seabed below, where it is reflected, i.e. produces an echo. The echo travels back through the water to the transducer, where it is received. The distance to the seabed can be worked out, if the speed of sound in water is known and the time between the transmission pulse leaving the transducer and the echo being received, the 'go and return time', is measured.

To measure the go and return time, the transducer transmits a pulse of ultrasound at the same time as a clock is started ($t = 0$). If the speed of sound in water is c and the depth is d, then the pulse reaches the seabed at time $t = d/c$. The returning echo also travels at speed c and takes a further time d/c to reach the transducer, where it is received. Hence, the echo arrives back at the transducer after a total go and return time $t = 2d/c$. Rearranging this equation, the depth d can be calculated from

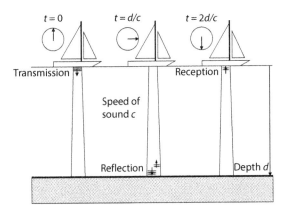

Figure 1.2 Measurement of water depth using the pulse-echo principle. A short burst or pulse of ultrasound is transmitted at time $t = 0$. It takes time d/c to travel to depth d, where c is the speed of sound. The echo arrives back at the source at time $2d/c$. The depth d can be worked out from the go and return time $2d/c$ if the speed of sound c is known.

$d = ct/2$. Thus, the system calculates the target range d by measuring the arrival time t of an echo, assuming a fixed value for the speed of sound c (usually 1540 m s^{-1} for human tissues).

In this example, only one reflecting surface was considered, i.e. the interface between the water and the seabed. The water contained no other interfaces or irregularities, which might generate additional echoes. When a pulse travels through the tissues of the body, it encounters many interfaces and scatterers, all of which generate echoes. After transmission of the short pulse, the transducer operates in receive mode, effectively listening for echoes. These begin to return immediately from targets close to the transducer, followed by echoes from greater and greater depths, in a continuous series, to the maximum depth of interest. This is known as the pulse-echo sequence.

Image formation

The time of arrival of echoes after transmission increases with target depth. As the sequence of echoes arrives, it is used to form a line in the image, the brightness of the line at each depth changing with the amplitude or strength of the echo. Each pulse-echo sequence results in the display of one line of information on the B-mode image. The two-dimensional B-mode image is formed from a large number of B-mode lines. A complete B-mode image, such as that in Figure 1.1, is made up typically of 100 or more B-mode lines.

Figure 1.3 illustrates the formation of a B-mode image. During the first pulse-echo sequence, an image line is formed, say on the left of the display (Line 1). The beam is then moved to the adjacent position. Here a new pulse-echo sequence produces a new image line, with a position on the

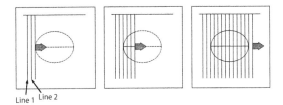

Figure 1.3 Formation of a two-dimensional B-mode image. The image is built up line-by-line as the beam is stepped along the transducer array.

display corresponding to that of the new beam (Line 2). The beam is progressively stepped across the image with a new pulse-echo sequence generating a new image line at each position. Figure 1.4 is a B-mode image of a section through the head of a fetus from an early ultrasound scanner. The image is made up of approximately 160 individual B-mode lines, which can be seen in the image. One complete sweep of the 160 lines takes approximately 1/30th of a second. This means that about 30 complete images could be formed in each second, allowing real-time display of the B-mode image. That is, the image is displayed with negligible delay as the information is acquired, rather than being recorded and then viewed, as with a radiograph or CT scan. Note that the individual B-mode lines in Figure 1.4 are not visible in the more up-to-date image of Figure 1.1, which has been improved by image processing.

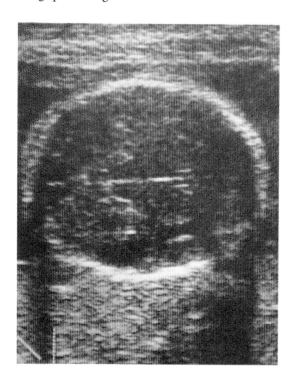

Figure 1.4 Individual B-mode lines can be seen in this image of a fetal head from an early ultrasound scanner. The image is formed from approximately 160 B-mode lines. (Reproduced from Evans JA et al. 1987. *BMUS Bulletin*, 44, 14–18, by kind permission of the British Medical Ultrasound Society.)

B-MODE FORMATS

The B-mode image, just described, was produced by a linear transducer array, i.e. a large number of small transducer elements arranged in a straight line (see Chapter 3). The ultrasound beams, and hence the B-mode lines, were all perpendicular to the line of transducer elements, and hence parallel to each other (Figure 1.5a). The resulting rectangular field of view is useful in applications where there is a need to image superficial areas of the body at the same time as organs at a deeper level.

Other scan formats are available to suit the application. For example, a curvilinear transducer (Figure 1.5b) gives a wide field of view near the transducer and an even wider field at deeper levels. This is also achieved by the trapezoidal field of view (Figure 1.5c). Curvilinear and trapezoidal fields of view are widely used in obstetric scanning to allow imaging of more superficial targets, such as the placenta, while giving the greatest coverage at the depth of the baby. The sector field of view (Figure 1.5d) is preferred for imaging of the heart, where access is normally through a narrow acoustic window between the ribs. In the sector format, all of the B-mode lines are close together near the transducer and pass through the narrow gap, but they diverge after that to give a wide field of view at the depth of the heart.

Transducers designed to be used internally, such as intravascular or rectal probes, may use the

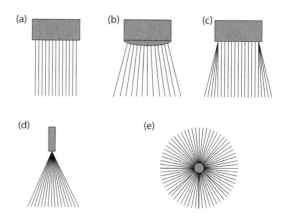

Figure 1.5 Scan line arrangements for the most common B-mode formats. These are (a) linear, (b) curvilinear, (c) trapezoidal, (d) sector and (e) radial.

radial format (Figure 1.5e) as well as sector and linear fields of view. The radial beam distribution is similar to that of beams of light from a lighthouse. This format may be obtained by rotating a single element transducer on the end of a catheter or rigid tube, which can be inserted into the body. Hence, the B-mode lines all radiate out from the centre of the field of view.

ACKNOWLEDGEMENT

The image in Figure 1.1 was obtained with the assistance of Hitachi Medical Systems, United Kingdom.

REFERENCES

Evans JA, McNay M, Gowland M, Farrant P. 1987. BMUS ultrasonic fetal measurement survey. *BMUS Bulletin*, 44, 14–18.

NHS England. 2017. *Diagnostic Imaging Dataset Statistical Release*. Leeds, UK: NHS England.

NHS Improvement. 2016. *Proposed National Tariff Prices: Planning for 2017/18 and 2018/19*. London: NHS Improvement.

2

Physics

KEVIN MARTIN AND KUMAR V RAMNARINE

INTRODUCTION

The formation of ultrasound images of internal anatomy involves a wide range of physical processes. These include the generation of ultrasound waves, the formation of ultrasound beams, reflection, scattering, diffraction and attenuation of ultrasound waves and the processing of echoes to create a recognisable image. It is a matter of good fortune that these processes work within human tissues in a way that allows the formation of images with a resolution of the order of 1 mm to the full depth of the larger organs of the human body such as the liver. However, safe and optimal use of ultrasound imaging systems requires good management. The clinical user must be able to make informed choices in the way that he or she uses the ultrasound imaging system. This requires an understanding of the physical processes involved in creating the image. As with any imaging system, ultrasound imaging systems are not without their limitations and pitfalls. The user must be aware of the limitations and the physical processes by which image artefacts arise and be able to recognise them for what they are without being misled.

The aim of this chapter is to explain the important physical processes involved in forming the ultrasound image and which lead to limitations and artefacts of ultrasound images. Detailed descriptions of how images are formed are given in Chapters 3 and 4. The limitations and artefacts of ultrasound images are described in detail in Chapter 5.

WAVES

What is a wave? We live in a world that contains waves in many different forms, e.g. sound waves, radio waves, light waves, and ocean waves. The Mexican wave is a simple example that demonstrates an important aspect of waves. The person at one end of a long line of people raises his or her arms briefly. As the arms of the first person are lowered, the next person in the line performs the same action. This action is continued along the line of people, resulting in the familiar Mexican wave. The movement in this case is a short disturbance that travels along the line as the action of one person triggers the same response from the next. No one moves out of position. Each person raises and lowers their arms in turn and the wave travels due to the interaction between neighbours.

Another familiar wave example is the formation of ripples on the surface of a pond caused by a stone being thrown into the water (Figure 2.1a). In this case, the movement of the water at any point is repetitive, rather than being a single disturbance, and the ripples extend some distance across the pond. Water displaced by the stone causes a local change in the height of the water, which causes a change in height in the water immediately adjacent to it and so on. Hence a wave travels out from the point of entry of the stone. The surface of the water at each point in the pond (Figure 2.1b) goes up and down like a weight on the end of a spring, giving rise to the oscillating nature of the wave. Again, it is only the disturbance which travels across the pond, and not the water. Energy is transported by the wave across the pond from the stone to the shore.

Radio waves and light waves are electromagnetic waves. They can travel not only through some solids, liquids and gasses, but also through the vacuum of space. The waves that we are concerned with in this text are mechanical waves, which travel through a deformable or elastic medium such as a solid or liquid. A mechanical disturbance can travel through the medium due to its elastic properties.

Transverse and longitudinal waves

Waves may be classed as transverse or longitudinal waves. In the case of the Mexican wave previously described, the movement of the arms was up and down, while the direction of travel of the wave was horizontal. This is an example of a transverse wave, as the direction of movement of the arms was transverse to the direction of travel of the wave. The ripples on the pond are also a transverse wave, as the water surface moves up and down as the ripples travel along.

Shear waves are also classed as transverse waves. Shear waves travel mainly in solid materials. The local disturbance or displacement in this case is a shearing action, where, for example, one part of the material is pushed sideways with respect to that immediately underneath. Shear waves may also propagate over short distances in tissue and can yield useful information on the elastic properties of tissues, as described in Chapter 14.

The sound waves used to form medical images are longitudinal waves which propagate through a physical medium (usually tissue or liquid). Here, the particles of the medium oscillate backwards and forwards along the direction of propagation of the wave (see Figure 2.2). Where particles in adjacent regions have moved towards each other, a

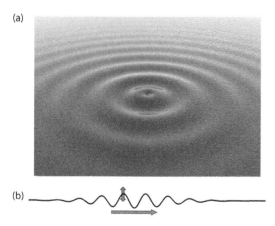

Figure 2.1 Ripples on the surface of a pond. (a) Ripples on the surface of a pond travel out from the point of entry of a stone. (b) Only the disturbance travels across the pond. The water surface simply goes up and down.

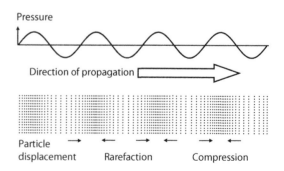

Figure 2.2 In a longitudinal wave, particle motion is aligned with the direction of travel, resulting in bands of high and low pressure.

region of compression (increased pressure) results, but where particles have moved apart, a region of rarefaction (reduced pressure) results. As in the transverse wave case, there is no net movement of the medium. Only the disturbance and its associated energy are transported.

Sound waves as illustrated in Figure 2.2 arise due to exchange of energy of motion of the molecules of the material (kinetic energy) with potential energy stored due to elastic compression and stretching of the material. The movement of the molecules can be modelled as shown in Figure 2.3. Here the mass of a small element of the material is represented by the black ball and the compressibility of the elastic material by two springs. In Figure 2.3a, there is no wave and the element is in its rest position and the springs relaxed. As the element of the material is displaced from its rest position (Figure 2.3b), the material is compressed on one side and stretched on the other, building up an increasing restoring

force and potential energy in the pair of springs. The element is then accelerated back towards its rest position by the forces in the springs, gaining kinetic energy as it goes. The kinetic energy carries it past the rest position, compressing and stretching the material in the other direction (Figure 2.2c). The result is that the element oscillates backwards and forwards. As in the simple waves previously described, interaction of the element with other neighbouring elements causes the disturbance and its associated energy to travel through the material and a travelling longitudinal wave is produced.

The most familiar sound waves are those that travel in air from a source of sound, e.g. a musical instrument or a bell, to the human ear. The surface of a bell vibrates when it is struck. The oscillating motion of the surface pushes and pulls against the air molecules adjacent to it. Neighbouring air molecules are then set in motion, which displace their neighbours and so the disturbance travels through the air as a sound wave. When the sound wave reaches the listener's ear, it causes the eardrum to vibrate, giving the sensation of sound. Energy from the bell is transported by the wave to the eardrum, causing it to move.

FREQUENCY, SPEED AND WAVELENGTH

Frequency

When the bell above is struck, its surface vibrates backwards and forwards at a certain frequency (number of times per second). An observer listening to the sound at any point nearby will detect the same number of vibrations per second. The frequency of the wave is the number of oscillations or wave crests passing a stationary observer per second (Figure 2.4) and is determined by the source of the sound wave. Frequency is normally given the symbol f and has units of hertz (1 Hz = 1 cycle per second). Sound waves with frequencies in the approximate range from 20 Hz to 20 kHz can be detected by the human ear. At frequencies above 15–20 kHz, the hearing mechanism in the human ear is unable to respond to the sound wave. Hence, sound waves with frequencies above approximately 20 kHz are referred to as ultrasound waves.

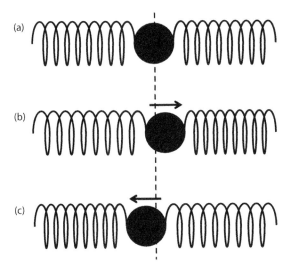

Figure 2.3 **(a)** The displacement of molecules by a longitudinal wave in an elastic material can be modelled by a localised mass (the black ball) and two springs. **(b)** As the mass is displaced from its rest position, the material is compressed on one side and stretched on the other, building up an increasing restoring force. **(c)** The mass is accelerated back by the springs and carried past its rest position by its kinetic energy, compressing and stretching the material in the other direction. The mass continues to oscillate backwards and forwards.

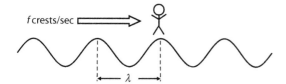

Figure 2.4 The frequency f of a wave is the number of wave crests passing a given point per second. The wavelength λ is the distance between wave crests.

Speed

Sound waves can travel through solids, liquids and gasses, but as will be shown in more detail later, the speed at which a sound wave travels is determined by the properties of the particular medium in which it is travelling. The speed of sound is normally given the symbol c and has units of m s^{-1} (metres per second). Examples are the speed of sound in air (330 m s^{-1}) and water (1480 m s^{-1}).

Wavelength

The ripples on the surface of the pond in Figure 2.1 show a regular pattern, in which the wave crests are all a similar distance apart. This distance between consecutive wave crests, or other similar points on the wave, is called the wavelength. Wavelength is illustrated in Figure 2.4. It is normally given the symbol λ (lambda) and has units of metres or millimetres.

A wave whose crests are λ metres apart and pass an observer at a rate of f per second must be travelling at a speed of $f \times \lambda$ metres per second:

$$\text{The speed of sound} \quad c = f\lambda$$

However, the form of this equation might suggest that the speed of sound can be chosen by choosing suitable values for f and λ. In physical terms, this is not the case and it is more informative to rearrange the equation to give a definition of the wavelength λ:

$$\text{Wavelength} \quad \lambda = \frac{c}{f}$$

This is because the frequency of a sound wave is determined by the source which produces it and the speed of the sound wave is determined by the material it is travelling through. The wavelength is the result of the combination of these two properties. That is, a wave from a source of frequency f, travelling through a medium in which the speed of sound is c, has a wavelength λ.

For example, a sound wave from a 30 kHz source travelling through water ($c \approx 1500$ m s^{-1}) has a wavelength of 50 mm, whereas a wave from the same source travelling through air ($c = 330$ m s^{-1}) has a wavelength of about 10 mm.

Phase

As a sound wave passes through a medium, the particles are displaced backwards and forwards from their rest positions in a repeating cycle. The pattern of displacement of the particles with time can often be described by a sine wave (Figure 2.5a). This pattern of displacement is as would be seen in the height of a rotating bicycle pedal when viewed from behind the bicycle. A complete cycle of the pedal height corresponds to a complete 360° rotation. The height of the pedal at any point in the cycle is related

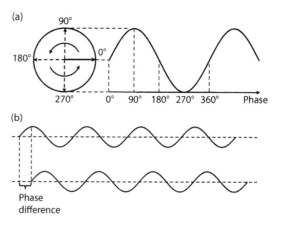

Figure 2.5 **(a)** During the passage of a sound wave, particles of the medium oscillate back and forth about their mean positions. The pattern of particle displacement often describes a sine wave. Phase describes the position within a cycle of oscillation and is measured in degrees. **(b)** Two waves of the same frequency and amplitude can be compared in terms of their phase difference.

to the angle of the pedal when viewed from the side. The phase of the pedal is its position within such a cycle of rotation and is measured in degrees. For example if a position of horizontal (zero height) and to the rear is defined as a phase of 0°, a phase of 90° will correspond to the pedal being in the vertical position where its height is at a maximum. At 180°, the pedal is horizontal and forwards with a height of zero. The height reaches its minimum value at 270° when the pedal is vertically down.

Two waves of the same frequency may differ in terms of their phase and can be compared in terms of their phase difference, measured in degrees (Figure 2.5b). Phase difference is an important concept when waves are added together, as described later in this chapter.

PRESSURE, INTENSITY AND POWER

As explained earlier, a sound wave passing through a medium causes the particles of the medium to oscillate back and forth along the direction of propagation (i.e. longitudinally). The longitudinal motion of the particles results in regions of compression and rarefaction so that at each point in the medium the pressure oscillates between maximum and minimum values as the wave passes. The difference between this actual pressure and the normal ambient pressure in the medium is called the *excess pressure* (see Figure 2.6), which is measured in pascals (Pa), where 1 Pa equals 1 N m^{-2} (the newton N is a measure of force). When the medium is compressed, the excess pressure is positive. When the medium undergoes rarefaction, the pressure is less than the normal ambient pressure, and so the

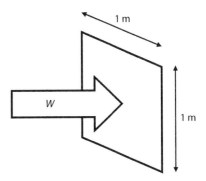

Figure 2.7 Intensity is the power W in watts flowing through unit area, e.g. 1 W m^{-2}.

excess pressure is negative. The amplitude of the wave may be described by the peak excess pressure, the maximum value during the passage of a wave. In practice, the ambient pressure is usually quite small compared to the excess pressure and the excess pressure is referred to simply as the pressure in the wave.

As the sound wave passes through the medium, it transports energy from the source into the medium. The rate at which this ultrasound energy is produced by the source is given by the ultrasound power. Energy is measured in joules (J) and power is measured in watts (W). One watt is equal to one joule per second, i.e. $1 \text{ W} = 1 \text{ J s}^{-1}$.

The ultrasound produced by the source travels through the tissues of the body along an ultrasound beam, and the associated power is distributed across the beam. As is discussed later in this chapter, the power is not distributed evenly across the beam, but may be more concentrated or intense near the centre. The intensity is a measure of the amount of power flowing through unit area of the beam cross section. Intensity is defined as the power flowing through unit area presented at 90° to the direction of propagation (Figure 2.7). Intensity I is measured in W m^{-2} or mW cm^{-2}.

As one might expect intuitively, the intensity associated with a wave increases with the pressure amplitude of the wave. In fact intensity I is proportional to p^2.

SPEED OF SOUND

As previously described, a sound wave travels or propagates through a medium due to the

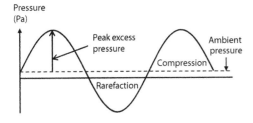

Figure 2.6 The pressure in the medium alternates between compression and rarefaction as the wave passes. The excess pressure in the sound wave is the pressure above or below the ambient pressure.

interaction between neighbouring particles or elements of the medium. The longitudinal movement of a particle and its associated energy are passed on to a neighbouring particle so that the wave propagates through the material. The speed of propagation of a sound wave is determined by the properties of the medium it is travelling in. In gases (e.g. air) the speed of sound is relatively low in relation to values in liquids, which in turn tend to be lower than values in solids.

The material properties which determine the speed of sound are density and stiffness. Density is a measure of the weight of a standard volume of material. For example, bone has a higher density than water; a 1 cm cube of bone weighs almost twice as much as a 1 cm cube of water. Density is normally given the symbol ρ (rho) and is measured in units of kilograms per cubic metre (kg m^{-3}). Bone has a density of 1850 kg m^{-3}; water has a density of 1000 kg m^{-3}, which is the same as 1 gram per cubic centimetre.

Stiffness is a measure of how well a material resists being deformed when it is squeezed. This is given by the pressure required to change its thickness by a given fraction. The pressure or stress (force per unit area) applied to the material is measured in pascal. The fractional change in thickness, or strain, is the ratio of the actual change in thickness to the original thickness of the sample. As strain is a ratio, it has no units of measurement. Hence the stiffness k, which is the ratio of stress to strain, is measured in units of pascal. Further details of stiffness can be found in Chapter 14.

The overall mass of a uniform material is distributed continuously throughout its bulk. However, during the passage of a wave, small elements of the material move back and forth and their movement can be modelled by a line of small discrete masses separated by springs which model the stiffness of the material, as shown in Figure 2.8. This simple model can be used to explain how the density and stiffness of a material determine its speed of sound. In Figure 2.8, the small masses m model the mass of a small element of a material of low density and the large masses M a small element of a material of high density. The masses are linked by springs which model the stiffness of the material. So springs K model a material of high stiffness and springs k a material of low stiffness. In Figure 2.8a,

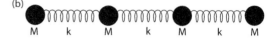

Figure 2.8 The speed of sound in a medium is determined by its density and stiffness, which can be modelled by a series of masses and springs. **(a)** A material with low density m and high stiffness K has a high speed of sound. **(b)** A material of high density M and low stiffness k has a low speed of sound.

the small masses m are linked by springs of high stiffness K, modelling a material with low density and high stiffness. In Figure 2.8b, the large masses M are linked by springs of low stiffness k, modelling a material of high density and low stiffness.

In Figure 2.8a, a longitudinal wave can be propagated along the row of small masses m by giving the first mass a momentary push to the right. This movement is coupled to the second small mass by a stiff spring K causing it to accelerate quickly to the right and pass on the movement to the third mass and so on. As the masses are light (low density), they can be accelerated quickly by the stiff springs (high stiffness) and the disturbance travels rapidly.

In Figure 2.8b, a momentary movement of the first large mass M to the right is coupled to the second mass by a weak spring k (low stiffness). The second large mass will accelerate relatively slowly in response to the small force from the weak spring. Its neighbours to the right also respond slowly so that the disturbance travels relatively slowly.

Hence, low density and high stiffness lead to high speed of sound whereas high density and low stiffness lead to low speed of sound. Mathematically this is expressed in the following equation:

$$\text{Speed of sound } c = \sqrt{\frac{k}{\rho}}$$

Table 2.1 shows typical values for the speed of sound in various materials, including several types of human tissue, water and air. It can be seen that the speed of sound in most tissues is similar to that in water, the speed of sound in bone is much

Table 2.1 Speed of sound in human tissues and liquids

Material	c (m s^{-1})
Liver	1578
Kidney	1560
Amniotic fluid	1534
Fat	1430
Average tissue	1540
Water	1480
Bone	3190–3406
Air	333

Source: Duck FA. 1990. *Physical Properties of Tissue—A Comprehensive Reference Book*. London: Academic Press.

Table 2.2 Wavelengths used in diagnosis

f (MHz)	λ (mm)
2	0.77
5	0.31
10	0.15
15	0.1

higher and that in air is much lower. As seen, the density of bone is almost twice that of tissue, which might be expected to lead to a lower speed of sound. However, the stiffness of bone is more than 10 times greater than that of tissue, leading to higher speed of sound. Gases, such as air, have much lower speed of sound than tissue. Although gases have low density, they have very low stiffness (high compressibility), leading to relatively low speed of sound compared to liquids and solids.

The similarity of speed of sound values in human soft tissues and water is explained by the fact that the density and compressibility of these materials is dependent on their molecular composition and short-range molecular interactions, rather than their long-range structure (Sarvazyan and Hill 2004). Most human soft tissues contain 70%–75% water and so behave in a similar manner at this level. The speed of sound in homogenised tissue is almost identical to that in its normal state. Fat does not have high water content and its speed of sound is significantly different from the average in soft tissues.

The most important point to note from Table 2.1 is that the values for the speed of sound in human soft tissues are rather similar. In fact they are sufficiently similar that the B-mode image-forming process can assume a single, average value of 1540 m s^{-1} without introducing significant errors or distortions in the image. All the values shown (with the exception of fat) are within 5% of this average value and are not much different from the value in water. As discussed later, the different speed of sound in fat can lead to imaging artefacts.

FREQUENCIES AND WAVELENGTHS USED IN DIAGNOSIS

The ultrasound frequencies used most commonly in medical diagnosis are in the range 2–15 MHz, although frequencies up to 40 MHz may be used in special applications and in research. The wavelengths in tissue which result from these frequencies can be calculated using the equation given earlier, which relates wavelength λ to the frequency f and speed c of a wave:

$$\lambda = \frac{c}{f}$$

Assuming the average speed of sound in soft tissues of 1540 m s^{-1}, values of λ at diagnostic frequencies are as shown in Table 2.2.

The wavelengths in soft tissues which result from these frequencies are within the range 0.1–1 mm. As discussed later in this chapter and in Chapter 5, the wavelength of the ultrasound wave has an important influence on the ability of the imaging system to resolve fine anatomical detail. Short wavelengths give rise to improved resolution, i.e. the ability to show closely spaced targets separately in the image.

REFLECTION OF ULTRASOUND WAVES

Reflection of ultrasound waves is a fundamental process in the formation of ultrasound images. In Chapter 1, a B-mode image was described as being constructed from echoes, which are generated by reflection of ultrasound waves at tissue boundaries and scattering by small irregularities within tissue. Reflection as described here is similar to reflection of light by a mirror and occurs where the interface extends to dimensions much greater than the wavelength of the ultrasound wave. This type of reflection is also referred to as specular reflection. Reflections occur at tissue boundaries where there is a change in acoustic properties of the tissue. Specifically, reflection occurs at an interface between two tissues of different acoustic impedance (see the next section). When an ultrasound wave travelling through one type of tissue encounters an interface with a tissue with different acoustic impedance, some of its energy is reflected back towards the source of the wave, while the remainder is transmitted into the second tissue.

Acoustic impedance

The acoustic impedance of a medium z is a measure of the response of the particles of the medium in terms of their velocity, to a wave of a given pressure:

$$\text{Acoustic impedance } z = \frac{p}{v}$$

where p is the local pressure and v is the local particle velocity. Acoustic impedance defined in this way is known as the specific acoustic impedance as it refers to local values of pressure and particle velocity. Acoustic impedance is analogous to electrical impedance (or resistance R), which is the ratio of the voltage V applied to an electrical component (the electrical driving force or pressure) to the resulting electrical current I which passes through it (the response), as expressed in Ohm's law:

$$R = V/I.$$

The acoustic impedance of a medium is again determined by its density ρ and stiffness k. It can be explained in more detail, as with the speed of sound, by modelling the medium as a row of small or large masses m and M linked by weak or stiff springs k and K as shown in Figure 2.9. In this case, however, the small masses m are linked by weak springs k, modelling a material with low density and low stiffness (Figure 2.9a). The large masses M are linked by stiff springs K, modelling a material with high density and stiffness (Figure 2.9b).

Figure 2.9 The acoustic impedance of a medium is determined by its density and stiffness, which can be modelled by a series of discrete masses and springs. **(a)** A material with low density m and stiffness k has low acoustic impedance. **(b)** A material with high density M and stiffness K has high acoustic impedance.

In Figure 2.9a, if a given pressure (due to a passing wave) is applied momentarily to the first small mass m, the mass is easily accelerated to the right (reaching increased velocity) and its movement encounters little opposing force from the weak spring k. This material has low acoustic impedance, as particle movements within it (in terms of velocity) in response to a given pressure are relatively large. In the second case (see Figure 2.9b), the larger masses M accelerate less in response to the applied pressure (reaching lower velocity) and their movements are further resisted by the stiff springs. Particle velocity (the response) in this material is lower for a given applied pressure and it has higher acoustic impedance. The two examples might be compared at a simple level to foam and concrete. A high jumper landing on a foam mattress (soft and light) will find that it yields to his body much more readily than concrete (solid and heavy).

The acoustic impedance z of a material is given by:

$$z = \sqrt{\rho k}$$

The equation shows that acoustic impedance increases with both density and stiffness. By combining this equation with that for the speed of sound given earlier, it can be shown also that:

$$z = \rho c$$

Acoustic impedance defined in this way is referred to as the characteristic acoustic impedance

Table 2.3 Values of characteristic acoustic impedance

Material	z (kg m^{-2} s^{-1})
Liver	1.66×10^6
Kidney	1.64×10^6
Blood	1.67×10^6
Fat	1.33×10^6
Water	1.48×10^6
Air	430
Bone	6.47×10^6

as it is given in terms of the macroscopic properties of the material. Acoustic impedance z has units of kg m^{-2} s^{-1}, but the term *rayl* (after Lord Rayleigh) is often used to express this unit.

Table 2.3 gives values of characteristic acoustic impedance z for some common types of human soft tissue, water, air and bone. The table shows that values of z in most human soft tissues and water are very similar. This is to be expected as most soft tissues and water have similar densities and as seen earlier, similar speeds of sound. The acoustic impedance for air is very small compared to soft tissues as its density and speed of sound are both much lower. For bone, which has high density and high speed of sound, z is about four times greater than for soft tissues.

Reflection

When a sound wave travelling through one medium meets an interface with a second medium of different acoustic impedance, some of the wave is transmitted into the second medium and some is reflected back into the first medium. The amplitudes of the transmitted and reflected waves depend on the change in acoustic impedance. If the change in acoustic impedance is large, the ultrasound wave will be strongly reflected back into the first medium with only weak transmission into the second. If the change in acoustic impedance is small, then most of the ultrasound energy will be transmitted into the second medium, with only weak reflection back into the first.

Figure 2.10a shows a sound wave travelling through a medium with acoustic impedance z_1, incident on an interface with a second medium

with acoustic impedance z_2. The incident wave is partially transmitted into medium 2 and partially reflected back into medium 1. At a macroscopic level, the characteristic acoustic impedance changes abruptly at the interface. At a microscopic level, the specific acoustic impedance, i.e. the ratio of particle pressure to particle velocity, must also change. However, the velocity of particles and the local pressure cannot change abruptly at the interface, without disruption of the medium; they must be continuous across the interface. It is this apparent mismatch that results in the formation of a reflected wave. The total wave pressure and velocity at the interface in medium 1, including those of the reflected wave, can then be equal to those of the transmitted wave in medium 2. In Figure 2.10a, the particle pressure and velocity in the incident wave are p_i and v_i, the particle pressure and velocity in the reflected wave are p_r and v_r and the particle pressure and velocity in the transmitted wave are p_t and v_t. To maintain continuity across the interface, the following conditions must hold:

$$p_t = p_i + p_r$$

$$v_t = v_i + v_r$$

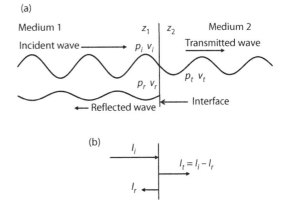

Figure 2.10 **(a)** The total particle pressure and total particle velocity must be continuous across an interface where there is a change in acoustic impedance. This requirement results in the formation of a reflected wave which travels back into the first medium. **(b)** The intensity transmitted across an interface is the incident intensity minus that reflected.

From the definition of specific acoustic impedance, we also have:

$$z_1 = \frac{p_i}{v_i} = \frac{p_r}{v_r}$$

and

$$z_2 = \frac{p_t}{v_t}$$

From these four equations, it can be shown that:

$$\frac{p_r}{p_i} = \frac{z_2 - z_1}{z_2 + z_1}$$

This ratio of reflected to incident pressure is commonly referred to as the amplitude reflection coefficient R_A of the interface. It is very important to ultrasound image formation as it determines the amplitude of echoes produced at boundaries between different types of tissue. It can be seen that the reflection coefficient is determined by the difference between the acoustic impedances of the two media divided by their sum. So if the difference in acoustic impedance is small, then the reflection coefficient is small and only a weak echo is produced. If the difference in acoustic impedance is large, e.g. if z_1 is a small fraction of z_2, then the reflection coefficient is large and a strong echo is produced.

Note that when the acoustic impedance in the first medium is greater than that in the second, i.e. z_1 is greater than z_2, the reflection coefficient is negative. This means that the reflected wave is inverted at the interface before travelling back into the first medium. Inversion of the reflected wave has an important role in the function of ultrasound transducers and their matching layers as explained in Chapter 3.

Table 2.4 shows values of amplitude reflection coefficient for some interfaces that might be encountered in the human body. As shown in Table 2.3, the characteristic acoustic impedances of common types of human soft tissue are very similar. For interfaces between these, the reflection coefficients, as calculated from the previous equation, are hence very small. For most soft tissue to soft tissue interfaces, the amplitude reflection coefficient is less than 0.01 (1%). This is another

Table 2.4 Amplitude reflection coefficients of interfaces

Interface	R_A
Liver-kidney	0.006
Kidney-spleen	0.003
Blood-kidney	0.009
Liver-fat	0.11
Liver-bone	0.59
Liver-air	0.9995

important characteristic for ultrasound imaging as it means that most of the pulse energy at soft tissue interfaces is transmitted on to produce further echoes at deeper interfaces. The amplitude reflection coefficient at a tissue-fat interface is about 10% due to the low speed of sound in fat.

As shown in Table 2.3, the characteristic acoustic impedance of air is very small ($430 \text{ kg m}^{-2}\text{ s}^{-1}$) compared to a soft tissue such as liver ($1.66 \times 10^6 \text{ kg m}^{-2}\text{ s}^{-1}$). At an interface between soft tissue and air, as might be encountered within the lungs or gas pockets in the gut, the change in acoustic impedance and the resulting reflection coefficient are very large. For such an interface, the reflection coefficient is 0.999 (99.9%), resulting in almost total reflection of the ultrasound wave and zero transmission into the second medium. No useful echoes can be obtained from beyond such an interface. For this reason, it is important to exclude air from between the ultrasound source (the transducer) and the patient's skin to ensure effective transmission of ultrasound. This also means that those regions of the body that contain gas, such as the lungs and gut, cannot be imaged effectively using ultrasound.

At an interface between soft tissue (characteristic acoustic impedance $1.66 \times 10^6 \text{ kg m}^{-2}\text{ s}^{-1}$) and bone (characteristic acoustic impedance $6.47 \times 10^6 \text{ kg m}^{-2}\text{ s}^{-1}$), the change in acoustic impedance is still quite large, giving rise to an amplitude reflection coefficient of approximately 0.5 (50%), making it difficult to obtain echoes from beyond bony structures such as ribs. Note that the reflection coefficient is not related to the frequency of the wave; it is determined only by the change in z at the interface between the two media.

So far, reflection has been described in terms of the amplitude reflection coefficient, which relates the amplitude of the reflected wave to that of the incident wave and was defined in terms of pressure. Reflection coefficient can also be described in terms of the intensity reflection coefficient, which is the ratio of the intensities of the reflected (I_r) and incident waves (I_i). As intensity is proportional to pressure squared, the intensity reflection coefficient R_I is given by:

$$\frac{I_r}{I_i} = R_I = R_A^2 = \left(\frac{z_2 - z_1}{z_2 + z_1}\right)^2$$

As described earlier in this chapter, intensity is a measure of the power, or rate of energy flow, through unit cross-sectional area. At the interface, the energy flow of the incident wave in terms of its intensity must be conserved and is split between the transmitted wave and the reflected wave, as shown in Figure 2.10b.

Hence, the incident intensity $I_i = I_t + I_r$.

This equation can be rearranged to give the transmitted intensity:

$$I_t = I_i - I_r$$

The intensity transmission coefficient T_i is the ratio of the intensity of the transmitted wave I_t to that of the incident wave I_i.

That is:

$$T_i = \frac{I_t}{I_i}$$

As stated, energy flow at the interface must be conserved and it can be shown that $T_i = 1 - R_i$. For example, if 0.01 (1%) of the incident intensity is reflected, then the other 0.99 (99%) must be transmitted across the boundary.

Law of reflection

In the description of reflection so far, it has been assumed that the interface is large compared to the wavelength of the wave and that the incident wave approaches the boundary at 90° (normal incidence). Under these circumstances, the reflected and transmitted waves also travel at 90° to the interface. In clinical practice, the incident wave

may approach the interface at any angle. The angle between the direction of propagation and a line at 90° to the interface (the normal) is called the angle of incidence θ_i (which has been 0° so far) as shown in Figure 2.11a. Similarly, the angle between the direction of the reflected wave and the normal is called the angle of reflection θ_r.

For a flat, smooth interface, the angle of reflection is equal to the angle of incidence, $\theta_r = \theta_i$. This is referred to as the law of reflection.

As discussed in Chapter 5, reflection at strongly reflecting interfaces can lead to a number of image artefacts.

(a)
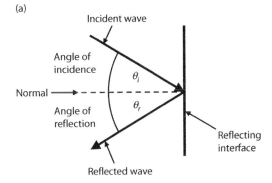

(b)
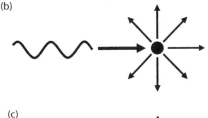

(c)
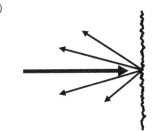

Figure 2.11 Ultrasound waves are reflected at large interfaces and scattered by small targets. (a) At a large, smooth interface, the angle of reflection is equal to the angle of incidence. (b) Small targets scatter the wave over a large angle. (c) A rough surface reflects the wave over a range of angles.

SCATTERING

Reflection, as just described, occurs at large interfaces such as those between organs, where there is a change in acoustic impedance. Within the parenchyma of most organs (e.g. liver and pancreas), there are many small-scale variations in acoustic properties, which constitute very small-scale reflecting targets (of size comparable to or less than the wavelength). Reflections from such very small targets do not follow the laws of reflection for large interfaces. When an ultrasound wave is incident on such a target, the wave is scattered over a large range of angles (Figure 2.11b). In fact, for a target which is much smaller than the wavelength, the wave may be scattered uniformly in all directions. For targets of the order of a wavelength in size, scattering will not be uniform in all directions but will still be over a wide angle.

The total ultrasound power scattered by a very small target is much less than that for a large interface and is related to the size d of the target and the wavelength λ of the wave. The scattered power is strongly dependent on these dimensions. For targets which are much smaller than a wavelength ($d \ll \lambda$), scattered power W_s is proportional to the sixth power of the size d and inversely proportional to the fourth power of the wavelength, i.e.:

$$W_s \propto \frac{d^6}{\lambda^4} \propto d^6 f^4$$

This frequency dependence is often referred to as Rayleigh scattering. As shown in the equation, the scattered power can also be expressed in terms of the frequency f of the wave. As f is proportional to $1/\lambda$, the scattered power W_s is proportional to the fourth power of the frequency.

Organs such as the liver contain non-uniformities in density and stiffness on scales ranging from the cellular level up to blood vessels, resulting in scattering characteristics which do not obey such simple rules over all frequencies used in diagnosis. The frequency dependence of scattering in real liver changes with frequency over the diagnostic range (3–10 MHz). The scattered power is proportional to f^m, where m increases with frequency from approximately 1 to 3 over this range (Dickinson 1986).

The way in which the various tissue types scatter ultrasound back to the transducer has a major influence on their appearance in the ultrasound image, with stronger backscatter leading to a brighter image. For example, liver and spleen scatter more strongly than brain (Duck 1990). The collagen content of tissue has a major influence on backscatter. Increased collagen, due to repair processes such as in myocardium following infarction, increases the backscatter. Backscatter from blood is very weak compared to most tissues, leading to a dark appearance in the ultrasound image. This is because the red blood cells, which are the main contributors to the backscattered signal, are much smaller than the wavelength (8 micron diameter compared to millimetre wavelengths) and hence act as Rayleigh scatterers. Also, the acoustic impedance mismatch between the red cells and blood fluid is low. Note that contrast agents (Chapter 13) are comparable in size to blood cells but scatter strongly due to the high acoustic impedance mismatch between the encapsulated gas and surrounding blood. Tissues such as muscle and tendon have long-range structure in one direction, i.e. along the muscle fibres, and do not scatter ultrasound uniformly in all directions. Along the direction of the fibres, such tissues may behave rather like a large interface, whereas across the fibres they may scatter ultrasound over a range of angles. Consequently, the appearance of muscle in an ultrasound image may change with the relative orientations of the ultrasound beams and the muscle fibres.

There are two important aspects of scattering to note for ultrasound imaging. First, the ultrasonic power scattered back to the transducer by small targets is small compared to that from a large interface, so the echoes from the parenchyma of organs such as the liver are relatively weak.

Second, as ultrasound is scattered over a wide angle by small targets, their response, and hence their appearance in the image, does not change significantly with the angle of incidence of the wave. Liver parenchyma looks similar ultrasonically regardless of the direction from which it is imaged. In contrast, the appearance of large interfaces is strongly dependent on the angle of incidence of the ultrasound beam, as described in Chapter 5.

DIFFUSE REFLECTION

The description of reflection provided assumed a perfectly flat, smooth interface. Some surfaces within the body may be slightly rough on the scale of a wavelength and reflect ultrasound waves over a range of angles, an effect similar to scattering from small targets. This type of reflection is known as diffuse reflection (Figure 2.11c).

REFRACTION

In the description of reflection presented, the angle of the reflected wave at a large interface was the same as that of the incident wave, but there was assumed to be no transmitted wave. Where the speed of sound is the same on both sides of the interface, the transmitted wave will carry on in the same direction as the incident wave. However, where there is a change in the speed of sound from the first medium to the next (and the angle of incidence is not 90°), the direction of the transmitted wave is altered due to refraction.

Refraction is commonly observed with light waves. For example, when an underwater object is looked at from the air above, light from the object is refracted as it emerges into the air, causing the apparent position of the object to be displaced from its real position (Figure 2.12a). As the speed of light in air is higher than that in water, the light emerging from the water is refracted away from the normal. The underwater object appears to an observer as though it is in line with the emerging beam of light, rather than in its true position. Refraction of ultrasound waves at boundaries where the speed of sound changes can also cause displacement of the image of a target from its true relative position in the patient, as described in Chapter 5.

Refraction can be explained as shown in Figure 2.12b. A wave front AB, travelling in medium 1 (propagation speed c_1), arrives at an interface with medium 2 (propagation speed c_2) where $c_2 > c_1$. The angle of incidence of the approaching wave is θ_i. The edge of the wave at A then passes into medium 2, where it travels more quickly than the edge at B, which is still in medium 1. By the time the wave edge at B has travelled to the interface (at point D), a wave front from point A will have travelled a greater distance to point C. The new wave

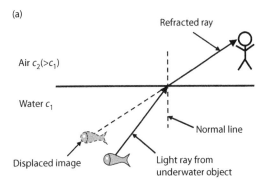

(a)

Refracted ray

Air $c_2 (> c_1)$

Water c_1

Normal line

Displaced image

Light ray from underwater object

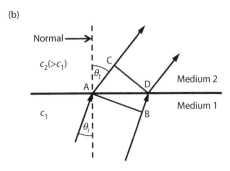

(b)

Normal

$c_2 (> c_1)$

θ_t

C

A

D

Medium 2

c_1

B

Medium 1

θ_i

Figure 2.12 **(a)** When a ray of light emerges at an angle from water into air, the direction of the ray is altered due to refraction. The image of an underwater object viewed via the light ray is displaced. Refraction occurs because the speed of light in air is greater than that in water. **(b)** The incident wave front AB arrives at the interface. The edge of the wave front at A passes into medium 2 first, where it travels more quickly than the edge at B (which is still in medium 1). By the time the wave edge at B has travelled to the interface (at point D), a wave front from point A has travelled a greater distance in medium 2 to point C. The new wave front CD is hence deviated away from the normal.

front CD is hence deviated away from the normal at angle θ_t. When the wave crosses an interface where the speed of sound increases, the angle to the normal also increases. Conversely, when the wave experiences a reduction in the speed of sound as it crosses the interface, the angle to the normal also decreases. The relationship between the angles θ_i, θ_t, c_1 and c_2 is described by Snell's law:

$$\frac{\sin\theta_i}{\sin\theta_t} = \frac{c_1}{c_2}$$

Snell's law shows that the angles of the incident and transmitted waves are the same when the speeds of sound in the two media are the same. The change in the direction of propagation of the wave as it crosses the boundary increases with increasing change in the speed of sound, i.e. larger changes in the speed of sound give rise to stronger refraction effects.

ATTENUATION

When an ultrasound wave propagates through soft tissue, the energy associated with the wave is gradually lost so that its intensity reduces with distance travelled, an effect known as attenuation. The way in which intensity falls with increasing distance is illustrated in Figure 2.13a. The pattern is that of an exponential decay curve, where the rate of decrease is rapid at first but becomes more gradual as the intensity reduces. The attenuation curve follows the simple rule that the fractional decrease in intensity is the same for every centimetre travelled into the tissue.

For example, if the intensity is reduced by a factor of 0.7 in the first centimetre, it will be reduced again by 0.7 in the second centimetre and again by 0.7 in the third centimetre. So if the intensity at the surface starts off at a value of 1.0 (relative intensity units), it will be reduced to 0.7 after 1 cm, 0.49 after 2 cm ($0.7 \times 0.7 = 0.49$) and to 0.34 after 3 cm ($0.7 \times 0.7 \times 0.7 = 0.34$) and so on.

The attenuation over a given distance in centimetres can be calculated more conveniently if the fractional decrease in intensity over each centimetre is expressed in decibels (dB). The decibel is used to express a factor or ratio on a logarithmic scale. As the rate of attenuation per centimetre in decibels is related to the logarithm of the ratio (see Appendix A), the attenuation over each centimetre travelled can be added rather than multiplied, and we can then measure attenuation in decibels per centimetre. A ratio of 0.7 is equal to 1.5 dB, so for the graph in Figure 2.13a the intensity is reduced by 1.5 dB for each centimetre of travel. After 1 cm, it is reduced by 1.5 dB, after 2 cm it is reduced by 3 dB (1.5 + 1.5) and after 3 cm it is reduced by 4.5 dB (1.5 + 1.5 + 1.5). If the attenuation curve in Figure 2.13a is plotted on a decibel scale as in

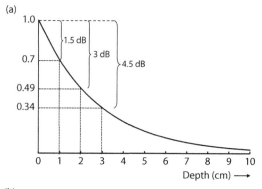

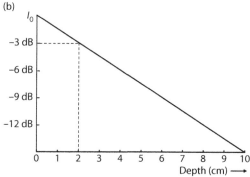

Figure 2.13 (a) The ultrasound intensity is attenuated (reduced) by the same fraction for each unit of distance travelled into the medium. This equates to attenuation by the same number of decibels (dB) for each unit of distance. **(b)** If the intensity is plotted on a decibel scale, the variation with depth becomes a straight line as it is attenuated by the same number of decibels for each centimetre (cm) of depth.

Figure 2.13b, the attenuation curve becomes a straight line, showing that the intensity decreases by the same number of decibels for each centimetre of depth. The rate at which the intensity of the wave is attenuated (in dB cm^{-1}), is referred to as the attenuation coefficient.

In Figure 2.13b, the starting intensity is labelled as I_0. The intensity at each depth is then plotted in relation to this value. The convention, as shown on the vertical axis, is to state the number of decibels by which the intensity has been reduced from I_0 as a negative number of decibels. For example, at a depth of 2 cm, the intensity is given as −3 dB with respect to I_0. The minus sign indicates that the intensity at that depth is 3 dB less than the starting value. An ultrasound wave can be attenuated by several mechanisms as it travels through tissue. The most important mechanism is absorption, in which ultrasound energy is converted into heat (Parker 1983). In most diagnostic systems, ultrasound propagates in the form of a beam, as described later in this chapter. The attenuation of practical interest is the rate at which ultrasound intensity in the beam decreases with distance. As well as absorption, the intensity in the beam may be reduced due to scattering of ultrasound out of the beam and to divergence or spreading of the beam with distance.

Absorption

Absorption is the process by which ultrasound energy is converted into heat in the medium. When an ultrasound wave passes through a medium, its particles move backwards and forwards in response to the pressure wave, as described earlier. At low frequencies, the particles are able to move in step with the passing pressure wave, and energy associated with the motion of the particles is effectively passed back to the wave as it moves on. However, the particles of the medium cannot move instantaneously and at high frequencies may be unable to keep up with the rapid fluctuations in pressure. They are unable to pass back all of the energy associated with their movements to the passing wave as they are out of step and some energy is retained by the medium, where it appears as heat. Absorption is likely to be strongest at frequencies which excite natural modes of vibration of the particular molecules of the medium as it is at such frequencies that they are most out of step with the passing wave. Understanding of the mechanisms and vibrational modes which lead to absorption of ultrasound is still imperfect, but absorption has been found to be strongly dependent on tissue composition and structure. For example, tissues with high collagen content such as tendons and cartilage show high absorption, whereas those with high water content show lower absorption. Water and liquids such as urine, amniotic fluid and blood have low absorption and low attenuation. Estimates of the contribution of absorption to attenuation are variable, but for many tissues absorption is the dominant loss mechanism.

Dependence on frequency

Attenuation of ultrasound by biological tissues increases with frequency. The attenuation coefficient of most tissues, when expressed in $dB\ cm^{-1}$, increases approximately linearly with frequency, i.e. doubling the frequency will increase the attenuation by approximately a factor of 2. Hence, for most tissues, it is possible to measure ultrasound attenuation in $dB\ cm^{-1}\ MHz^{-1}$. This allows the total attenuation of an ultrasound pulse to be calculated easily from the frequency and the distance travelled. For example, if a tissue attenuates by $0.7\ dB\ cm^{-1}\ MHz^{-1}$, then a 5 MHz ultrasound wave, after travelling a distance of 10 cm, will be attenuated by $5\ MHz \times 10\ cm \times 0.7\ dB\ cm^{-1}\ MHz^{-1}$ $= 35\ dB$.

Note that this attenuation is experienced by the transmitted pulse as it travels into tissue and by the echoes as they return to the transducer. Hence, for the example just given, echoes from a depth of 10 cm will be smaller than those received from tissues close to the transducer by 70 dB.

Table 2.5 shows measured values of attenuation for some human tissues expressed in units of $dB\ cm^{-1}\ MHz^{-1}$. Values are typically in the range $0.3–0.6\ dB\ cm^{-1}\ MHz^{-1}$ for soft tissues but much lower for water (and watery body fluids). Attenuation in bone is very high and does not increase linearly with frequency as in the case of soft tissues. Bone and other calcified materials are effective barriers to propagation of ultrasound. As was seen earlier, ultrasound is strongly reflected at the surface of bone. Ultrasound which is transmitted into the bone is then rapidly attenuated. Any

Table 2.5 Values of attenuation for some human tissues

Tissue	Attenuation ($dB\ cm^{-1}\ MHz^{-1}$)
Liver	0.399
Brain	0.435
Muscle	0.57
Blood	0.15
Water	0.02
Bone	22

echoes returned from beyond the bone surface will be too weak to detect.

The increase in attenuation with frequency is an important feature of ultrasound which has a major impact on its use in forming an image and leads to a classic trade-off that the operator must make. Attenuation of ultrasound transmitted into tissue and echoes returned to the transducer increases with depth so that echoes from the deepest parts of an organ are weak and difficult to detect. Attenuation is also higher at higher frequencies. Hence for imaging large or deep organs, a low frequency (3–5 MHz) must be used to be able to detect echoes from the most distal parts. High frequencies (10–15 MHz) can only be used to image relatively small, superficial targets (e.g. thyroid) as they are attenuated more rapidly. The short wavelengths associated with high-frequency ultrasound enable improved resolution of image detail. Low frequencies lead to poorer resolution. The operator must choose the optimum frequency for each particular application. This is a compromise that ensures that the best image resolution is obtained, while allowing echoes to be received from the required depth.

ULTRASOUND BEAMS

The description of ultrasound wave propagation so far has concentrated mainly on the properties of the wave and how it is affected by the medium in the direction of propagation. The principles outlined apply regardless of the sideways extent of the wave, i.e. in the transverse direction. It is clear from the outline of B-mode image formation given in Chapter 1 that to be able to define the origin of an echo in the imaged cross section, the extent of the wave in the transverse direction has to be very limited. That is, the wave must propagate along a narrow corridor or beam. This section describes the formation and properties of ultrasound beams.

Interference of waves

So far, we have considered only a single wave propagating through a medium. When two or more waves from different sources propagate through the same medium, they interfere with each other. That is, the effects of each individual wave are added at each point in the medium. This is to be

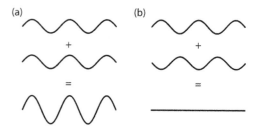

Figure 2.14 The effects of two waves travelling through the same medium are added, i.e. the waves interfere with each other. **(a)** Waves with the same frequency and phase interfere constructively (add). **(b)** Waves with opposite phase interfere destructively (cancel).

expected, as the pressure at a point in the medium will be the sum of the pressures on it from the different waves.

Figure 2.14 shows the simple case of two waves with the same frequency and amplitude propagating in the same direction. In Figure 2.14a, the waves are in phase or in step with each other and the peaks of the two waves coincide, as do the troughs. In this case the resulting wave amplitude is twice that of the individual waves. This case is referred to as constructive interference, as the resulting amplitude is greater than that of both the individual waves.

In Figure 2.14b, the waves are in anti-phase; that is the peaks of one wave coincide with the troughs of the other. Hence, at each point in the medium the particles are being pushed in one direction by the first wave and by the same pressure in the opposite direction by the second wave. The resulting pressure on the particles is zero and the effects of the waves cancel out. This case is referred to as destructive interference.

Diffraction

When a source generates a sound wave, the way in which the wave spreads out as it moves away from the source is determined by the relationship between the width of the source (the aperture) and the wavelength of the wave.

If the aperture is smaller than the wavelength, the wave spreads out as it travels (diverges), an effect known as diffraction. This is rather like the wave on the pond, spreading out from the point of entry of a small stone as shown in Figure 2.1. For a sound wave from a small point source inside a medium, the wave spreads out as an expanding sphere (a spherical wave) rather than a circle as on a water surface. The small scattering targets within tissue described earlier effectively act as sources of such spherical waves.

If the width of the source is much greater than the wavelength of the wave, the waves are relatively flat (plane) rather than curved and lie parallel to the surface of the source. Such waves travel in a direction perpendicular to the surface of the source with relatively little sideways spread, i.e. in the form of a parallel-sided beam.

These two different cases of curved waves from a small source and plane waves from a large source can be linked by considering the large source to be made up of a long row of small sources, as shown in Figure 2.15.

Each of the small sources generates a sound wave of the same frequency and amplitude and all are in phase with each other. The curved waves from each propagate outwards and the parts of the curve which are parallel to the surface of the source align to form plane waves. The other, non-parallel parts of the curved waves tend to interfere destructively and cancel out. The diagram shows

only four point sources for clarity, but in practice the large source is considered as a continuous series of small sources. This view of the generation of a plane wave from a large plane source is put to practical use in forming ultrasound beams from rows of small sources (array transducers) as described in Chapter 3.

Practical sources of ultrasound beams

An ideal ultrasound beam would be one that is very narrow throughout its whole length. This would allow fine detail to be resolved at all imaged depths. However, as indicated previously, a narrow source (in relation to the wavelength) produces a beam that is narrow near the source but diverges quickly. A source that is wide in relation to the wavelength produces a parallel beam which is too wide to resolve fine detail. Practical ultrasound sources must compromise between these two extremes to give the optimum combination of narrow beam width and minimal divergence.

Plane disc source

One example of a practical source of ultrasound is a plane disc transducer, also referred to as a plane circular piston source. The surface of this source is a flat disc and it is assumed that all parts of the surface move backwards and forwards exactly in phase and with the same amplitude. The surface of the disc source can be considered to be made up of many small elements, each of which emits a spherical wave. The pressure amplitude at each point in the beam is determined by the sum of the spherical waves from all of the elements (Figure 2.16). The different path lengths from the various elements to the summing point mean that each of the spherical waves has a different phase when it arrives. At some points, this results in overall constructive interference, giving rise to an amplitude maximum. At other points, the overall effect is destructive and a minimum is formed. At points close to the source, the path lengths can be different by many wavelengths.

The basic shape of the ultrasound beam produced by a plane disc transducer is illustrated in

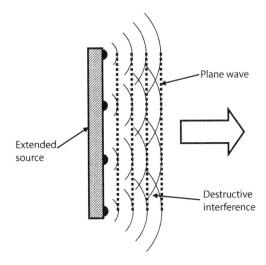

Figure 2.15 An extended source can be considered as a row of small point sources (only four points shown for clarity). Spherical waves from each interfere to form a series of wave fronts.

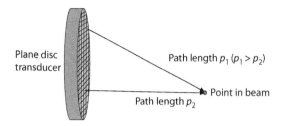

Figure 2.16 The surface of the disc source can be considered to be made up of many small elements, each of which emits a spherical wave. The pressure amplitude at each point in the beam is determined by the sum of the spherical waves from all of the elements. The different path lengths, from the various elements to the summing point, mean that each of the spherical waves has a different phase when it arrives.

Figure 2.17a. To a first approximation, it can be divided into two parts. These are:

1. The near field, which is roughly cylindrical in shape and has approximately the same diameter as the source.
2. The far field, which diverges gradually.

Figure 2.17b shows the real distribution of pressure amplitude within a beam from a plane disc source. Within the near field, the pressure amplitude of the wave is not constant everywhere, but shows many peaks and troughs due to constructive and destructive interference. As the source is circular, these pressure variations have circular symmetry. That is, the peaks and troughs are in the form of rings centred on the beam axis. The end of the near field is defined as being the distance from the source at which the maximum path length difference is $\lambda/2$. This distance, the near-field length (NFL), is given by the equation:

$$NFL = \frac{a^2}{\lambda}$$

where a is the radius of the source. Figure 2.17c shows the variation in pressure amplitude along the central beam axis. It can be seen that the pressure amplitude reaches a final maximum value at the end of the near field.

In the far field, destructive interference does not occur in the central lobe of the beam, as the path

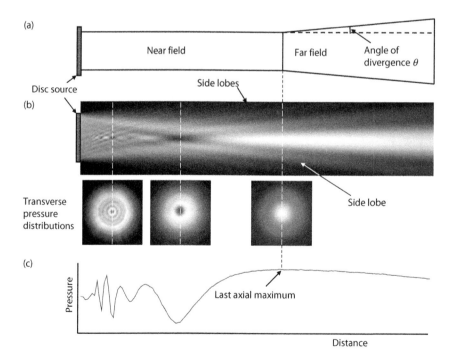

Figure 2.17 The ultrasound beam from a plane disc source consists of a near field, in which the pressure distribution is complex, and a far field, in which it is more uniform.

length differences are all less than $\lambda/2$. The resulting beam structure is relatively simple, showing a maximum value on the beam axis, which falls away uniformly with radial distance from the axis. The intensity along the beam axis in the far field falls approximately as the inverse square law, i.e. proportional to $1/z^2$, where z is the distance from the transducer. The beam diverges in the far field. The angle of divergence θ is the angle between the beam axis and the edge of the central lobe of the beam (where the intensity falls to zero) and is given by the equation:

$$\sin\theta = 0.61\frac{\lambda}{a}$$

When the aperture a is similar in size to the wavelength λ, the near field is short and the beam diverges rapidly in the far field, i.e. θ is large. Figure 2.18a shows an example for the case of a 3 MHz wave in a medium with speed of sound c of 1500 m s^{-1}. The wavelength λ is 0.5 mm. If the source radius is also 0.5 mm, then the near field is short (0.5 mm) and the angle of divergence is large (37°). In Figure 2.18b, the source radius is 10 mm (20λ) leading to a NFL of 200 mm and an angle of divergence in the far field of 1.7°. However, the beam is approximately 20 mm wide. So when a is large compared to λ, the near field is long and there is little divergence in the far field, i.e. θ is small. Hence, the beam is almost parallel, but too wide to image fine detail.

For the plane disc source, the optimum beam shape is achieved when the radius is about 10λ–15λ, i.e. about 20–30 wavelengths in diameter. Figure 2.18c shows the example of a radius of 7.5 mm (15λ) leading to a NFL of 112.5 mm and an angle of divergence of 2.3° in the far field. At higher ultrasound frequencies, the wavelength is shorter, but the ideal beam shape is still obtained for a source radius of about 10λ–15λ. For example, a 10 MHz source ($\lambda = 0.15$ mm) with a radius of 2 mm (13λ) has a NFL of 27 mm and an angle of divergence in the far field of 2.6°. Hence,

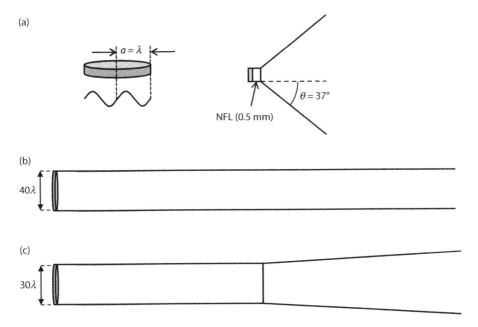

Figure 2.18 The shape of the ultrasound beam from a simple disc source is determined by the relationship between the width of the source and the wavelength. (a) Assume $f = 3$ MHz and $c = 1500$ m s^{-1} ($\lambda = 0.5$ mm). If the radius $a = \lambda$, then the NFL = 0.5 mm and the angle of divergence $\theta = 37°$. (b) Assume the radius $a = 10$ mm (20λ). Then NFL = 200 mm and $\theta = 1.7°$. (c) Assume the radius $a = 7.5$ mm (15λ). Then NFL = 112.5 mm and $\theta = 2.3°$. A radius of about 10λ–15λ (i.e. diameter $\approx$ 20λ–30λ) gives a good compromise between beam width and shape.

increased frequency allows the source diameter and the beam width to be scaled down while maintaining the beam shape in terms of low divergence in the far field. High frequencies, e.g. 10 MHz are used to image small, superficial anatomical features such as the eye or thyroid, giving finer resolution of the small anatomical details. As attenuation of ultrasound increases with frequency, lower frequencies must be used for larger organs such as the liver to allow penetration to the deepest parts, leading to a wider beam and poorer resolution.

This description of a beam from a disc source relates to what is called the 'main lobe' of the beam. The angle of divergence θ defines the edge of the main lobe in the far field where destructive interference causes a minimum to be formed in the pattern of interference at that angle. At increasing angles to the main lobe greater than θ, alternate maxima and minima are formed, as can be seen in Figure 2.17b. The regions containing these maxima are referred to as side lobes. Side lobes are weaker than the main lobe but can give rise to significant echoes if they are incident on a strongly reflecting target adjacent to the main lobe, resulting in acoustic noise in the image. Manufacturers normally design their transducers to minimise side lobes. This can be done by applying stronger excitation to the centre of the transducer than at the edges, a technique known as apodization. Apodization reduces the amplitude of side lobes but leads to an increase in the width of the main lobe.

This description of the beam from a plane disc-shaped source assumes that the transmitted wave is continuous and hence contains only a single frequency. For imaging purposes, the source must produce a short burst or pulse of ultrasound, which gives distinct echoes from interfaces. As described later, a short pulse contains energy at a range of frequencies rather than just one, each of which produces a slightly different beam. These are effectively added together in the pulsed beam, resulting in smearing out of the pressure variations compared to those in the continuous-wave beam shown in Figure 2.17. Also, at points close to the source, the time of arrival of short pulses from the middle and the edge of the source may be so different that the pulses do not overlap in time. Hence

some of the interference effects seen with a continuous wave do not occur.

Beams from array transducers

The description given of ultrasound beam shapes assumes a circular source of ultrasound, giving rise to a beam with circular symmetry. The assumption of this simple symmetry is convenient in outlining the basic properties of ultrasound beams. As discussed in the next chapter, ultrasound imaging systems typically use transducers which consist of a long array of narrow rectangular elements. The active aperture along the direction of the array is variable and often much greater than the array width. The essential features of the ultrasound beam, such as the near field, far field and last axial maximum are still present. However, the scales of these features along and across the array are different, leading to a more complex beam structure.

Figure 2.19 shows the pattern of array elements in a typical array transducer. The width of the array is determined by the length of the elements and is typically 30λ across, giving a fixed beam shape in that direction similar to that shown in Figure 2.18c. Along the array (the scan plane), the aperture is changed according to the depth of the tissue region being imaged. In this plane, the dimensions of features such as the near field change according to the aperture used. Hence, the beam shape is different in the two planes. The width of each element (along the array) is typically of the order of 1.3λ (see Chapter 3) so that individual strips produce a beam in the scan plane similar to that in Figure 2.18a. In the scan

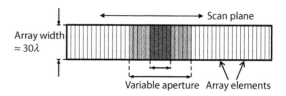

Figure 2.19 In an array transducer, the width of the array is typically fixed by the length of the array elements at approximately 30λ. Along the array (the scan plane), the aperture is variable according to the depth of the tissue region being imaged. Hence the dimensions of features such as near-field length are different in the two planes.

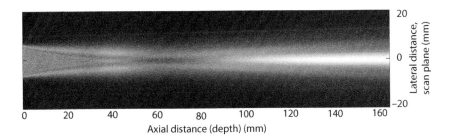

Figure 2.20 A simulation of the beam in the scan plane of an array transducer with a square aperture (15 × 15 mm). The aperture in the scan plane is made up of 24 active elements. The ultrasound wave has a frequency of 3 MHz and is pulsed rather than continuous. (Image courtesy of Dr Elly Martin, University College London, United Kingdom.)

plane, each element acts rather like the imaginary elements of the plane circular disc described earlier. When activated together to form the required aperture, their individual contributions interfere to form a beam in the plane of the array.

Figure 2.20 shows a simulation of the beam in the scan plane of an array transducer where the aperture along the array is the same size as that across the array, i.e. a square aperture. The ultrasound frequency is 3 MHz, giving a wavelength of approximately 0.5 mm. In the scan plane, 24 elements positioned at intervals of 1.3λ are activated at the same time, giving an aperture that is 15 mm wide. The ultrasound wave in this case is pulsed rather than continuous. It can be seen that the main features of the beam are similar to those for the continuous wave beam shown in Figure 2.17b, but the pattern of pressure variations near to the source is smoothed out.

FOCUSING

For imaging purposes, a narrow ultrasound beam is desirable as it allows closely spaced targets to be shown separately in the image. For a practical source of a given frequency, the aperture is chosen to give the best compromise between beam width and beam divergence.

A significant improvement to the beam width can be obtained by focusing. Here, the source is designed to produce wave fronts which are concave rather than flat, as shown in Figure 2.21a. Each part of the concave wave travels at right angles to its surface, so that the waves converge towards a point in the beam, the focus, where the beam achieves

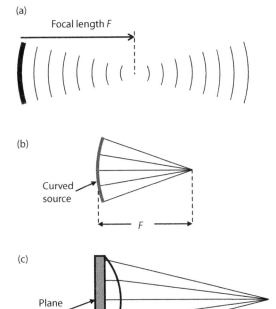

Figure 2.21 Focusing of ultrasound beams. (a) The wave fronts from a curved source converge towards a focus. (b) Focusing can be achieved by using a curved source. (c) Focusing can also be achieved by adding an acoustic lens to a plane source.

its minimum width. Beyond the focus, the waves become convex and the beam diverges again, but more rapidly than for an unfocused beam with the same aperture and frequency. The distance from the source to the focus is the focal length F.

For a single-element source, focusing can be achieved in one of two ways. These are by use of (1) a curved source or (2) an acoustic lens. The curved source (Figure 2.21b) is manufactured with a radius of curvature of F and hence produces curved wave fronts which converge at a focus F cm from the source. A flat source can be focused by attaching an acoustic lens to the front face of the source. The lens produces curved wave fronts by refraction at its outer surface in a similar manner to focusing in an optical lens (Figure 2.21c). The convex acoustic lens is made from material which has a lower speed of sound than tissue. Wave fronts from the source pass into the lens at normal incidence and are undeviated. On arrival at the interface between the lens and tissue, the increase in the speed of sound causes the direction of propagation to be deviated away from the normal and a converging wave front is formed. A cylindrical acoustic lens is used on the face of an array transducer to achieve fixed focusing at right angles to the image plane. As described in Chapter 3, focusing is achieved in the scan plane of array transducers by manipulating the timing of transmissions from individual elements in the active aperture.

To achieve effective focusing, the focus must be within the near field of the equivalent unfocused beam. The focusing effect on the beam is strongest when the focus is in the first half of the near field. Here, the waves converge rapidly to a very narrow beam width at the focus and then diverge rapidly again beyond that point. Focusing is weaker when the focus lies within the second half of the near-field length (Kossoff 1979). Commercial ultrasound imaging systems always use some degree of focusing and hence operate in the near field of the equivalent unfocused source.

For a focused beam, the beam width W_F at the focus is given by the equation:

$$W_F = k \frac{F}{D} \lambda$$

This equation shows that for a given wavelength λ, the beam width W_F is proportional to the ratio of focal length F to aperture diameter D, often referred to as the f-number. When the f-number is small (e.g. 2), the beam width at the focus W_F is small, i.e. focusing is strong. Also,

W_F decreases as λ is reduced, i.e. as frequency is increased.

For a circular source, the constant k is determined by the level at which the beam width at the focus is measured. If it is measured between points on each side of the beam at which the intensity has fallen to half of the intensity at the centre of the beam, i.e. the -3 dB level, k is equal to 1. When measured between -12 dB points, the value of k is 2, i.e. the beam width is doubled. For a rectangular source, such as an array transducer, the equation (with k = 2) gives a measure of the beam width between points on each side of the beam where the intensity falls to zero (Angelson et al. 1995).

Figure 2.22a is a focused version of the simulated beam shown in Figure 2.20. The aperture is square with 15 mm sides and the pulse frequency is 3 MHz. In this simulation, a focus has been applied at a depth of 60 mm (f-number of 4). In the region between approximately 45 and 80 mm depth, the focused beam is much narrower than the unfocused version and more intense. The equation given earlier in this section predicts that the beam width in the focal region should be 4 mm, i.e. much narrower than the unfocused beam in that region. In the simulation shown in Figure 2.22b, the focal depth is again 60 mm, but the aperture has been increased to 22.5 mm (f-number of 2.7) giving stronger focusing. It can be seen that the beam converges more rapidly towards the focus and diverges more rapidly afterwards. The focal region is shorter than in the previous case and the beam is narrower in this region. The predicted beam width at the focus is 2.7 mm.

The simulations in Figure 2.22 show that when the f-number is low, e.g. 2.7, the focusing action is strong, i.e. there is a large reduction in beam width, but the length of the focal zone, referred to as the depth of focus, is quite short. For a higher f-number, e.g. 4, the focusing action is weaker. That is, there is less reduction in the beam width, but the depth of focus is greater. Hence, weaker focusing is associated with more moderate reductions in beam width and a longer focal zone. The depth of focus d_F can be estimated from the following equation:

$$d_F = 4\lambda \left(\frac{F}{D} \right)^2$$

(a)

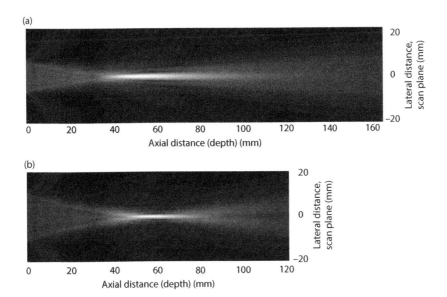

(b)

Figure 2.22 **(a)** A focused version of the simulated beam shown in Figure 2.20. The aperture is 15 mm wide and the focal distance is 60 mm (f-number of 4). The beam width in the focal region is 4 mm and the depth of focus is 32 mm. **(b)** In this simulation, the focal depth is 60 mm, but the aperture is increased to 22.5 mm (f-number of **2.7**) giving stronger focusing. The focal region is shorter (approximately 14 mm) and the beam is narrower (**2.7** mm) at the focal depth. (Images courtesy of Dr Elly Martin, University College London, United Kingdom.)

This equation shows that the depth of focus increases rapidly with increases in f-number. Doubling the f-number gives a fourfold increase in depth of focus. Depth of focus also increases with the wavelength. The equation predicts depths of focus of 32 and 14 mm, respectively, for the beams in Figure 2.22a and b.

To obtain a beam that is narrow over its whole length, ultrasound imaging systems often use several transmissions along the same beam axis, but focused to different depths to obtain a composite beam with an extended focal region. For array transducers, it is common to increase the aperture width in the scan plane as the focal depth is increased so that the f-number remains constant.

ULTRASOUND PULSE

As described, a B-mode image is formed from echoes produced by reflection from interfaces and scattering from smaller targets. To produce a distinct echo which corresponds to a particular interface, ultrasound must be transmitted in the form of a short burst or pulse. To allow echoes

from closely spaced interfaces to be resolved separately, the pulse must be short. A typical ultrasound pulse consists of a few cycles of oscillation at the nominal frequency of the wave, as illustrated in Figure 2.23. The amplitude of the

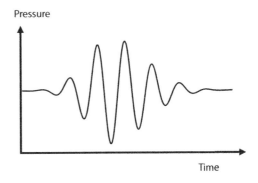

Figure 2.23 The pressure waveform of a typical ultrasound pulse consists of a few cycles of oscillation at the nominal frequency of the wave. The amplitude of the oscillation in the pulse increases at the leading edge, reaches a peak and then decreases more slowly in the trailing edge.

oscillation in the pulse increases at the leading edge, reaches a peak, and then decreases more slowly in the trailing edge.

Frequency and bandwidth

When a source of ultrasound is excited continuously to produce a continuous wave whose pressure varies as a pure sine wave, the wave has a specific frequency. A graph of wave amplitude against frequency would show a single value at that frequency and zero at all other frequencies as in Figure 2.24a. A continuous wave, which does not have the form of a pure sine wave, contains other frequencies in addition to the main frequency of the wave. For a continuous wave, these other frequencies are multiples of the main frequency known as harmonics. For example, the sound of a continuous note on a violin is closer to a sawtooth wave than a sine

wave and contains a series of harmonics in addition to the fundamental frequency as shown in Figure 2.24b.

Figure 2.25 shows how such a sawtooth wave can be constructed from a range of harmonics of the fundamental frequency. The upper wave is a pure sine wave of frequency f. The series of sine waves shown beneath are harmonics at frequencies of $2f$, $3f$, $4f$ and $5f$. However, the amplitudes of the harmonics have been reduced by the same multiples. So the amplitude at frequency $2f$ is divided by 2, that at frequency $3f$ is divided by 3 and so on. When the series is added together, the result is a waveform which is approximately triangular in

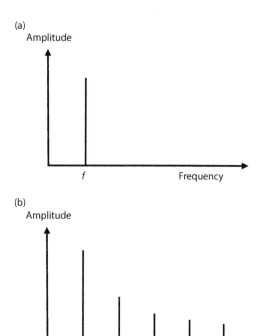

Figure 2.24 **(a)** For a pure sine wave of frequency f, a graph of amplitude against frequency shows a single value of amplitude at f. **(b)** A sound wave from a violin is shaped like a sawtooth and contains a series of harmonic frequencies in addition to the fundamental frequency.

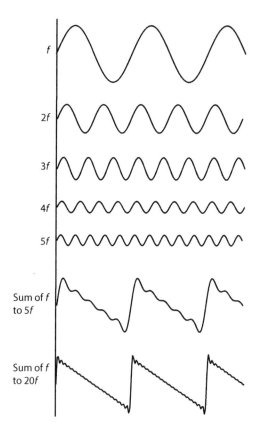

Figure 2.25 A continuous sawtooth waveform can be constructed by adding together a series of sine waves which are multiples or harmonics of the fundamental frequency f. The amplitude of each harmonic is reduced by the same number used to multiply the fundamental frequency. A sharper sawtooth is obtained by extending the series to higher harmonics.

form. The set of harmonics is known as a Fourier series. By extending the series to a frequency of $20f$, it can be seen that a much sharper sawtooth wave is obtained. Hence, to produce a continuous waveform which contains sharp transitions, it is necessary to include a wide range of harmonic frequencies. The high-frequency components are required to produce the sharp edges of the sawtooth.

PULSE SPECTRUM

The ultrasound pulse shown in Figure 2.23 is not a continuous sine wave. Its amplitude changes during the pulse and it lasts for only a few cycles of oscillation. From the foregoing discussion, it must contain frequencies other than the nominal frequency of the pulse. However, as the pulse waveform is not continuous, the other frequencies are not all multiples of the nominal frequency. A graph of amplitude against frequency for such a pulse shows a continuous range of frequencies, rather than discrete harmonics, as shown in Figure 2.26. However, as for the continuous sawtooth wave, rapid changes in the shape of the pulse require a wide range of frequency components. A short pulse

contains a wide range or spectrum of frequencies, whereas a long pulse contains a relatively narrow spectrum of frequencies. The range of frequencies is referred to as the bandwidth (BW) of the pulse. The bandwidth is related to the pulse duration T by the simple equation:

$$BW = \frac{1}{T}$$

So if the pulse duration is halved, the bandwidth is doubled. Figure 2.26a shows the waveform of a long 3 MHz pulse. The graph of amplitude against frequency beneath shows that the bandwidth is about 1 MHz. The 3 MHz pulse shown in Figure 2.26b is much shorter and its bandwidth is more than twice that of the long pulse. A short pulse gives precise time resolution and hence distance resolution but contains a wide spectrum of frequencies.

NON-LINEAR PROPAGATION

In the description of propagation of sound waves given earlier in this chapter, the wave propagated with a fixed speed determined by the properties of

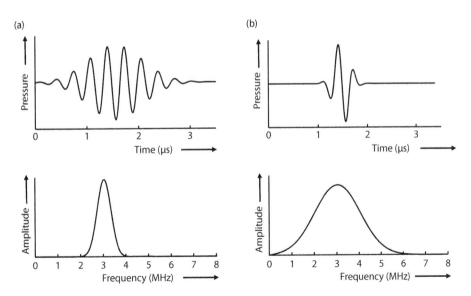

Figure 2.26 The range of frequencies contained within a pulse (its bandwidth) is determined by its duration or number of cycles of oscillation. A long pulse has a narrow bandwidth. A pulse which is half as long has twice the bandwidth. **(a)** A long pulse, i.e. one containing many cycles of oscillation, has a relatively narrow bandwidth. **(b)** A short pulse, containing few cycles of oscillation has a wider bandwidth.

the medium. It was also assumed that there was a linear relationship between the amplitude of the wave at the source and the amplitude elsewhere in the beam. Adding two waves with the same amplitude and phase resulted in a wave with twice the amplitude. This description of linear propagation is a good approximation to reality when the amplitude of the wave is small.

At high-pressure amplitudes (>1 MPa), this simple picture breaks down and non-linear propagation effects become noticeable (Duck 2002; Humphrey 2000). The speed at which each part of the wave travels is related to the properties of the medium and to the local particle velocity, which enhances or reduces the local speed. At high-pressure amplitudes, the medium becomes compressed, resulting in an increase in its stiffness and hence an increase in the speed of sound. In addition, the effect of particle velocity becomes significant. In the high-pressure (compression) parts of the wave, particle motion is in the direction of propagation, resulting in a slight increase in phase speed, whereas in the low-pressure (rarefaction) parts of the wave, motion is in the opposite direction and the phase speed is slightly reduced. As the wave propagates into the medium, the compression parts of the wave gradually catch up with the rarefaction parts. In a plane wave, the leading edges of the compression parts of the wave become steeper and may form a 'shock' front, an instantaneous increase in pressure. In diagnostic beams (Figure 2.27a shows

a real diagnostic pulse propagating in water), the compression parts become taller and narrower, while the rarefaction parts become lower in amplitude and longer (Duck 2002).

As described earlier in relation to continuous waves, abrupt changes in a waveform, such as that in a sawtooth wave, require the inclusion of harmonic frequencies in the pulse spectrum. However, as the waveform in Figure 2.27a is a pulse and not a continuous wave, the harmonics appear as repeats of the pulse spectrum around the fundamental frequency rather than the narrow spectral lines of a continuous wave, as shown in Figure 2.27b. The pulse spectrum that might be expected from an undistorted pulse is shown centred on the fundamental frequency f_0. Repeats of this spectrum appear centred at frequencies of $2f_0$ and $3f_0$. The frequency $2f_0$ is known as the second harmonic, $3f_0$ as the third harmonic and so on. Non-linear propagation results in some of the energy in the pulse being transferred from the fundamental frequency f_0 to its harmonics. As the pulse travels further into the medium, the high-frequency components are attenuated more rapidly than the low-frequency components and the pulse shape becomes more rounded again as the overall amplitude is reduced.

HARMONIC IMAGING

The changes in the pulse spectrum due to non-linear propagation are put to good use in harmonic

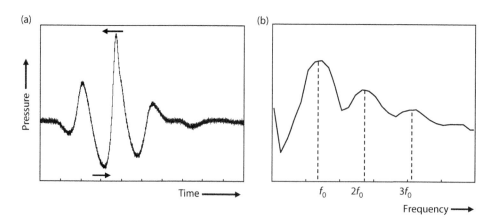

Figure 2.27 The pulse waveform and spectrum for a high-amplitude pulse. (a) At high pressure amplitudes, the pulse waveform becomes distorted due to nonlinear propagation. (b) The spectrum of the distorted pulse contains significant components at multiples of the centre frequency (harmonics).

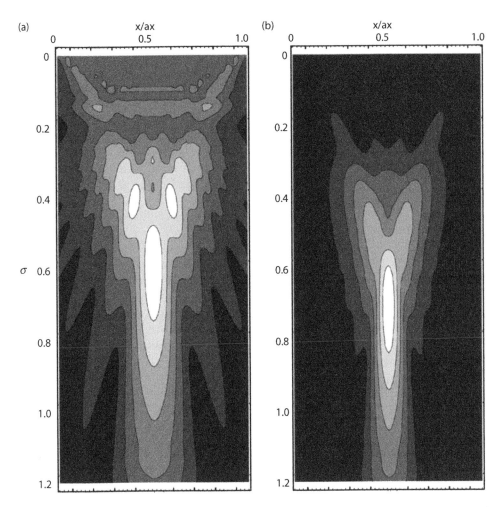

Figure 2.28 Contour beam plots for the fundamental **(a)** and second harmonic **(b)** frequencies. The lower amplitude regions of the fundamental beam, including the side lobes, are suppressed in the second harmonic beam. The resulting harmonic beam is also narrower, improving resolution. (Reprinted from *Comptes Rendus de l'Académie des Sciences-Series IV-Physics*, 2[8], Averkiou MA., Tissue harmonic ultrasonic imaging, 1139–1151, Copyright 2001, with permission from Elsevier.)

imaging, as described in Chapter 4. In harmonic imaging, a pulse is transmitted with fundamental frequency f_0 but, due to non-linear propagation the echoes returned from within the tissues contain energy at harmonic frequencies $2f_0$, $3f_0$, etc. The imaging system ignores the frequencies in the fundamental part of the spectrum and forms an image using only the second harmonic $(2f_0)$ part of the pulse (Tranquart et al. 1999; Desser et al. 2000; Averkiou 2001). The effective ultrasound beam which this produces, the harmonic beam, is narrower than the conventional beam and suppresses

artefacts such as side lobes (Figure 2.28). This is due to the fact that non-linear propagation and hence the formation of harmonics occurs most strongly in the highest-amplitude parts of the transmitted beam, i.e. near the beam axis. Weaker parts of the beam such as the side lobes and edges of the main lobe produce little harmonic energy and are suppressed in relation to the central part of the beam. Harmonic imaging can also reduce other forms of acoustic noise such as the weak echoes due to reverberations and multiple path artefacts, as described in Chapter 5.

QUESTIONS

Multiple Choice Questions

Q1. Ultrasound in clinical use has frequencies in the range:
 a. 2–50 Hz
 b. 2–50 kHz
 c. 2–50 MHz
 d. 2–50 GHz
 e. 2–50 THz

Q2. Ultrasound waves:
 a. Involve particle motion in the same direction as wave travel
 b. Involve no motion of the particles
 c. Are longitudinal waves
 d. Are shear waves
 e. Involve particle motion transverse to the direction of wave motion

Q3. The speed of sound in tissue assumed by manufacturers is:
 a. 1480 m/s
 b. 1540 m/s
 c. 1580 m/s
 d. 1620 m/s
 e. 1660 m/s

Q4. The power of an ultrasound beam is:
 a. The energy of the wave at a particular location
 b. The peak amplitude of the wave
 c. The frequency of the wave
 d. The distance between consecutive peaks
 e. The amount of beam energy transferred per unit time

Q5. Acoustic impedance is the product of:
 a. Density and speed of sound
 b. Density and wavelength
 c. Density and frequency
 d. Frequency and wavelength
 e. Frequency divided by wavelength

Q6. Reflection of ultrasound at a boundary between two tissues may occur when there is a mismatch in:
 a. Speed of sound
 b. Density
 c. Acoustic impedance
 d. Frequency
 e. Wavelength

Q7. Ultrasound gel is needed between the transducer and the patient because:
 a. It helps the patient feel relaxed
 b. If there was no gel there would be an air gap and no ultrasound would be transmitted into the patient
 c. It helps cool the transducer
 d. It helps the transducer slide over the patient's skin
 e. It ensures that the wavelength is correct

Q8. Specular reflection occurs when the object size is:
 a. The same as the frequency
 b. The same as the wavelength
 c. Greater than the acoustic impedance
 d. Much greater than the wavelength
 e. Greater than the frequency

Q9. The amplitude reflection coefficient at a typical tissue-tissue interface:
 a. Is dependent on the frequency
 b. Is less than 1%
 c. Is about 50%
 d. Results in high attenuation
 e. Is greater than for a tissue-air interface

Q10. Scattering from within tissues such as the liver:
 a. Is independent of the frequency
 b. Depends on the angle on incidence
 c. Cannot be seen on an ultrasound image
 d. Is much weaker than reflections from large interfaces
 e. Is independent of the angle of incidence

Q11. Which of the following statements is true concerning refraction?
 a. It occurs when there is a change in speed of sound between two tissues
 b. It occurs when there is no difference between the speed of sound between two tissues
 c. There is no deviation of the ultrasound beam when there is refraction
 d. There is always a 20° deviation of the ultrasound beam when there is refraction
 e. It causes image loss due to attenuation

Q12. Concerning attenuation of ultrasound waves:
 a. There is loss of energy from the beam due to attenuation
 b. There is change in the speed of sound

c. Attenuation is due only to absorption of ultrasound

d. The attenuation in soft tissues is proportional to the transmit frequency

e. The attenuation in soft tissues is inversely proportional to the transmit frequency

Q13. Diffraction is:

a. The change in direction due to changes in the speed of sound

b. The increase in amplitude of waves at the focus

c. The spread of waves from a source which is similar in extent to the wavelength

d. The change in direction when a wave encounters an object

e. The loss of energy due to absorption from the beam

Q14. If two waves overlap and are in phase, the amplitude at the point where they overlap is:

a. Larger

b. Smaller

c. The same

d. Ten times bigger

e. Minus 3 dB

Q15. If two waves overlap and are in anti-phase, the amplitude at the point where they overlap is:

a. Larger

b. Smaller

c. The same

d. Ten times bigger

e. Minus 3 dB

Q16. Which of the following is true concerning ultrasound beam formation?

a. The ultrasound source can be considered to be made of many small elements

b. The amplitude at a location x, y, z is made up of the sum of wavelets from all of the elements

c. The elements must be square

d. The wavelets always combine in an additive manner giving rise to larger amplitudes

e. The wavelets always combine in a subtractive manner giving rise to smaller amplitudes

Q17. The beam from a plane disc source consists of:

a. No main lobe or side lobes

b. Just a main lobe

c. A main lobe plus several side lobes

d. A main lobe plus two side lobes

e. Just side lobes

Q18. A plane disc source that is two wavelengths across will produce a beam:

a. That is two wavelengths wide at all depths

b. With a long near field and little divergence in the far field

c. With a short near field and little divergence in the far field

d. With a short near field and rapid divergence in the far field

e. That is 20 wavelengths long

Q19. An ultrasound beam from a single element disc source can be focused by:

a. Increasing the frequency

b. Placing an acoustic lens on the face of the transducer

c. Increasing the wavelength

d. Shaping the face of the transducer to be curved

e. Making the face of the transducer flat

Q20. Non-linear propagation is concerned with:

a. Decrease in frequency of the fundamental

b. Steepening in the shape of the ultrasound pulse

c. Increase in frequency of the fundamental

d. Generation of harmonics at multiples of the fundamental frequency

e. Scattering of ultrasound by blood

Short-Answer Questions

Q1. Explain the term *acoustic impedance* and state the tissue properties which determine its value.

Q2. What is the difference between acoustic impedance and acoustic absorption?

Q3. A medium attenuates ultrasound at a rate of 0.7 dB cm^{-1} MHz^{-1}. A target at a depth of 5 cm below the transducer is imaged with a 3 MHz ultrasound pulse. By how many decibels will the echo from the target be attenuated compared to a similar target at a depth of 1 cm?

Q4. Explain why bone and gas limit the areas of clinical application of ultrasound.

Q5. Two tissue types have speed of sound and density of (a) 1580 m s^{-1} and 1.05×10^3 kg m^{-3} and (b) 1430 m s^{-1} and 0.93×10^3 kg m^{-3}. Calculate the amplitude reflection coefficient for a large interface between them.

Q6. Explain the term *refraction* and how it might affect ultrasound images.

Q7. A plane disc transducer, with a diameter of 1.5 cm, is driven at 3 MHz to produce a continuous-wave beam in tissue with a speed of sound of 1500 m s^{-1}. Calculate the near-field length of the beam and its angle of divergence in the far field.

Q8. Explain how focusing of an ultrasound beam can be achieved.

Q9. How is the beam width in the focal region related to the focal length, aperture and wavelength? For the beam in question 7, estimate the beam width at the focus if a lens with a focal length of 6 cm is added to the transducer.

Q10. Describe the principle of harmonic imaging and explain how it can reduce some types of acoustic noise in B-mode images.

REFERENCES

Angelson BAJ, Torp H, Holm S et al. 1995. Which transducer array is best? *European Journal of Ultrasound*, 2, 151–164.

Averkiou MA. 2001. Tissue harmonic ultrasonic imaging. *Comptes Rendus de l'Académie des Sciences-Series IV-Physics*, 2(8), 1139–1151.

Desser TS, Jedrzejewicz T, Bradley C. 2000. Native tissue harmonic imaging: Basic principles and clinical applications. *Ultrasound Quarterly*, 16, 40–48.

Dickinson RJ. 1986. Reflection and scattering. In: Hill CR (Ed.), *Physical Principles of Medical Ultrasonics*. Chichester: Ellis Horwood.

Duck FA. 1990. *Physical Properties of Tissue—A Comprehensive Reference Book*. London: Academic Press.

Duck FA. 2002. Nonlinear acoustics in diagnostic ultrasound. *Ultrasound in Medicine and Biology*, 28, 1–18.

Humphrey VF. 2000. Nonlinear propagation in ultrasonic fields: Measurements, modeling and harmonic imaging. *Ultrasonics*, 38, 267–272.

Kossoff G. 1979. Analysis of focusing action of spherically curved transducers. *Ultrasound in Medicine and Biology*, 5, 359–365.

Parker KJ. 1983. Ultrasonic attenuation and absorption in liver tissue. *Ultrasound in Medicine and Biology*, 9, 363–369.

Sarvazyan AP, Hill CR. 2004. Physical chemistry of the ultrasound---tissue interaction. In: Hill CR, Bamber JC, ter Haar GR. (Eds.), *Physical Principles of Medical Ultrasonics*. Chichester: John Wiley & Sons, Ltd.

Tranquart F, Grenier N, Eder V, Pourcelot L. 1999. Clinical use of ultrasound tissue harmonic imaging. *Ultrasound in Medicine and Biology*, 25, 889–894.

Transducers and beam forming

TONY WHITTINGHAM AND KEVIN MARTIN

INTRODUCTION

The basic principles of B-mode scanning were introduced in Chapter 1. The way the beam is formed and manipulated to interrogate tissues and organs is now described in more detail.

The transducer is the device that actually converts electrical transmission pulses into ultrasonic pulses and, conversely, ultrasonic echo pulses into electrical echo signals. The simplest way to interrogate all the scan lines that make up a B-mode image is to physically move the transducer so that the beam is swept through the tissues as the pulse-echo cycle is repeated. This was the original method used, but it has been superseded by electronic scanning methods which use multi-element array transducers with no moving parts. Array transducers allow the beam to be moved instantly from one position to another along the probe face, and also allow the shape and size of the beam to be changed to suit the needs of each examination. The echo signal sequences received by individual array elements are stored temporarily to allow a range of processing methods to be applied.

The beam-former is the part of the scanner that determines the shape, size and position of the interrogating beams by processing electrical signals to and from the transducer array elements. In transmission, it generates the electrical signals that drive the active transducer elements, and in reception it combines the echo sequences received by all the active transducer elements into a single echo sequence. The beam-former allows the beam to be moved and steered in different directions to ensure that all tissue areas within the scan plane are interrogated.

The echo signals received at the transducer must also undergo other processes such as amplification and digitization before being used in the formation of the displayed B-mode image. These processes are the subject of Chapter 4. However, we begin with a description of the classic beam-forming methods used with array transducers.

WHICH BEAM DO YOU MEAN?

Before discussing transducers and beam forming further, it is helpful to consider the idea of beams a little more. In Chapter 2, the shape and size of the beam transmitted by a simple disc transducer were discussed. In most imaging techniques, ultrasound is transmitted in short pulses. The 'transmit beam' then represents the 'corridor' along which the pulses travel. The lateral extent of the pulse at a given depth is determined by the width of the beam at that depth.

It is also possible to talk of a 'receive beam', which describes the region in which a point source of ultrasound must lie if it is to produce a detectable electrical signal at the receiving transducer. In the case of a simple disc transducer, since the same transducer is used for both transmission and reception (at different times, of course), the transmission beam has an identical shape and size to the receive beam. In other words, those points in the transmission beam that have the greatest intensity will also be the points in the receive beam where a point source would produce the greatest electrical signal at the transducer. However, in the case of array transducers, the combination of transducer elements used for transmission is usually different from that used for reception, and so the two beams are different in shape and size.

Figure 3.1 shows a beam from an array transducer. This is moved electronically along the length of the transducer face, so that its axis sweeps out a flat scan plane, which defines the image cross section obtained from the patient. The plane that passes through the beam at right angles to the scan plane is the elevation plane. The shape and size of the beam in the scan plane are different from those in the elevation plane, and it is necessary to be clear which plane is being considered. The beam width in the scan plane determines the lateral resolution of the scanner, whereas that in the elevation plane defines the 'slice thickness', and hence the acoustic noise of the image and, to some extent, the sensitivity of the scanner. The transmitting and receiving apertures of array transducers are generally rectangular rather than circular. This results in the beams having rectangular cross sections close to the transducer, becoming roughly elliptical towards the focal region and beyond. These topics are discussed in later sections under the headings of the different types of transducers. Lateral resolution and slice thickness are also discussed more fully in Chapter 5.

COMMON FEATURES OF ALL TRANSDUCERS AND TRANSDUCER ELEMENTS

All transducers or transducer elements have the same basic components: a piezoelectric plate, a matching layer and a backing layer, as shown in Figure 3.2. Usually, there is also a lens, but in array transducers it is usual for one large lens to extend

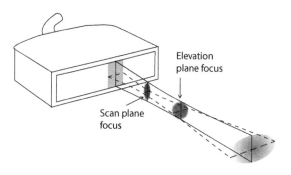

Figure 3.1 The rectangular beam aperture of an array transducer produces a beam with non-circular cross sections. In this example, the focal beam width in the vertical (elevation) direction is wider and at a greater range than that in the horizontal (scan) plane.

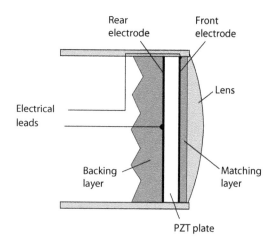

Figure 3.2 The basic component elements in an imaging ultrasound transducer.

across all the transducer elements. The number, size, shape and arrangement of transducer elements vary according to the transducer type and application, but these are addressed in detail later, as each transducer type is discussed.

Piezoelectric plate

The actual ultrasound-generating and detecting component is a thin piezoelectric plate. Piezoelectric materials expand or contract when a positive or negative electrical voltage is applied across them and, conversely, generate positive or negative voltages when compressed or stretched by an external force. Some piezoelectric materials, such as quartz, occur naturally but the piezoelectric material normally used for transducers in medical imaging is a synthetic polycrystalline (i.e. made up of many crystalline grains) ceramic material: lead zirconate titanate (PZT). Various types of PZT are available and are chosen according to the properties required, such as high sensitivity, or the ability to cope with large acoustic powers (Szabo and Lewin 2007). One advantage of PZT is that, in its original powder form, it can be formed into curved or flat shapes as desired, and then fired to make it rigid.

After firing, the way that the positively charged lead, zirconium and titanium ions and negatively charged oxygen ions are arranged in each unit cell of PZT means that each cell contains a net positive charge and a net negative charge of equal magnitude a small distance apart. A pair of opposite but equal electric charges a fixed distance apart like this is known as an electric dipole. The orientation of this dipole is constrained to lie in one of three mutually perpendicular directions determined by the cell structure. Each crystalline grain of PZT is divided into a number of regions, known as domains, each of which is polarized, i.e. all the cell dipoles within a domain are oriented in the same direction. The orientations of the cell dipoles in adjacent domains are at $\pm 90°$ or $\pm 180°$ to this direction, as indicated in diagram (a) of Figure 3.3. At this stage, the polycrystalline PZT plate as a whole has no net polarization because of the diversity of dipole directions within each grain.

In order to make the PZT plate behave as a piezoelectric, it must be polarized. This is achieved by heating it to above 200°C and applying a strong electric field across it, a process known as 'poling'. This causes the dipoles within each domain to switch to the allowed direction that is closest to the direction of the poling electric field. Once cooled, nearly all the dipoles in a given grain are 'frozen' in their new orientations, giving the whole PZT plate a net polarization that is parallel to the poling field, as shown in diagram (b) of Figure 3.3.

After being polarized, the PZT plate acts as a piezoelectric. If opposite faces of the plate are

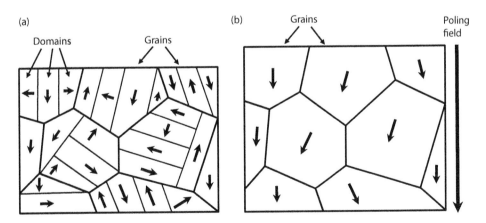

Figure 3.3 **(a)** Before poling the PZT wafer, the polarizations of the domains are mixed within each randomly orientated grain, resulting in no net polarization and, hence, no piezoelectric effect. **(b)** Poling results in uniform polarization within, and partial alignment of, the domains, leaving a net polarization and piezoelectric effect.

squeezed together, all the positive charges throughout the plate are moved slightly closer to one face and all the negative charges are moved slightly closer to the other. This creates a positive charge at one face and a negative charge at the other, manifested as a voltage difference across the plate. Mechanically pulling opposite faces apart leads to all the internal positive and negative charges moving in the opposite directions, resulting in a voltage difference of the opposite polarity. If, instead, a voltage, having the same polarity as the dipoles, is applied across the plate, the opposite charges within every unit cell are pushed towards each other, reducing the thickness of every cell and of the crystal as a whole. If a voltage of the opposite polarity is applied across the plate, the opposite charges in every cell will be pulled apart, increasing the thickness of every cell and of the plate as a whole.

A modified form of PZT, known as 'composite PZT' is commonly used to give improved bandwidth and sensitivity. This material is made by cutting closely spaced narrow channels (kerfs) through a solid plate of PZT and filling them with an inert polymer (Figure 3.4). The resulting PZT columns each vibrate more efficiently along their length than does a slab of PZT. Also, the lower density of the polymer lowers the overall density of the plate, and

hence its acoustic impedance. Both of these effects make the composite PZT plate more efficient at converting electrical into mechanical energy, and vice versa. As explained in the section 'Developments in Transducer Technology', this improves the sensitivity and bandwidth of the transducer.

Note that, as each grain in the PZT (or compound PZT) plate is orientated differently, some of the grain dipoles are aligned more parallel to the poling direction than others. This incomplete alignment of all the dipoles limits the piezoelectric performance of the polycrystalline PZT, a shortcoming that is addressed by the recent development of single-crystal PZT (see 'Developments in Transducer Technology').

The thin plate of PZT (or compound PZT) has conductive layers applied to both faces, forming electrodes, to which electrical connections are bonded. In order to transmit an ultrasonic pulse, a corresponding oscillating voltage is applied to the electrodes, making the PZT plate expand and contract at the required frequency (Figure 3.5). The back-and-forth movements of the front face send an ultrasonic wave into the patient's tissues. In reception, the pressure variations of returning echoes cause the plate to contract and expand. The voltage generated at the electrodes is directly proportional to the pressure variations, giving an electrical version (the echo signal) of the ultrasonic echo.

The thickness of the plate is chosen to be half a wavelength at the intended centre frequency of the transmit pulse as this will make it expand and contract most strongly when an alternating voltage at this frequency is applied. This strong vibration at a particular frequency, known as 'half-wave

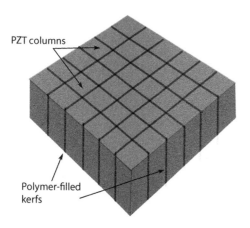

Figure 3.4 Composite PZT is made by cutting closely spaced narrow channels (kerfs) through a solid plate of PZT and filling them with an inert polymer. The resulting columns of PZT vibrate more strongly along their length than does a slab of PZT across its thickness. Also, the lower density of the polymer lowers the overall density of the plate, and hence its acoustic impedance.

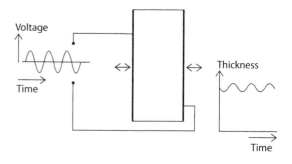

Figure 3.5 Changes in voltage across a PZT plate produce corresponding changes in thickness.

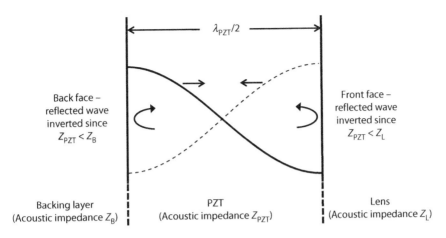

Figure 3.6 Half-wave resonance occurs in a plate of PZT. The wave reverberates back and forth between the front and back faces and travels one wavelength on each round trip, returning in phase with the original wave.

resonance', occurs because a wave reverberating back and forth across a plate of this thickness will have travelled exactly one wavelength when it arrives back at its starting position (Figure 3.6). All of the reverberations will, therefore, be in phase and will sum together to produce a wave of large amplitude. Note that as the acoustic impedance outside of the plate is less than that of PZT, the reverberating wave is inverted as it is reflected at the front and back faces (see 'Reflection' in Chapter 2). Also note that the wavelength here is calculated using the speed with which sound propagates between the two flat faces of the PZT plate. In array probes, discussed later, the plate is made up of PZT elements separated by narrow grooves filled with inert material (kerfs) having a lower speed of sound. Consequently the average speed across such a plate is lower than that in PZT.

Backing layer

PZT has advantages as a transducer material in that it is efficient at converting electrical energy to mechanical energy, and vice versa, and is relatively easy to machine or mould to any required shape or size. However, it has a significant disadvantage in that, even in the compound PZT form, it has a characteristic acoustic impedance that is many times higher than that of the soft tissue. If the front face of the PZT plate were to be in direct contact with the patient, a large fraction of the ultrasound

wave's power would be reflected at the PZT–tissue interface (see 'Reflection' in Chapter 2). If nothing were to be done about this, the resonance within the PZT plate, as previously referred to, would be very marked and vibrations would continue long after the applied driving voltage had finished. Such 'ringing', as it is called, would result in an overly long pulse. As explained in Chapter 2, this is another way of saying that the transducer would have a narrow bandwidth.

This unwanted 'ringing' can be much reduced by having a backing (damping) layer behind the PZT, made of a material with both high characteristic acoustic impedance and the ability to absorb ultrasound. If the impedance of the backing layer were identical to that of PZT, all the sound energy would cross the boundary between the PZT and the backing layer, and none would be reflected back into the PZT. Once in the backing layer the sound would be completely absorbed and converted into heat. This would eliminate ringing, but would be at the expense of sensitivity, as some of the energy of the electrical driving pulse and also of the returning echo sound pulses would be wasted as heat in the backing layer. In modern practice, a backing layer with impedance somewhat lower than that of the PZT is used. This compromise impedance is chosen to give a useful reduction in ringing without lowering sensitivity too much. The remaining ringing is removed by using matching layers, as discussed next.

Matching layer(s)

Apart from the ringing problem, the fact that only a small fraction of the wave's power would be transmitted through the front PZT-patient interface means there is also a potential problem of poor sensitivity. In order to overcome this, at least one 'impedance matching layer' is bonded to the front face of the PZT. A single matching layer can increase the transmission across the front face to 100%, provided that two important conditions are met. First, the matching layer should have a thickness equal to a quarter of a wavelength. Second, it should have an impedance equal to $\sqrt{(z_{PZT} \times z_T)}$, where z_{PZT} is the impedance of PZT and z_T is the impedance of tissue. The explanation behind this remarkable achievement is that the sound reverberates back and forth repeatedly within the matching layer, transmitting a series of overlapping pulses into the patient that align peak to peak with each other and hence sum, as illustrated in Figure 3.7, to give a large-amplitude output. At the same time, another series of pulses, also aligned peak to peak with each other, is sent back into the PZT, cancelling out the wave originally reflected at the PZT–matching layer interface. The reason

why there is peak-to-peak alignment within both these series of pulses after a round trip of just half a wavelength across the matching layer and back is that the pulses are inverted on reflection from the matching layer–patient interface, since the acoustic impedance changes from higher to lower at this interface (see the discussion of reflection in Chapter 2).

The 100% transmission through the matching layer only occurs at one frequency, for which its thickness is exactly one-quarter of a wavelength. It is normal to choose the matching layer thickness to satisfy this condition for the centre frequency of the pulse, as there is more energy at this frequency than at any other. Smaller, but nevertheless worthwhile, improvements in transmission will occur at frequencies close to the centre frequency. However, at frequencies well removed from the centre frequency the matching layer will not be very effective. A −3 dB transducer bandwidth (Figure 3.8) can be defined, in a similar way to the pulse bandwidth, described in Chapter 2. For a transducer, it is the range of frequencies over which its efficiency, as a converter of electrical energy to sound energy or vice versa, is more than half its maximum. It is evident that, although the matching layer improves sensitivity at the centre frequency, it also acts as a frequency

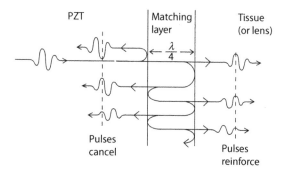

Figure 3.7 Quarter-wave matching layer. Reverberations within the plate produce multiple transmissions into the patient that reinforce each other to give a large-amplitude resultant pulse. The resultant of the multiple reflections back into the PZT cancel out the original (top) reflection back into the backing layer. Note that the width of the matching layer and the pulse sequences transmitted and reflected are not drawn here to the same distance scale. In practice the wavelengths in the PZT, matching layer and tissue (or lens) would all be different.

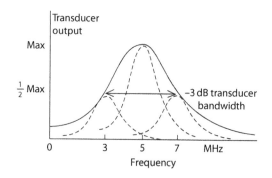

Figure 3.8 The −3 dB transducer bandwidth is the range of frequencies over which the output power for a given applied peak-to-peak voltage is more than half of the maximum. A multi-frequency probe must have a large bandwidth, so that it can transmit and receive pulses with several different centre frequencies, as indicated by the dashed pulse spectra within the transducer frequency response curve.

filter, reducing the bandwidth. For a transducer with a quarter-wave matching layer, a −3 dB bandwidth of about 60% of the centre frequency can be achieved. Thus, a 3 MHz transducer with a quarter-wave matching layer could have a −3 dB bandwidth of up to 1.8 MHz.

A large transducer bandwidth is crucial to good axial resolution, since the latter depends on a large pulse bandwidth (consistent with a short pulse, for example) and the pulse bandwidth cannot be more than that of the transducer producing it. The transducer bandwidth can be increased by using two, or more, matching layers, progressively reducing in characteristic impedance from the PZT to the patient's skin. The reason why this leads to a greater bandwidth is that the impedance change, and hence the reflection coefficient, at the PZT front face is less than that for the case of a single matching layer. This applies at all frequencies, so there is less difference in the performance at the centre frequency relative to that at other frequencies. Improvements in backing layer and multiple-matching-layer technologies have meant that −3 dB bandwidths greater than 100% of centre frequency are now available.

A large transducer bandwidth is also needed for harmonic imaging (Chapter 4) and other modern developments. It is also a prerequisite of a 'multi-frequency transducer'. This type of transducer allows the operator to select one of a choice of operating frequencies according to the penetration required. Whatever centre frequency is selected, short bursts of oscillating voltage at that frequency are applied across the PZT plate to produce the ultrasound transmission pulses. At the same time, the receiving amplifiers (Chapter 4) are tuned to that frequency. In order for a single transducer to be able to operate at three frequencies, say 3, 5 and 7 MHz, it would need to have a centre frequency of 5 MHz and a bandwidth of 4 MHz, which is 80% of the centre frequency (Figure 3.8). Note that the bandwidths of the pulses generated at the upper and lower frequencies must be less than that of the transducer itself – this means the axial resolution to be expected from a probe in multi-frequency mode will be less than that which would be possible if the whole transducer bandwidth were used to generate a full 5 MHz bandwidth pulse.

Lens

A lens is usually incorporated after the matching layer. If there is no electronic focusing, as for example in a single-element probe, the width of the beam is least and the receive sensitivity and transmission amplitude are greatest near the focus of the lens. In linear-array probes, focusing in the scan plane is achieved entirely by electronic means, and so a cylindrical lens, producing focusing only in the plane perpendicular to the scan plane (elevation plane), is used. In phased-array transducers (see later) the lens may have some curvature (focusing action) in the scan plane as well as in the elevation plane, in order to augment the electronic focusing in the scan plane.

DEVELOPMENTS IN TRANSDUCER TECHNOLOGY

The performance of the transducer in terms of its efficiency and bandwidth is critical to the overall performance of the ultrasound system. Transducer performance has been improved in recent years through the development of new piezoelectric materials and a new type of transducer known as a micro-machined ultrasonic transducer (MUT).

Single-crystal piezoelectric transducers

As mentioned when discussing the piezoelectric plate, one of the limitations of the ceramic piezoelectric material PZT is that during polarization, the granular structure limits the alignment of piezoelectric domains that can be achieved and hence the piezoelectric efficiency of the device (Figure 3.9a). Transducers are now commercially available which use alternative piezoelectric materials, grown as single crystals (Chen et al. 2006; Zhou et al. 2014). These include lead titanate (PT) doped with various other elements, such as lead, magnesium and niobium (PMN–PT) or lead, zinc and niobium (PZN–PT). The single crystal is produced by melting the constituent materials and drawing out a seed crystal very slowly under careful temperature control. The resultant crystal does not have any grain structure, so that when it is sliced into multiple wafers, these can be poled to have almost uniform

(a)

(b)

Grains

Poling
field

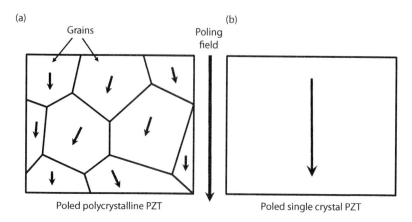

Poled polycrystalline PZT

Poled single crystal PZT

Figure 3.9 **(a)** The granular structure of polycrystalline PZT limits the alignment of piezoelectric domains that can be achieved during poling and hence the piezoelectric efficiency of the device. **(b)** The single crystal material has no grain boundaries and can be poled to give near perfect alignment of the dipoles, giving much greater transducer efficiency.

polarization, i.e. all the electric dipoles in line with the poling electric field (Figure 3.9b). These single-crystal wafers are much more efficient at converting the electrical energy supplied to the transducer into mechanical energy of vibration, and have a much larger response in terms of thickness change for a given applied voltage. They also lead to transducers with wider bandwidths since their greater efficiency means that they retain less unconverted energy and therefore ring for less time. This greater bandwidth improves axial resolution as well as being necessary for multi-frequency operation and harmonic imaging (see Chapter 4).

Micro-machined ultrasonic transducers

MUTs employ different principles of operation from those used for the PZT transducers described so far. Compared to conventional PZT transducers, MUTs have much better acoustic matching to tissue, obviating the need for a matching layer and giving up to 130% fractional bandwidth (Caronti et al. 2006). The wider bandwidth enables the use of short ultrasound pulses, giving improved axial resolution. They are fabricated using the well-established techniques employed in the manufacture of electronic integrated circuits (microchips), enabling reductions in production costs and close integration with electronics on the same silicon substrate (Khuri-Yakub

and Oralkan 2011; Qiu et al. 2015). The basic cells are much smaller than standard piezoelectric transducer elements, being typically of the order of a few 10 s to several hundred microns across (1 μm = 0.001 mm). To form a single element of an array transducer, many MUT cells are connected together in parallel (Figure 3.10).

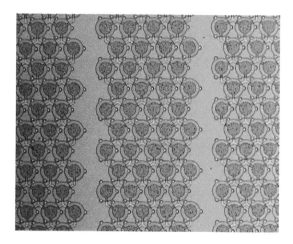

Figure 3.10 A portion of a 64-element one-dimensional CMUT array showing how individual CMUT cells are aggregated to form the required transducer element area. (Reprinted from *Microelectronics Journal*, 37, Caronti A, Caliano G, Carotenuto R et al., Capacitive micromachined ultrasonic transducer (CMUT) arrays for medical imaging, 770–777, Copyright 2006, with permission from Elsevier.)

The two principal types of MUT devices are the capacitive micro-machined ultrasonic transducer (CMUT) and the piezoelectric micro-machined ultrasonic transducer (PMUT). Although the basic CMUT and PMUT units have some similarities in construction, their principles of operation are fundamentally different.

The basic CMUT element is similar in structure to a small drum, with a fixed base above which is a thin flexible membrane separated by a vacuum gap (Figure 3.11). The membrane and the fixed base contain electrodes which form a capacitor. When a steady bias voltage is applied between the electrodes, electrostatic attraction causes the flexible membrane to be drawn towards the base. The attraction is opposed by the elastic restoring force in the membrane. An alternating voltage applied between the electrodes causes the membrane to vibrate. This mechanism can be used to generate an ultrasound wave in a medium in acoustic contact with the membrane. In reception, movement of the membrane due to the pressure changes of the incoming wave causes changes in the capacitance of the device. Under a steady bias voltage, the changes in capacitance give rise to a corresponding electrical current into the device, whose amplitude is related to the bias voltage and the frequency of the wave. This current may be detected and processed as an ultrasound echo signal.

For efficient operation in transmission and reception, a steady bias voltage must be applied between the electrodes of the CMUT, in addition to the oscillating drive voltage. Increasing the bias voltage causes a reduction in the gap between the membrane and the base. Increasing the bias voltage beyond a limit known as the collapse voltage causes the membrane to make contact with the base, resulting in electrical breakdown or mechanical damage to the device. Unfortunately, the greatest degree of membrane movement for a given change in voltage is obtained when the membrane is very close to the base. That is, the greatest efficiency is obtained by operating the device near the collapse voltage (Caronti et al. 2006). One successful commercial realisation protects against this occurrence by employing a small structure beneath the membrane (Otake et al. 2017). This allows the device to be operated very close to the collapse voltage and thus achieve a wide amplitude of membrane movement.

The basic PMUT element also consists of a small drum with a membrane suspended over a narrow vacuum gap. However, the membrane in this case consists of a thin plate of piezoelectric material with electrodes on each surface bonded to a passive, elastic layer (Figure 3.12). Application of a voltage between the electrodes causes the thickness of the piezoelectric plate to change but this, in turn, causes the lateral dimension of the plate to change in the opposite sense. For example, an expansion of the plate thickness causes a contraction of its lateral dimension. The differential expansion between the piezoelectric layer and the passive layer causes the membrane to bend, producing a vertical deflection. In transmission, application of an alternating voltage leads to vibration of the membrane. In reception, the deflection of the membrane due to the incoming pressure wave causes lateral changes in dimension which produce an electrical signal at the electrodes.

An operating bandwidth of 2–22 MHz has been achieved in a single commercially available

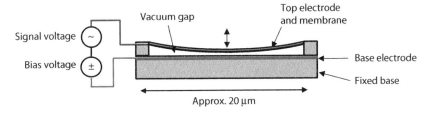

Figure 3.11 The basic CMUT element is like a small drum, with a fixed base above which is a thin flexible membrane separated by a vacuum gap. If an alternating voltage is applied between electrodes in the membrane and in the fixed base, electrostatic attraction between the two causes the membrane to vibrate. Individual CMUT cells are typically about 20 μm across.

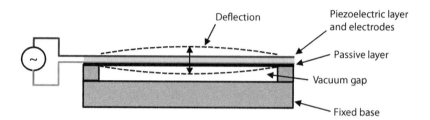

Figure 3.12 The basic PMUT element contains a thin piezoelectric plate with electrodes on each surface, bonded to a passive, elastic layer. Application of a voltage between the electrodes causes changes in the lateral dimension of the electrode causing the plate to bend. Application of an alternating voltage leads to vibration of the membrane. In reception, the deflection of the membrane due to the incoming pressure wave causes lateral changes in dimension which produce an electrical signal at the electrodes.

CMUT transducer (Tanaka 2017). CMUT transducers have much potential for use in harmonic imaging and high-frequency small-scale transducers for intravascular imaging. There is also the potential for the manufacture of two-dimensional (2D) matrix-array transducers for three-dimensional (3D) volumetric imaging using this technology (Khuri-Yakub and Oralkan 2011). Small prototype PMUT arrays have been demonstrated in this application. Dausch et al. (2014) constructed 5 MHz rectangular arrays with 256 and 512 elements. These were integrated into an intra-cardiac catheter for *in vivo* real-time 3D imaging.

LINEAR-ARRAY AND CURVILINEAR-ARRAY TRANSDUCERS (BEAM-STEPPING ARRAYS)

A common type of transducer is the linear array, or its curved version – the curvilinear array. Linear arrays offer a rectangular field of view (Figure 3.13) that maintains its width close up to the transducer

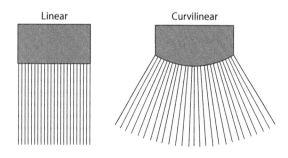

Figure 3.13 Linear and curvilinear scan formats.

face and are therefore particularly suitable when the region of interest extends right up to the surface (e.g. neck or limbs). Curvilinear arrays work in the same way as linear arrays, but differ in that the array of elements along the front face forms an arc, rather than a straight line. They share the same benefit of a wide field of view at the surface, but have the additional advantage that the field of view becomes wider with depth. They are therefore popular for abdominal applications, including obstetrics. However, in order to maintain full contact, it is necessary to press the convex front face slightly into the patient. This makes the linear array more suitable than the curvilinear array for applications where superficial structures, such as arteries or veins, should not be deformed, or where the skin is sensitive. An answer to this problem is offered by trapezoidal (virtual curvilinear) arrays, discussed later.

From the outside, a linear-array transducer appears as a plastic block designed to fit comfortably into the operator's hand with a rubber lens along the face that makes contact with the patient (Figure 3.14). Behind the lens is a matching layer, and behind this is a linear array of typically 128 regularly spaced, narrow, rectangular transducer elements, separated by narrow barriers (kerfs), made of an inert material, usually a polymer or epoxy. Some linear-array transducers have as many as 256 elements, but cost considerations and fabrication difficulties mean that 128 is a more common number. Note that these numbers are chosen, rather than say 200 or 100, since they are 'round' numbers in binary terms, and hence more convenient for digital control and processing.

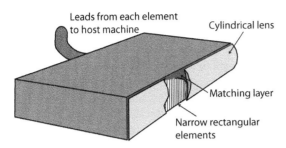

Figure 3.14 Cut-away view of a linear-array transducer, showing the elements, matching layer and lens.

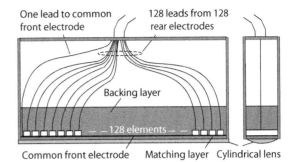

Figure 3.15 Section through a linear-array transducer. For clarity, the sub-dicing of each element is not shown.

The width of each array element is typically about 1.3 wavelengths, being a compromise that gives a reasonably wide array of say about 85 mm (= 128 × 1.3 × 0.5 mm) at 3 MHz, and hence a usefully wide field of view, while still allowing the elements to be narrow enough to radiate over a wide range of angles in the scan plane (Chapter 2). The longer side of each element determines the width of the beam in the elevation direction, and has a typical value of about 30 wavelengths. This means the weakly focused cylindrical lens gives a reasonably narrow elevation beam width at all depths. Since all transducer dimensions are proportional to wavelength, high-frequency transducers are smaller than low-frequency ones. Thus, assuming 128 elements, a 3 MHz (wavelength = 0.5 mm) transducer might typically have a lens face measuring about 85 by 15 mm, with each element being about 0.65 mm wide, whereas a 7.5 MHz (wavelength = 0.2 mm) transducer will have a lens face measuring about 35 by 6 mm, with each element being about 0.25 mm wide.

The front electrodes of all the elements are usually connected together, so they share a common electrical lead. However, the rear electrode of each element is provided with a separate electrical lead (Figure 3.15), allowing the signals to and from each element to be individually processed by the beam-former. In practice, each element is usually further 'sub-diced' into two or three even narrower elements. This is done because otherwise each element would be approximately as wide as it is thick and an undesirable resonant vibration across the element width would accompany, and take energy from, the desired thickness vibration.

Note that although the typical element width of 1.3 wavelengths might seem very different to the element thickness of 0.5 wavelengths, the former is for a wave in tissue, while the latter is for a wave in PZT. Since the speed of sound, and hence the wavelength, in PZT is two to three times higher than it is in tissue, the two dimensions are, in fact, similar. The mechanical sub-dicing does not affect the number of electrically addressable elements, since the rear electrodes of the two or three sub-diced elements making up the original element are connected together and share a single lead.

Active group of elements

In order to interrogate a particular scan line, an 'active group' of adjacent transducer elements, centred on the required scan line, is used. While that scan line is being interrogated, all the other elements in the probe are disconnected and idle. (Note that this may not be case the when line multiplexing is used, see 'Time-Saving Techniques for Array Transducers' later.) First, a pulse is transmitted, using say the central 20 elements of the group. This pulse travels along the transmit beam, centred on the scan line. As soon as the pulse has been transmitted, elements in a different combination, still centred on the scan line, act together as a receiving transducer, defining the receive beam. The number of elements used for reception is initially less than that used for transmission, but this number is progressively increased as echoes return from deeper and deeper targets until it eventually exceeds that used for transmission (see 'Scan Plane

Dynamic Focusing and Aperture in Reception' later). Both the transmit and receive beams can be focused, or otherwise altered, by controlling the signals to or from each of the elements in the active group, as described later.

Once all echoes have been received from one scan line, a new active group of elements, centred on the next scan line, is activated. This is achieved by dropping an element from one end of the old group and adding a new one at the other end (Figure 3.16). This advances the centre of the active group, and hence the scan line, by the width of one element. The new scan line is then interrogated by a new transmit and receive beam, centred on that line. The process is repeated until all the scan lines across the field of view have been interrogated, when a new sweep across the whole array is commenced.

Beam shape control in the scan plane

SCAN PLANE FOCUSING IN TRANSMISSION

Since the cylindrical lens does nothing to reduce the beam width in the scan plane, an electronic method of focusing must be provided if good lateral resolution is to be achieved. This is controlled by the operator, who sets the transmit focus at the depth for which optimum lateral resolution is desired. This ensures that the transmission beam is as narrow as possible there (the receive beam must also be narrow there, but this is considered next). Usually an arrowhead or other indicator alongside the image indicates the depth at which the

transmit focus has been set. Pulses from all the elements in the active group must arrive at the transmit focus simultaneously in order to concentrate the power into a narrow 'focal zone'. However, the distance between an element and the focus, which lies on the beam axis passing through the centre of the group, is slightly, but crucially, greater for the outer elements of the group than for more central elements. Pulses from elements further from the centre of the active group must, therefore, be transmitted slightly earlier than those nearer the centre (Figure 3.17). The manufacturer builds a look-up table into the machine for each possible choice of transmission focus depth available to the operator. These tell the controlling computer the appropriate 'early start' for each element. At points outside the required focal zone, the individual pulses from different elements arrive at different times, producing no more than weak acoustic noise.

SCAN PLANE DYNAMIC FOCUSING AND APERTURE IN RECEPTION

Focusing in reception means that, for each scan line at any given moment, the scanner is made particularly sensitive to echoes originating at a specified depth (the receive focus) on the scan line. This also results in the receive beam being narrowed near this focus, further improving lateral resolution. In order for the sensitivity to be high for an echo coming from the receive focus, the echo signals produced by all transducer elements in the active group must contribute simultaneously to the resultant electronic echo signal. As in the case

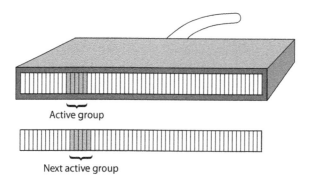

Active group

Next active group

Figure 3.16 The active group is stepped along the array by dropping an element from one end and adding a new one to the other. In reality, the active group would contain at least 20 elements rather than the five shown here.

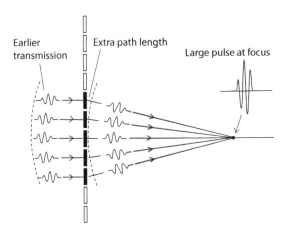

Figure 3.17 Creating a transmission focus for a linear-array transducer. In order to form a large-amplitude pulse at the focus, pulses from all elements must arrive there at the same time. This is achieved by transmitting slightly earlier from elements that are farther from the centre of the group.

of transmission focusing, allowance must be made for the fact that the distance between the required focus and a receiving element is greater for elements situated towards the outside of the group than for those near the centre. This is done by electronically delaying the electrical echo signals produced by all transducer elements except the outermost, before summing them together (Figure 3.18). The delays are chosen such that the sum of the travel time as a sound wave (from the focus to a particular

element) plus the delay imposed on the electrical echo signal is the same for all elements. This means that the imposed electronic delays are greater for elements closer to the centre of the active group, for which the sound wave travel times are least. In this way, the echo signals are all aligned in phase when they are summed and a large amplitude signal is obtained for echoes from the desired receive focus, but only a weak-summed signal (acoustic noise) results from echoes from elsewhere. This technique for focusing in reception is widely referred to as the delay and sum method.

In practice, focusing in reception is controlled automatically by the machine, with no receive focus control available for the operator. This is because the ideal depth for the receive focus at any time is the depth of origin of the echoes arriving at the transducer at that time. This is zero immediately after transmission, becoming progressively greater as echoes return from deeper and deeper targets. Since the time needed for a two-way trip increases by 13 μs for every additional 1 cm of target depth, the machine automatically advances the receive focus at the rate of 1 cm every 13 μs. The continual advancement of the receive focus to greater and greater depths gives rise to the name 'dynamic focusing in reception'. In fact, high-performance machines advance the receive focus in several hundred tiny steps (as many as one for each image pixel down a scan line), during the echo-receiving interval after each transmission. The 'effective receive beam' (Figure 3.19) consists

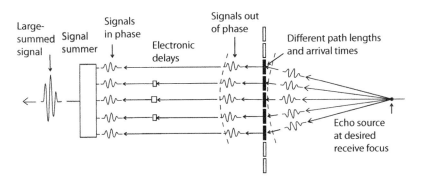

Figure 3.18 The delay and sum method for creating a receive focus for a linear-array transducer. In order to obtain a large echo signal from a target at a desired receive focus, contributions from all elements must arrive at the signal summer at the same time. This is achieved by electronic delays that are greater for elements closer to the centre of the group.

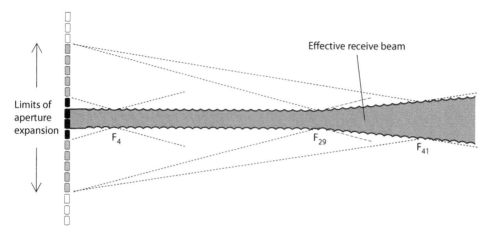

Figure 3.19 Dynamic focusing and aperture in reception. The machine automatically changes the delays so that the receive focus advances at the rate of 1 cm every 13 µs. At the same time the aperture is expanded, so that the width of the beam at all the foci remains constant (up to the 29th focus in this example). If the aperture stops expanding, the beam widths at deeper foci become progressively greater. The scalloped lines enclosing all the focal zones indicate the effective receive beam.

of a sequence of closely spaced focal zones, and is therefore narrow over a wide range of depths, not just at a single focal zone.

At the same time as the receive focus is advanced, the number of elements in the active receive group is increased. The reason for this comes from the fact (Chapter 2) that the beam width at the focus is inversely proportional to the transducer aperture. It is, therefore, desirable that the active receiving group has as many elements (as large an aperture) as possible. However, there is no benefit in using a large group when receiving echoes from superficial targets, since elements far from the centre of the group would not be able to receive echoes from them; these targets would be outside the individual receive beams of the outer elements (Figure 3.20). Such elements would be able to receive echoes from deeper targets, but they would contribute nothing but noise for echoes from close targets. Thus the maximum number of elements it is worth including in the beam increases with time after transmission, in proportion to the depth of the reception focus. This means the beam width in the successive focal zones remains fairly constant, keeping lateral resolution as uniformly good as possible at all depths.

Some machines limit the number of elements in the receiving group, on cost grounds, to a maximum of about 30. This means a constant receive beam width is only maintained up to the depth at which the aperture expansion is stopped (Figure 3.19). At greater depths, the beam becomes wider, and lateral resolution becomes noticeably worse. In some more sophisticated (expensive)

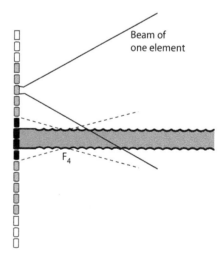

Figure 3.20 An element can only contribute usefully to the receive active group if the target lies in the individual beam of that element. Here, elements outside the four elements nearest the scan line cannot receive echoes from focal zone F_4 or nearer. The maximum useful aperture for the active group increases as the depth of the receive focus increases.

machines, however, the receiving group continues to expand until all the elements in the array are included. Such machines can maintain good lateral resolution to much greater depths.

SCAN PLANE APODIZATION

Another beam-forming process, known as 'apodization', can also be employed. In transmission, this involves exciting the elements of the active group non-uniformly in order to control the intensity profile across the beam. Usually the inner elements are excited more than the outer elements, resulting in the reduction of side lobe amplitude and the extension of the focal zone. However, as these benefits are at the expense of a broadening of the main lobe (Figure 3.21), a compromise is necessary and this is one judgement in which there is no common view among manufacturers. Apodization of the receive beam can be achieved by giving different amplifications to the signals from each element. The receive beam apodization can be changed dynamically to control side lobe characteristics as the receive focus is advanced.

SCAN PLANE MULTIPLE-ZONE FOCUSING

Further improvement in lateral resolution, albeit at the expense of frame rate, is possible by sub-dividing each scan line into two or more depth zones and

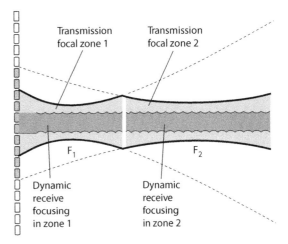

Figure 3.22 Multiple-zone focusing. The operator has selected two focal zones (F_1 and F_2). Targets lying between the transducer and a point about halfway between the two foci are interrogated with a transmission pulse focused at F_1. Targets beyond the halfway point are interrogated by transmitting another pulse along the same scan line, but focused at F_2. The heavy and light scalloped lines indicate the 'effective transmission beam', and the 'effective receive beam', respectively.

interrogating each zone with a separate transmission pulse, focused at its centre (Figure 3.22). For example, the operator might select transmission foci

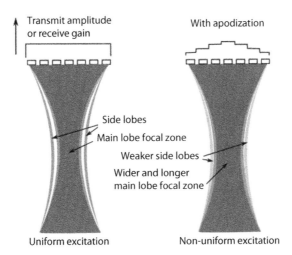

Figure 3.21 Apodization of a focused beam. By exciting outer elements less than those in the centre, side lobes can be suppressed and the focal zone extended. However, the width of the main lobe is increased. Non-uniform amplification of echoes from different elements can achieve similar changes in the receive beam.

at two different depths – F1 and F2. These would be indicated by two arrowheads or other focus indicators down the side of the image. One pulse would be transmitted with a focus at F1, and echoes from depths up to about halfway between F1 and F2 would be captured. Then a second pulse would be transmitted with a focus at F2, and echoes from all greater depths would be captured. The greater the number of transmission focal zones, the greater the depth range over which the 'effective transmission beam' is narrow. Unfortunately, the greater the number of focal zones, the longer is spent on each scan line, and so the lower the frame rate.

When using multiple-transmission focal zones, other transmission parameters such as centre frequency, pulse length and shape, aperture and apodization may all be optimized independently for each of the focal zones. These changes can take account of the fact that pulses sent out to interrogate deeper regions will experience greater attenuation of the high frequencies in their spectra.

Grating lobes

Grating lobes can occur with any transducer having regularly spaced elements, such as a linear or curvilinear array or a phased array (discussed later). They are weak replicas of the main beam,

at substantial angles (up to 90°) on each side of it (Figure 3.23a). They are named 'grating lobes' after the analogous phenomenon that occurs when a light beam passes through a grating of closely and regularly spaced narrow slits (a diffraction grating) to form a series of new deflected beams on each side of the original beam. Grating lobes contribute spurious echoes (acoustic noise) and effectively widen the beam in the scan plane, degrading both lateral resolution and contrast resolution.

Let us initially consider an unfocused transducer array, with an echo arriving from a distant target at an angle to the main beam. The pulse will reach the various transducer elements at slightly different times. As the angle considered increases, the difference in arrival times at a pair of adjacent elements will increase. If the distance between adjacent elements is large enough, an angle will exist for which this time difference is a full wave period (Figure 3.23b). Thus when, say, the first peak in the pulse is arriving at the more distant element of the pair, the second peak in the pulse will be arriving at the nearer element. When summed in the beamformer, the coincidence of these two peaks (and of other peaks and troughs in the pulse) will lead to the electrical pulses from the two transducer elements reinforcing each other. The regular element spacing means the same thing happens for every pair of

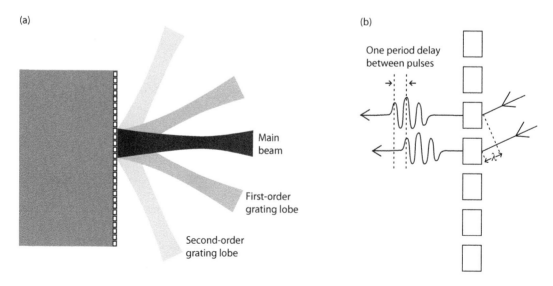

(a)

Main beam

First-order grating lobe

Second-order grating lobe

(b)

One period delay between pulses

Figure 3.23 **(a)** Grating lobes are weak replicas of the main beam, at angles of up to 90°. The greater the angle from the straight ahead direction, the weaker is the grating lobe. **(b)** The first grating lobe occurs at that angle for which the arrival times of an echo at adjacent elements differ by one period.

adjacent elements in the receive aperture, so that a large-amplitude electronic echo pulse (grating lobe signal) is produced when electrical pulses from all the elements in the active group are combined.

Depending on the spacing of the elements and the number of cycles in the pulse, second-order or even third-order grating lobe pairs may exist outside the first grating lobes. For these, the pulse arrives at one element two or three periods, respectively, ahead of that at its neighbouring element. However, for an Nth-order grating lobe to exist, there must be at least N cycles in the pulse; otherwise, there could be no constructive overlap between the signals from two elements. Clearly, the longer pulses and continuous waves used for Doppler techniques are more likely than the two to three cycle pulses used for imaging to produce such high-order grating lobes. In all cases, the greater the angle at which a grating lobe occurs, the weaker it will be, since each element is less efficient at transmitting or receiving sound waves in directions at large angles to the straight ahead direction.

The explanation provided has been for reception of echoes, but similar arguments apply in transmission by considering the arrival of pulses from pairs of adjacent elements at a distant point in the scan plane. They also apply to a focused array transducer when reception from or transmission to a target in the focal plane is considered. This is because the focusing delays applied to the signals from and to each transducer element compensate for the different distances between the different elements and the target.

The smaller the centre-to-centre distance between elements, the larger is the angle needed to produce the difference of one period needed for the first grating lobe. If this distance is less than half a wavelength, even a pulse arriving at an angle of 90° would produce a time difference less than half a period. This would mean there could be no overlap between the first peak in the electrical pulse from one element and the second peak in the electrical pulse from the nearer adjacent element. Consequently, there can be no grating lobes, if the centre-to-centre distance between elements is half a wavelength or less.

Applying this rule to clinical linear-array probes with 128 or so elements, grating lobes are always to be expected, since the centre-to-centre distance between elements is then typically about 1.3 wavelengths. For the relatively few probes with 256 elements, and thus a centre-to-centre distance of about 0.65 wavelengths, grating lobes will still occur but will lie at much greater angles to the intended beam. This results in much weaker grating lobes, due to the fall-off in transmission and reception efficiency with angle, as previously explained.

Slice thickness

Since the transmit and receive beams have a non-zero width in the elevation direction, echoes may be received from targets situated close to, but not actually in, the intended scan plane. Such echoes will contribute acoustical noise and will, therefore, tend to limit penetration and contrast resolution. In effect the image is the result of interrogating a slice of tissue, rather than a 2D plane. At any particular depth, the thickness of the slice is equal to the width of the beam in the elevation direction. The slice thickness is least at the depth at which the cylindrical lens is focused, and hence this is the depth where least acoustic noise can be expected (Figure 3.1). It is also the depth at which the greatest sensitivity can be achieved. The variation of sensitivity with depth depends on the depth at which the operator sets the scan plane transmission focus, but if this coincides with the elevation focus, the beam will be at its narrowest in both dimensions and the sensitivity at that depth will be particularly high.

Slice thickness may be improved at other depths by the use of multi-row arrays. Multiple rows are created by dividing each of the 128 or so strip elements of a one-dimensional array into, typically, three or five separate sections. In the simplest implementation, the 1.25D array, the multiple rows are used simply to expand the aperture from one row to say five as target depth increases. In the 1.5D array (Figure 3.24), electronic delays are applied to the signals to and from each row, in order to reduce beam width in the elevation direction, using the electronic focusing techniques (such as transmission focusing and dynamic focusing in reception) described for the scan plane. The focal length in the elevation plane can thus be changed automatically to match the scan plane focal length. The names used (1.25D and 1.5D) are intended to distinguish transducers with slice thickness control from full

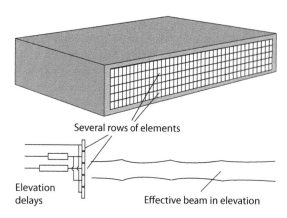

Figure 3.24 Multiple-row (1.5D) linear-array probe. Transmission and receive focusing techniques used for beam forming in the scan plane can be used to control focusing in the elevation plane. This means a narrower slice thickness and hence less acoustic noise.

2D arrays, with equal numbers of elements in both directions, which are discussed later under 3D imaging.

Strongly convex curvilinear-array transducers

Curvilinear arrays can be made with such tight curvature that their field of view becomes sector shaped (Figure 3.25). The advantages of a sector format include a small 'acoustic window' at the body surface with an increasingly wide field of

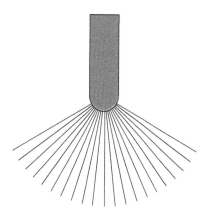

Figure 3.25 A strongly convex curvilinear array offers many of the advantages of a sector scan format for linear-array systems.

view at depth. Phased-array scanning systems (discussed next) are particularly well suited to sector scanning, but strongly convex curvilinear arrays allow manufacturers of linear-array systems to offer sector scanning transducers, without having to build in the specialized electronics that phased-array transducers require. However, the convexity of the curvilinear transducer face means that it is less suitable in situations where the flat face of a phased-array transducer is needed.

The maximum useful size of the active element group of a curvilinear array is more limited than in a linear array employing the same-sized elements. Consequently, beam width in the focal zone is greater, and lateral resolution is poorer, than in a comparable linear array. This is primarily due to the fact that, as the number of active elements is increased, the outermost elements point more and more away from the centre line of the group (scan line), until eventually they cannot transmit or receive in that direction at all. Also, the longer paths between the outer elements and a receive focus mean that the problems of providing compensating delays are more challenging.

PHASED-ARRAY TRANSDUCERS (BEAM-STEERING ARRAYS)

The phased-array transducer produces a 'sector' scan format in which the scan lines emanate in a fan-like formation from a point in the centre of the transducer face (Figure 3.26). As with all types of scanner, each scan line represents the axis of a transmit-receive beam.

The phased-array transducer is constructed in a similar way to the linear-array transducer. There are typically 128 rectangular transducer elements sharing a common lead to all front electrodes, and an individual lead to each rear electrode, as well as matching and backing layers (Figure 3.27). A lens provides fixed weak focusing in the elevation direction, and in some cases a modest degree of focusing in the scan plane to augment the electronic focusing in that plane. However, compared to a linear array, the transducer array is much shorter in the scan plane direction, with an overall aperture of typically 30 wavelengths square. The individual elements are much narrower (typically half a wavelength), and one advantage of this is

Phased-array transducers 55

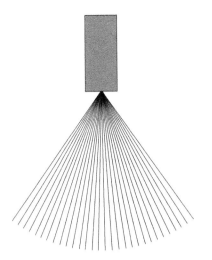

Figure 3.26 Sector scan format of a phased-array transducer.

that they do not need to be sub-diced, as is the case for the wider elements in a linear array. Unlike the linear array, which uses a different 'active group' of elements to interrogate each scan line, all the elements in the phased array are used to form the transmit and receive beams for every scan line. Since the dimension of the array and use of a fixed lens to provide focusing in the elevation direction are the same as for a linear array, the two types of transducer give similar slice thickness.

Electronic beam steering and focusing in the scan plane

Similar delay and sum techniques to those previously described for linear-array transducers are used to

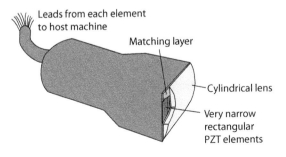

Figure 3.27 Cut-away view of the elements, matching layer and lens of a phased-array transducer.

achieve focusing of the transmit and receive beams in the scan plane. However, as well as being focused in the scan plane, the beams must also be steered by up to ±45°. The principle behind beam steering is really just an extension of that used for focusing. In fact, it follows automatically from arranging for the transmission focus and the multiple-receive foci to all lie on an oblique scan line.

SCAN PLANE FOCUSING (AND STEERING) IN TRANSMISSION

As described previously for linear-array transducers, focusing in transmission requires that pulses from all the elements arrive simultaneously at the transmission focus. The early starts needed by each element can be pre-calculated by the manufacturer for each possible position of the transmission focus along the various scan lines (Figure 3.28). The fact that the transmission focus on a particular scan line is not directly in front of the transmitting elements is of little consequence to the focusing process, provided the transmission beams of the individual elements diverge sufficiently to allow the sound from every element to reach it. The use of very narrow elements ensures this, since, as described in Chapter 2, a

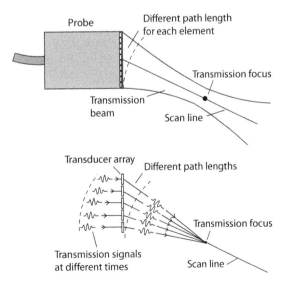

Figure 3.28 Creating a transmission focus for a phased-array transducer. The principle is the same as for a linear-array transducer, except that the focus lies on a scan line that is generally oblique.

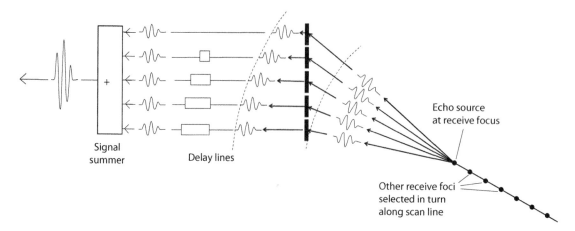

Figure 3.29 Creating a receive focus for a phased-array transducer. The principle is the same as for a linear-array transducer, except that the receive foci lie along a scan line that is generally oblique.

very narrow element will have an extremely short near field and a far field with a very large angle of divergence.

SCAN PLANE FOCUSING (AND STEERING) IN RECEPTION

Similarly, in reception, carefully pre-selected electronic delays are used to ensure that an echo from a desired receive focus takes the same time to reach the signal summer, irrespective of which element is considered (Figure 3.29). As for a linear array, dynamic focusing is used in reception, so again there is no receive focus control available to the operator. Since all the receive foci lie along a scan line that is generally at some angle to the probe's axis, it is necessary that each element can receive from ('see') any receive focus. Again, this is ensured by the very narrow width of the elements.

Other techniques for improving lateral resolution, such as apodization and multiple-zone focusing in transmission, are also used in phased-array systems. These techniques are identical to those already described for linear-array systems.

Variation in image quality across the field of view

When using a phased-array transducer, the user should always angle the probe so that any region of particular interest is in the centre of the field of view. This is where beam deflections are least and,

as discussed below, where beam widths are least and signal-to-noise ratio is highest. It will, therefore, be where the best lateral resolution, the highest sensitivity and the best contrast resolution are obtained.

DEPENDENCE OF BEAM WIDTH AND SENSITIVITY ON ANGLE

A particular artefact of phased arrays is that the width of the beam, measured at its focus, increases with increasing steering angle (Figure 3.30). Hence

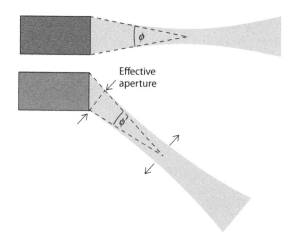

Figure 3.30 The beam from a phased-array transducer becomes wider as the angle of deflection increases. This is because the angular width of the transducer, as seen from the focus, becomes less (angle $\varphi' <$ angle φ).

lateral resolution becomes poorer towards the sides of the sector-shaped field of view. The width of a strongly focused beam at its focus is inversely proportional to the ratio of transducer aperture to focal length (Chapter 2). Another way of expressing this would be to say the beam width becomes smaller if the angular width of the transducer, as 'seen' from the focus, is large. Just as a door or window looks wider when viewed from directly in front than from somewhat to the side, so the angular width of the transducer aperture is greatest when seen from a point on a scan line at right angles to the probe face, and is less when seen from a scan line steered to a large angle.

Another problem is that, because the individual elements are most efficient when transmitting in, or receiving from, directions close to the 'straight ahead' direction, sensitivity decreases with steering angle.

GRATING LOBES

In general, the close spacing of the elements in phased-array probes tends to reduce the seriousness of grating lobes compared to those in linear arrays. In view of the 'half-wavelength' criterion given earlier when discussing grating lobes for linear arrays, it might be thought that grating lobes should be impossible for phased arrays, since their element widths are less than half a wavelength. However, wavelength here refers to that of the centre frequency of the pulse, and it should be remembered that a typical wide-bandwidth pulse will have significant energy at frequencies much higher than this. Such higher frequencies have shorter wavelengths and so may not satisfy the half-wavelength condition. Thus weak side lobes at these higher frequencies do, in fact, occur.

Furthermore, such grating lobes will grow stronger at large steering angles (Figure 3.31). As mentioned previously, the individual transducer elements are most efficient when transmitting in, or receiving from, directions close to the 'straight ahead' direction. Since beam steering deflects the grating lobes as well as the intended beam, steering the main lobe to one side will also steer the grating lobe that is following behind it more towards the straight ahead direction. This grating lobe will therefore strengthen and generate more acoustic noise, at the same time as the deflected main lobe is weakening.

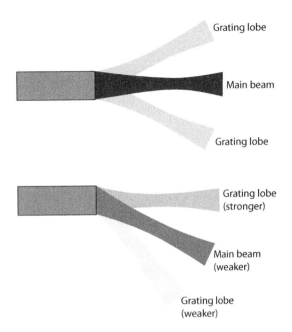

Figure 3.31 Grating lobes from a phased-array transducer are generally weak. As the deflection of the main beam increases, some grating lobes point more ahead and so become stronger. At the same time the main beam becomes weaker.

HYBRID BEAM-STEPPING/BEAM-STEERING TRANSDUCERS

Some scanning techniques involve steering the beams from a linear- or curvilinear-array transducer away from the normal straight ahead direction. One example is in a duplex linear-array system (see 'Mixed-Mode Scanning', later) where the Doppler beam is deflected to reduce the angle it makes with the direction of blood flow in a vessel lying parallel to the skin surface.

Beam steering in linear arrays is achieved using the combined focusing and steering technique that was described earlier for phased-array scanners. However, the relatively large width of transducer elements in linear-array transducers compared to those in phased arrays leads to greater problems. As the beam is progressively deflected there is greater reduction in sensitivity and greater acoustic noise from grating lobes. Despite these problems, a number of hybrid scan formats have been developed for linear arrays, in which both beam stepping and beam steering are used.

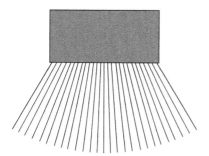

Figure 3.32 A trapezoidal scanning format is similar to that of a curvilinear transducer, but with the practical advantage of a flat transducer face.

Trapezoidal (virtual curvilinear) scanning

Some linear-array systems achieve a trapezoidal field of view by steering the scan lines situated towards the ends of the transducer progressively outwards (Figure 3.32). Such transducers provide the large field of view advantage of a curvilinear array, without the tissue compression problem that a convex front face generates.

Steered linear-array transducers

It is sometimes advantageous to be able to steer the whole field of view of a linear-array transducer to one side – to view blood vessels under the angle of the jaw, for example. A number of manufacturers, therefore, provide the option of steering all transmit and receive beams (i.e. all the scan lines) to the left or right, producing a parallelogram-shaped field of view (Figure 3.33).

Compound scanning

An extension of the steered linear-array technique is to superimpose several such angled views in a single 'compound' scan (Figure 3.34), such that the grey level of each pixel is the average of the values for that pixel in the several superimposed views. This technique, which is possible for both linear and curvilinear transducers, gives more complete delineation of the curved boundaries of anatomical features, since many such boundaries only give a strong reflection where

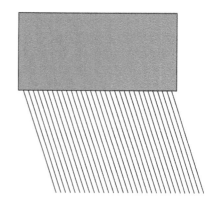

Figure 3.33 A linear-array transducer with beam steering.

the scan lines meet them at perpendicular incidence (see 'Specular Reflection' in Chapter 5). As illustrated in Figure 3.34a, the parts of a curved interface which are near horizontal are shown

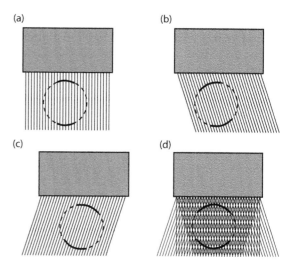

Figure 3.34 Compound scanning. (a) In an image with the beams in the straight ahead direction, boundaries that are approximately parallel to the probe face are clearly displayed as they are near perpendicular to the beams. (b and c) With the beams deflected to the right or left, boundaries that are at near perpendicular incidence to the beams are again most clearly displayed. (d) When the three images are compounded, images of the boundary from the three directions are averaged to give more extended delineation of the boundary. Compounding of the B-mode images also reduces speckle and noise.

most clearly when the beams are steered straight ahead, as incidence is near perpendicular. With the beams deflected to the right or left (Figures 3.34b and c), boundaries that are at near perpendicular incidence to the deflected beams are again more clearly delineated. When the three images are compounded (Figure 3.34d), the boundary is more completely delineated.

Compounding also results in reduced speckle (Chapter 5). The speckle pattern at any point in the image arises from the coherent addition of echoes from small scattering features within the sample volume. The deflection of the beams in compound scanning results in echoes from different sets of scatterers combining to form a different speckle pattern. When the images are compounded, the different speckle patterns at each point in the image are averaged, reducing the average grey level of the speckles and making the speckle pattern finer. Electronic noise will also tend to be reduced by the averaging inherent in the compounding process, due to its random nature. Echoes from genuine tissue structures are not reduced by the averaging process, so that the signal-to-noise ratio of the image is improved.

Compounding involves a loss of temporal resolution, since each displayed frame is the average of several sweeps. In common with the 'frame-averaging' noise-reduction technique described in Chapter 4, this introduces a degree of 'persistence' to the image. Thus, if each displayed image was the average of the previous nine different sweeps (views), and an organ cross section were to change instantly from A to B, nine frames would need to pass before all traces of the image of A were lost from the displayed image. Where necessary, a reasonable compromise in temporal resolution is possible by displaying the average of just the last three or so sweeps.

TIME-SAVING TECHNIQUES FOR ARRAY TRANSDUCERS

The scanning technique described so far is commonly referred to as 'line-by-line' scanning, since, for each image line, the transmit pulse must travel to the deepest part of the image and the echo return to the transducer before the next pulse is transmitted. The go and return time for each image line is determined by the speed of sound in the tissue and this leads to compromises between three competing qualities:

1. Temporal resolution
2. Size of the field of view
3. Image quality (e.g. lateral resolution, contrast resolution, dynamic range)

Improvements in any one of these must be at the expense of one or both of the other two (see Chapter 5). For example, a reduction in frame rate occurs if the maximum depth is increased (more time per scan line) or the width of the field of view is increased (more scan lines). Multiple-zone transmit focusing improves the range of depths over which good lateral resolution can be achieved, but this is at the expense of temporal resolution because of the need to remain longer on each scan line while interrogating several zones instead of just one. Compound scanning improves image quality through reduced speckle and acoustic noise and by better boundary delineation, but reduces temporal resolution. Some of the techniques described in Chapter 4, such as frame averaging and tissue harmonic imaging by pulse inversion, improve image quality, but are also at the expense of temporal resolution.

In order to help with these trade-offs, manufacturers have introduced beam-forming methods that make more efficient use of time. The time savings can be used to increase one or more of the above three qualities. The following techniques are examples of this.

Write zoom

This technique presents a full-screen real-time image of a restricted area of the normal field of view, selected by the user (Figure 3.35). Once selected, write zoom restricts the interrogation process to this operator-defined region of interest. One obvious way of using the consequent time savings would be to increase the frame rate. Alternatively, some or all of the time savings could be used to improve lateral resolution, both by narrowing the beam using multiple-zone focusing and by increasing the line density.

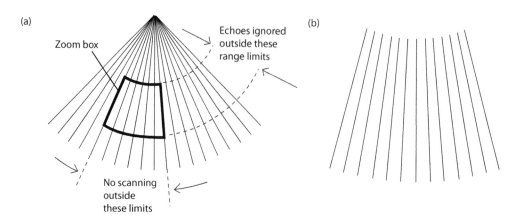

Figure 3.35 Write zoom. A 'zoom box' is defined by the operator **(a)**. Scanning is then restricted to this area **(b)**, allowing either a higher frame rate or improved lateral resolution by scanning with narrower, more closely spaced beams, or both.

Line multiplexing

The frame-rate penalty associated with using multiple-zone transmission focusing can be reduced by breaking the usual 'rule' that all echoes should have returned from one scan line before transmitting along the next line. Instead, interrogation of each line may be divided over several periods, interspersed with periods spent interrogating other lines. For example (Figure 3.36), when echoes have been received from the first (most superficial)

focal zone on scan line 1, rather than wait for the unwanted echoes from deeper structures to return before transmitting to the second focal zone on line 1, a transmission is sent from another active group centred on a well-removed scan line (e.g. scan line 50) to interrogate the first focal zone of that line. The choice of a distant scan line, in this case about halfway along the transducer, ensures that the unwanted echoes from one line do not reach the receiving element group on the other line. As soon as the echoes have arrived from the first zone on

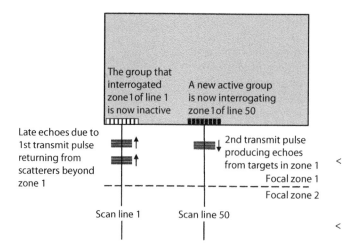

Figure 3.36 The diagram represents a moment soon after the part of line 1 lying within focal zone 1 has been interrogated. Rather than wait for unwanted late echoes from deeper structures to return before transmitting to focal zone 2, a transmission is sent from another active group centred on a well-removed scan line (e.g. scan line 50) to interrogate the first focal zone of that line.

line 50, a transmission might then be sent to the first focal zone on scan line 2, etc. Only when the first focal zone on all the scan lines has been interrogated are the second focal zones on all lines interrogated. These, too, are interrogated in the scan line sequence 1, 50, 2, 51, etc., to save time in the same way. Note that, although more than one transmission pulse is in flight at a time, echo reception does not take place on more than one line at a time.

Parallel beam forming in reception (multi-line acquisition)

This is an important and common technique, taken to much greater lengths in the next section. Since dynamic focusing is used in reception, the effective receive beam is narrower than the weakly focused transmit beam. In fact, it can be arranged for the transmit beam to accommodate two receive beams side by side (Figure 3.37). By sharing the same transmitted pulse, the echoes from targets located on two adjacent scan lines can be processed simultaneously ('in parallel') in

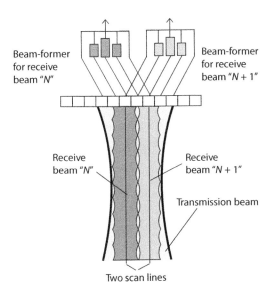

Figure 3.37 Parallel beam forming in reception saves time. One pulse is transmitted down a weakly focused transmission beam. In this example, two receive beam-formers act in parallel to simultaneously interrogate two scan lines lying within the transmission beam, using just that one transmission pulse.

two separate receive beam-formers. The different focusing delays for each receive beam may be in the form of hardware or software. In the former case, each element is connected to two physically separate beam-forming circuits. In the latter case, the two sets of delays are applied on a time-shared basis to digital samples of the echo signal from each element. For cardiac applications, where high frame rates can be important, as many as four scan lines may be interrogated in parallel. However, in order to accommodate four receive beams, an even broader transmit beam is required, further compromising lateral resolution.

RECENT DEVELOPMENTS IN BEAM FORMING FOR ARRAY PROBES

This section describes recent methods for improving the frame rate or the depth range over which good spatial and contrast resolution is obtained, or both. These techniques involve very rapid processing of large amounts of echo data, so their development has only been possible since the availability of large-scale rapid data storage and processing microchips. They all make use of the parallel beam-forming methods previously described. Rather than building up the image line by line, with one or more transmit pulses per line they produce each frame of a real-time B-mode image using a relatively small number of broad transmit pulses.

Plane wave techniques

Plane wave techniques involve transmitting plane wave pulses, commonly as wide as the field of view, in order to interrogate many, or all, of the scan lines simultaneously, using the parallel dynamic receive beam-forming technique described earlier. A key benefit is a high frame rate.

SINGLE PLANE WAVE IMAGING

In the simplest form, each frame consists of transmitting a single plane wave pulse, produced by firing all the elements in the transducer array at the same time. The unique benefit of this technique is the extremely high frame rate that can be achieved (several thousand frames/sec), since the interrogation of the entire field of view is completed within one transmit-receive cycle. However, this benefit comes at the cost of poor lateral resolution due to the extreme width of the transmit beam; recall that the transverse intensity profile across the transmit-receive beam is that of the transmit beam multiplied by that of the receive beam (see Chapter 5). Another consequence of the wide transmit beam is a high level of acoustic noise caused by echoes from scatterers well to the side of a receive beam axis (scan line) but, nevertheless, within the transmit beam. This acoustic noise leads to poor contrast resolution, penetration and sensitivity. Penetration and sensitivity are further reduced by the low amplitude of the transmit pulse, caused by the lack of transmit focusing.

Figure 3.38 illustrates the receive focusing on just two scan lines for clarity (lines 7 and 16 in this case) although, in practice, focusing is applied to all the scan lines simultaneously. It shows the situation at two moments in time, but, as in normal dynamic receive focusing, the focus advances steadily. In this way, beam-formed radio-frequency (RF) echo data from all points along every scan line in the frame are obtained and stored from a single transmission.

COHERENT PLANE WAVE COMPOUNDING

A more advanced form of plane wave scanning is known as 'coherent plane wave compounding'. This development allows the image quality at all depths to be much improved in terms of lateral resolution and signal-to-noise ratio, still with a very high frame rate although not as high as for the single plane wave technique just described (Montaldo et al. 2009). The field of view is interrogated several times (sometimes as many as 30 or 40 times) in every frame, each time by a plane wave pulse transmitted at a different direction to the array. These different directions are achieved by introducing a small delay between the firing of adjacent elements in the same way that beams from phased-array transducers are steered. The greater this delay, the greater is the deflection of the plane wave pulse from the straight ahead direction (see 'Phased-Array Transducers [Beam-Steering Arrays]'). As in the single plane wave technique, immediately after each transmission the RF echo data for all the scan lines within the transmit beam are obtained using parallel dynamic receive beam forming. This is illustrated in Figure 3.39 for one

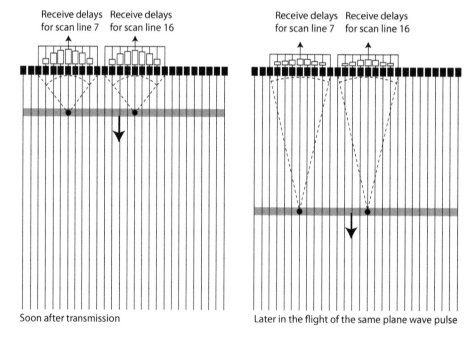

Receive delays for scan line 7   Receive delays for scan line 16

Soon after transmission

Receive delays for scan line 7   Receive delays for scan line 16

Later in the flight of the same plane wave pulse

Figure 3.38 Plane wave imaging. A plane wave pulse as wide as the field of view is transmitted. Parallel dynamic receive beam forming on all the scan lines completes one frame of the image for every transmission. Representative receive focus delays are shown for just two lines – lines 7 and 16.

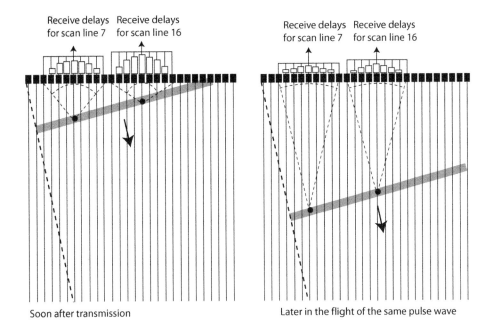

Receive delays Receive delays
for scan line 7 for scan line 16

Receive delays Receive delays
for scan line 7 for scan line 16

Soon after transmission

Later in the flight of the same pulse wave

Figure 3.39 Parallel dynamic receive beam forming for an angled transmit plane wave pulse. As the pulse advances, receive beam-formers for each scan line ensure the receive focus for that line advances in step with the depth of origin of echoes arriving back at the transducer. Representative receive focus delays are shown for just two lines – lines 7 and 16.

of the angled plane wave transmit pulses. Again, for clarity, the advancing receive foci are shown for just two scan lines (lines 7 and 16). At the end of all the transmissions in the frame, a final representative echo amplitude for every point on every scan line is calculated by coherently summing the N echoes for that point, where N is the number of transmissions in each frame. Coherent summation means that the N echoes are first delayed relative to each other to correct for differences in total propagation time to and from that point between the N different transmissions.

How this procedure improves signal-to-noise ratio and lateral resolution may be appreciated from a simple example, illustrated in Figure 3.40, where, for clarity, the number of transmissions per frame has been restricted to just three, i.e. $N = 3$. When the three RF echoes, one from each transmission (i–iii), from any particular point in the field of view, such as P, are coherently summed, the resulting final echo for P will have three times the amplitude that would result from a single transmission. This tripling of echo amplitude will apply only to echoes from within the relatively small crossover

region around P that is common to all three pulses (iv). The acoustic noise due to scatterers elsewhere in the field of view, on the other hand, will not be tripled in amplitude since the random nature of the peaks and troughs of noise means they do not align

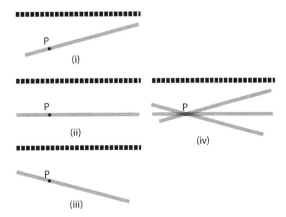

Figure 3.40 In this simple example, each point in the field of view, such as P, is insonated by just three plane wave pulses from different directions, each being transmitted after the echoes from the previous one have been stored.

when summed. This means that the signal-to-noise ratio of the final echo amplitude calculated for the scatterers at or close to P will be much better than was the case for the single plane wave technique. It also means that only scatterers at or close to P (i.e. within the crossover region) will contribute to the final echo amplitude for that point with a useful signal-to-noise ratio. In other words, the width of the crossover region determines the lateral resolution of the image.

The final echo obtained from a point P by this process is, in fact, almost identical to that which would be obtained from a concave transmit pulse converging to a focus at P. This is illustrated in Figure 3.41, where such a convergent pulse is considered to be approximated by three plane wave pulses A, B and C, arriving simultaneously at P from different directions (clearly, the greater the number of such plane transmit pulses at different angles, the closer the approximation to a converging pulse). The contribution to the final echo from P due to the echo produced by pulse A is almost the same as that which would have been produced by the part of the converging pulse labelled part a. Similarly, pulses B and C, respectively, contribute much the same as would parts b and c of the converging pulse. The difference is that each of the pulses A, B and C are transmitted at different times, not simultaneously as would be the case for parts a, b and c of the converging pulse. This

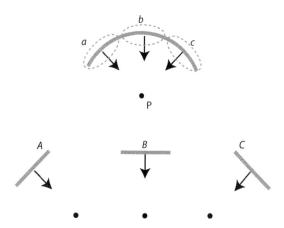

Figure 3.41 A concave pulse converging to a point can be considered as equivalent to a number of plane wave pulses converging to the same point from different directions.

process may, therefore, be correctly described as a form of transmit focusing but, because it is carried out after all the transmit-receive sequences have been completed, it is called 'retrospective transmit focusing'.

In line-by-line scanning, each scan line is interrogated in turn by a transmit beam focused to a specific depth chosen by the user. This produces an image with better signal-to-noise ratio and lateral resolution at this chosen depth than at other depths. The advantage of retrospective focusing is that it improves signal-to-noise ratio and lateral resolution at all points throughout the field of view that are insonated from a range of directions. This eliminates the need for a transmit focus control, although some manufacturers adopting this technique continue to provide one as an option. Retrospective focusing is also a feature of the synthetic aperture imaging methods described in the following section.

Plane wave imaging techniques were developed in the context of shear-wave imaging (see Chapter 14), an elastography technique designed to provide information on the stiffness of tissues (Bercoff 2011; Tanter and Fink 2014). Shear waves, produced by mechanical or acoustic vibration of the tissue, propagate through the tissue from the source of disturbance at a speed which is related to the stiffness of the tissues. By mapping the speed at which shear waves travel, an image of variations in tissue stiffness can be created. The shear wave velocity in soft tissues is in the range 1–10 m s^{-1}, so that the time to traverse a region 1 cm in width is of the order of 1–10 ms. In order to acquire several snapshots of the shear wave as it traverses this region, frame rates of several thousand hertz are required. This is impossible to achieve using previous ultrasound scanning techniques which, as previously noted, are able typically to produce frame rates up to about 60 Hz. The higher frame rates of plane wave imaging can also be used in Doppler and colour Doppler applications to give much improved performance in terms of velocity estimation and tissue discrimination (see Chapter 11).

Synthetic aperture imaging

Another technique for improving frame rates and gaining the benefit of retrospective transmit focusing is 'synthetic aperture imaging'. The term originates in radar imaging from a moving object such as an aeroplane or satellite (Figure 3.42), where several overlapping diverging pulses are transmitted in sequence as the object moves over the terrain of interest. The distance moved by the transmitter between the beginning and end of an imaging sequence defines the 'synthetic' transmit aperture. This effective aperture (2a) is much bigger than the transmitting antenna itself, leading to good lateral resolution according to the formula $w_F = F\lambda/a$, given in Chapter 2.

In the case of synthetic aperture diagnostic ultrasound, the sequence of transmissions is generated by an active group of adjacent elements that is stepped along a linear array. Apart from the increased frame rate that comes from transmitting fewer pulses in each frame, synthetic aperture techniques in diagnostic ultrasound have the inherent benefit of retrospective focusing, introduced in the section 'Coherent Plane Wave Compounding'. As mentioned there, this improves lateral resolution and signal-to-noise ratio at all depths, improving image uniformity compared to line-by-line scanning and relieves the user of the need to set the transmit focus at a particular depth of interest.

The basic form of synthetic aperture imaging involves transmitting a diverging wave from a single element, or small group of elements, at regularly spaced positions along the array. Each transmitted wave insonates most of, or in some cases the entire, field of view. Parallel receive beam forming, employing conventional delay and sum receive focusing, is used to obtain echoes from along all the scan lines. In this way, a set of low-resolution images is obtained, one for each transmit pulse (Figure 3.43). These low-resolution images, with the echoes in RF form, are compounded to produce a final high-resolution image.

A major problem with this basic type of synthetic aperture technique in diagnostic ultrasound imaging is that, because only one, or a small number of elements, is used in transmission, the amplitude of the transmitted ultrasound pulse is limited. This results in a low signal-to-noise ratio in the reconstructed images, as well as prohibiting the use of tissue harmonic imaging, since, as explained in Chapter 2, the latter requires transmit pulses of large amplitude.

For further reading see Misaridis and Jensen (2005), Nikolov et al. (2010) and Jensen et al. (2006), which also discuss the use of multiple transmit beams with different codes to further increase frame rate.

Synthetic aperture sequential imaging

A number of modern commercial imaging systems employ a form of synthetic aperture that overcomes the problem of limited transmit amplitude, mentioned previously (Bradley 2008; Mclaughlin 2012; Thiele et al. 2013), as well as reducing the amount of RF echo data to be stored and processed. In common with other synthetic aperture methods this method offers good lateral resolution and signal-to-noise ratio over a wide range of depths and makes the setting of a transmit focus at a particular depth of interest by the user unnecessary or, on some machines, an option. It involves stepping a focused, transmit beam along a linear array, the width of the beam being sufficient to span and simultaneously interrogate many scan lines using the parallel receive beam-forming technique described earlier. The distance by which the transmit beam is stepped between transmissions is more than the width of several elements, resulting in a higher frame rate than for line-by-line scanning. Crucially however,

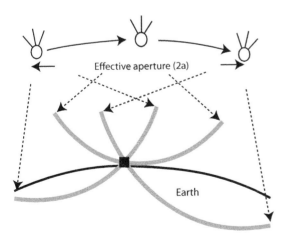

Figure 3.42 By transmitting from a range of positions a satellite creates a wide effective ('synthetic') aperture, resulting in excellent lateral resolution.

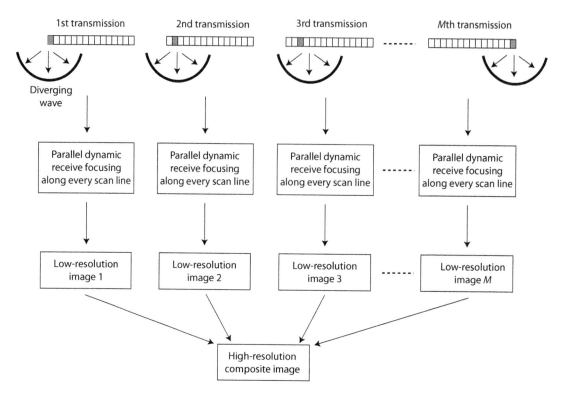

Figure 3.43 After each transmission-reception sequence a low-resolution image is formed. If *M* regularly spaced transmissions are used, a set of *M* low-resolution images is formed. A single high-resolution image is formed by coherently compounding these images.

the stepping distance is less than the width of the beam so that a number of scan lines are common to two or more adjacent beam positions. This results in each point on every scan line being interrogated several times, each time from a slightly different direction, as is explained now.

In the illustrative example of Figure 3.44, described by Thiele et al. (2013), each transmit beam spans six scan lines, which form the axes of six receive beams, operating in parallel, as shown in Figure 3.44a. As soon as echoes are receive beam-formed and stored from the six scan lines, the transmit beam is advanced by two elements along the array and echoes from six scan lines are again beam-formed and stored. Four of these scan lines are common to the previous transmission. The transmit beam is again advanced by two elements and again echoes are beam-formed and stored from six scan lines. Two of these, marked with dots in Figure 3.44b, are common to all three transmissions. These two lines will not be within

the following transmit beam, so interrogation of these two lines is now complete. For each of these lines there are now three sets of stored RF echoes which can be combined coherently to produce a final representative echo for every point along the line. Since the three transmit pulses responsible for the three stored echoes for each point arrive at that point from different directions, the final echo for each point benefits from retrospective transmit focusing. After each further transmission, final echoes for two more lines are formed. This means, for this illustrative example, that the image frame can be completed with half the number of transmissions that would be needed for line-by-line scanning, resulting in a twofold increase in frame rate. In comparison to the basic method of synthetic aperture, there is a considerable saving in echo data storage and processing requirements. Instead of the echo data for all the scan lines having to be stored and processed at the end of each frame, in the sequential technique, echo data for

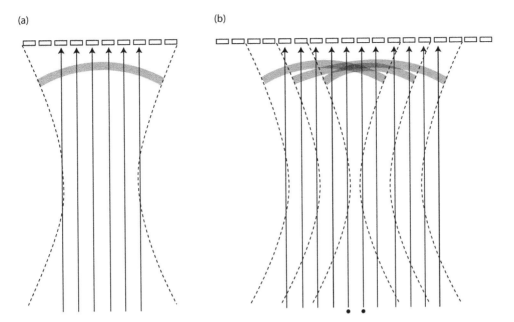

Figure 3.44 **(a)** A focused beam interrogates six scan lines in parallel. **(b)** When advanced along the array in steps of two elements, the two lines marked • are interrogated three times.

only three scan lines need to be stored and combined to form the final echo data for two lines.

In some commercial versions, a greater number of receive beams operate in parallel within each transmit beam and the transmit beam is advanced by more than two elements between transmissions, permitting greater increases in frame rate. For example, Thiele et al. (2013) describe a commercial version with a fourfold increase in frame rate, achieved by having 16 receive beams operating in parallel within each transmit beam, with a stepping distance of four scan lines between transmissions and each transmit beam overlapping the next by 12 scan lines.

MIXED-MODE SCANNING

Array scanners make it possible for the operator to highlight a specific scan line on a B-mode image and simultaneously generate a real-time M-mode scan, A-mode scan or Doppler spectrum for that line on the same display screen.

This is particularly useful in cardiological applications, where a phased-array transducer is commonly used because of its ability to fit between ribs. Here, the simultaneous display of an M-mode line and a real-time B-mode (2D) scan allows the operator to check that the M-mode line is placed, and remains, in the correct anatomical position (Figure 3.45). Although the two scans appear to be formed simultaneously, in fact the beam-former rapidly switches back and forth between B-mode and M-mode interrogations. After every few lines of B-mode interrogation, the beam is made to jump to the selected M-mode scan line for one transmission and echo acquisition sequence. It then jumps back to continue the B-mode scan for another few lines; then jumps back to the M-mode line, etc.

'Duplex' Doppler scanning is another example of mixed-mode scanning. Here, Doppler measurements are made of blood flow or tissue movements in a 'sample volume', whose position is indicated on a 'Doppler line' on the B-mode image (Chapter 9). This line can be set by the user to be either parallel to or at an angle to the image scan lines. When set at an angle to the scan lines, the beam-steering and focusing techniques described earlier (see 'Phased-Array Transducers [Beam-Steering Arrays]') are used to interrogate that line.

As explained in Chapter 9, the Doppler line must be interrogated at a high repetition frequency, much higher than would be possible even by jumping to the Doppler line after every B-mode

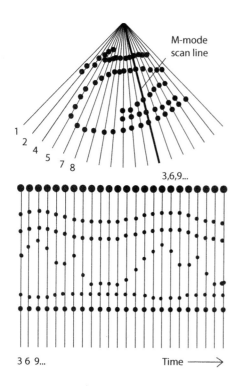

Figure 3.45 Mixed M-mode and B-mode scanning. The first two transmissions are directed along two lines of the B-mode. The third is transmitted along the M-mode scan line. The fourth and fifth pulses interrogate the next two B-mode scan lines, then the sixth interrogates the M-mode scan line again, etc. Thus the two scans proceed in parallel.

line. The Doppler line is therefore interrogated without interruption, except for very brief periods (about 20 ms) every second or so, as determined by the user, in which the machine performs one complete 'update' frame of the B-mode. Each of these B-mode update images is held frozen on the screen next to the ongoing Doppler display until it is automatically replaced by the next one. Note that the pulses transmitted along the Doppler line usually have a lower frequency and greater length than those transmitted along the imaging lines.

3D/4D TRANSDUCERS

The beam formation and scanning techniques described so far have all been designed to acquire echo information from a single cross section through the target tissues. The echo information

would then be processed and displayed as a real-time 2D, B-mode image (see Chapter 4). The same beam forming and scanning techniques can be extended to acquire echo information from a 3D volume of tissue. The resulting 3D volume data set can then be processed to create a number of alternative modes of display, as described in Chapter 12. Repetition of the 3D acquisition and display at a few hertz (refresh rates up to 20 Hz are possible) results in a real-time 3D display, referred to as four-dimensional (4D) display.

There are two commonly used approaches to the design of 3D/4D transducers. The design illustrated in Figure 3.46 incorporates an array transducer, in this case a curvilinear array, mounted on a swivel inside an enclosed bath of acoustic coupling fluid (linear-array transducers are also used). The transmit pulse and returning echoes pass through a thin acoustic window to acquire a 2D section from the adjacent tissues. A motor, coupled to the swivel by pulleys and a drive belt, rotates the transducer array so that the 2D scan plane is swept at 90° to the imaged plane, to interrogate and acquire echoes from a pyramidal volume of tissue.

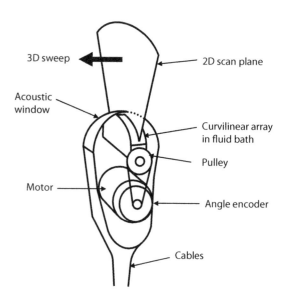

Figure 3.46 The mechanical 3D transducer contains a stepped array mounted on a swivel in an enclosed fluid bath beneath a thin acoustic window. The image plane is swept from side to side by a motor coupled via pulleys and drive belt. 3D volumes can be acquired at up to 5 Hz.

Figure 3.47 The matrix array transducer contains several thousand square elements (over 9,000 from one manufacturer) arranged in a 2D matrix. (Image courtesy of Philips Medical Systems.)

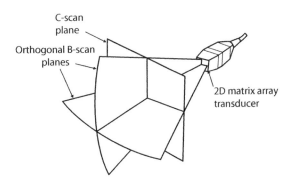

Figure 3.48 The 2D matrix-array transducer has a rectangular array of elements. It uses beam-steering techniques to sweep a 2D sector scan through a 3D volume or interrogate orthogonal B-scan planes. Volume data can be displayed as a series of B-scan planes or as a C-scan. 3D display modes are also widely used.

The orientation of each scan plane is measured by an angle encoder attached to the motor. A sequence of closely spaced, adjacent 2D sections is acquired, which constitutes a 3D volume data set. By rapidly repeating the sweeping motion of the array, a 4D image can be created. Due to the mechanical nature of the transducer movement and the line-by-line scanning of each slice, 3D acquisition rates are limited to a few volumes per second.

The spatial resolution in each 2D slice of the acquired data is subject to the same limitations as described earlier for linear-array transducers. That is, the lateral resolution in any of the scan planes of the moving transducer array is much better than that in the associated elevation planes, due to the lack of electronic focusing in the elevation direction. This leads to poor lateral resolution in 2D images of planes perpendicular to the original scan planes, reconstructed from the 3D volume data set.

Recent forms of 3D/4D scanners use a 2D matrix transducer array (Figure 3.47). This type of transducer contains several thousand square elements (over 9,000 in one case) arranged in a 2D matrix. Using beam-steering techniques as described earlier, the matrix array can be used to create a 2D sector scan in a single plane. By applying beam-steering techniques also in the orthogonal direction (elevation), the 2D sector can be swept sideways to describe a pyramidal volume (Figure 3.48). As the matrix of elements is square, it is possible to achieve similar resolution in the lateral (scan plane) and elevation directions by applying the same degree of electronic focusing

and aperture control to both planes. The stored volume data can be interrogated and displayed as selected 2D sector scans in any plane, including a C-scan, where the imaged section is parallel to the transducer face. Alternative modes of 3D display, including surface rendering, are described in detail in Chapter 12.

In order to achieve useful volume repetition rates, parallel beam formation must again be used. Commercial systems use relatively wide transmit beams, within each of which up to 64 receive beams (8 × 8) operate in parallel (Ustener 2008). By providing substantial overlap between the wide transmit beams, retrospective transmit beam focusing (as described previously for plane-wave and synthetic aperture techniques) can be used to achieve good lateral and elevation resolution, while maintaining useful volume rates of up to several tens of hertz (Bradley 2008).

MECHANICALLY SCANNED TRANSDUCERS

Mechanical scanning is usually used for special applications at frequencies above around 20 MHz. This is due to the difficulties of fabricating high-frequency linear-array and phased-array transducers, although array transducers operating at up to 50 MHz have been produced for small animal scanning (Foster et al. 2009). Mechanical scanners producing rectangular, or trapezoidal, fields of view may be used for high-frequency scanning of superficial sites, such as the eye and skin. Here, the transducer is driven back and forth inside an enclosed water bath at the end of a handheld probe. The transducer must move in a water-filled bath, rather than in air, since the latter would result in virtually zero transmission across both the transducer-air and the air-skin interfaces (Chapter 2). Part of the wall of this water bath is made from a thin plastic membrane, allowing transmission into the patient. Reverberations (Chapter 5) between the transducer and this wall (effectively the patient) can cause a noticeable artefact. Linear mechanical scanning is only practical for transducers operating at frequencies above 10 MHz or so, because of the vibration that heavier, lower-frequency transducers would produce when constantly reversing direction. Despite the general disadvantages that are commonly associated with the moving components and water baths of mechanical scanners – bulk, vibration, leakage, wear and tear, etc. – a general advantage of all mechanical scanners over linear- or phased-array transducers is that they do not suffer from grating lobes, and thus generate less acoustic noise.

ENDO-CAVITY TRANSDUCERS

Endo-cavity transducers are intended for insertion into a natural body cavity or through a surgical opening. A number of different types are represented in Figure 3.49. The ability to place the transducer close to a target organ or mass means that there is less attenuation from intervening tissue, which in turn means a higher frequency may be used and superior lateral and axial resolution obtained. The image distortions and artefacts due to any tissue heterogeneity or strongly reflecting or refracting interfaces between the transducer and the target are also reduced.

All the beam-forming techniques previously mentioned find application in endo-cavity transducers, the choice being determined primarily by the anatomical features and constraints of the particular application. Thus, a curvilinear array offers a field of view of an appropriate shape for trans-vaginal scanning. The wide field of view close to a linear array is suitable for imaging the prostate

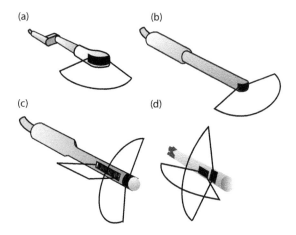

Figure 3.49 Examples of endo-transducers. **(a)** Curvilinear transducer for trans-vaginal scanning. **(b)** 'End-fire' curvilinear-array transducer for trans-rectal or trans-vaginal scanning. **(c)** 'Bi-plane' trans-rectal transducer with both a linear array and a curvilinear array – allowing both transverse and longitudinal scans of the prostate. **(d)** Trans-oesophageal transducer with two phased arrays set at right angles, giving two orthogonal cross sections of the heart.

from the rectum. Phased arrays give a wide field of view for visualizing the left side of the heart from a trans-oesophageal probe.

360° Mechanically scanned endo-probes

These rotate a transducer about the axis of a probe so that the beam sweeps through a 360° circle, in a similar way to a light beam sweeping around a lighthouse. Such probes consist of an outer tube within which is a rotating inner rod, bearing the outwardly pointing transducer. In order for the signals to and from the transducer to cross between the stationary and rotating parts, either slip-rings or a transformer arrangement are used. A disposable rubber sheath may be clamped to the outer tube which, when inflated with water, provides an offset between the rotating transducer and the patient. Applications of this type of probe include scanning the prostate from within the rectum, and imaging the bladder wall with a probe introduced into the bladder via the urethra. The technique can be used with low-frequency transducers, but because of the size and mass of the transducers such scanners are likely to be limited to low frame rates or involve manual rotation of the transducer.

High-frequency versions (e.g. 30 MHz) can be inserted via a catheter into a blood vessel in order to visualize the vessel wall. The transducer is attached to a rotating wire within a non-rotating outer cable. One difficulty with this method is that friction between the rotating wire and the outer cable causes the cable, and hence the transducer, to weave around within the blood vessel. The continual movement of the viewing point with respect to the target leads to difficulties in interpreting the images. The cylindrical-array probes discussed next do not have this problem and offer an alternative method for intra-luminal scanning.

Intra-luminal and intra-cardiac catheter probes using transducer arrays

Strongly convex transducers were discussed earlier as an exaggerated form of curvilinear array. The ultimate development of this idea is to curve a linear array so tightly that a complete cylindrical array is formed. Such transducer arrays offer an alternative to the mechanical 360° scanners mentioned previously (Figure 3.50a).

Tiny, high-frequency (e.g. 2 mm diameter, 30 MHz), cylindrical arrays, mounted on catheters,

can be inserted into a blood vessel. This gives a direct high-resolution image of the internal wall of the blood vessel. One construction method is to mount the transducer elements, their connecting leads and other electronic hardware on a flexible printed circuit and then roll this into the required final cylindrical form. Probes made this way have been designed as single-use, disposable devices.

Another type of probe designed for intra-cardiac imaging has a 64-element phased array (5–9 MHz) mounted on the side of a 3 mm diameter catheter (Figure 3.50b). The catheter can be introduced into the heart via the femoral or jugular vein (Proulx et al. 2005).

A problem facing the designers of catheter-mounted arrays is that it is not possible to accommodate 128 or so leads within the narrow catheter. Consequently, the arrays have fewer elements, and in some cases the 'synthetic-aperture' technique, described previously, is employed. This allows the number of leads to be a fraction of the number of array elements. Electronic switches are mounted next to the transducer array so that a given lead can be connected to one of several different elements as required. After each transmission, the echo sequence from the selected elements is digitized and stored. Several transmission-reception sequences are carried out along the same scan line, with the leads being connected to different elements for each one. When all the elements have been selected, all the stored echo sequences are summed together coherently, with each sequence being delayed relative to the others by the appropriate delay-time interval. These delays are just those that would have been used if the echo sequences from all the elements had been available at the same time.

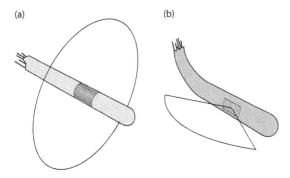

(a)   (b)

Figure 3.50 **(a)** Intra-luminal transducer with a cylindrical high-frequency (typically 30 MHz) array, providing 360° transverse images of blood-vessel walls. **(b)** Intracardiac phased-array transducer mounted on the end of a steerable catheter.

QUESTIONS

Multiple Choice Questions

Q1. The transducer has maximum output when the piezoelectric plate thickness is:
a. One-third the wavelength
b. Two-thirds the wavelength
c. Half the wavelength
d. Twice the wavelength
e. Three-quarters the wavelength

Q2. If the transducer did not have a damping backing layer, this would lead to:
a. Infinitely long pulses
b. Very short pulses
c. Pulses of zero duration in time
d. Ringing
e. Long pulses

Q3. The amplitude of pulses transmitted into the patient can be increased by:
a. Using composite PZT instead of PZT
b. One or more matching layers
c. Increasing the peak-to-peak voltage of the driving pulse applied across the PZT plate
d. Increasing the pulse repetition frequency
e. A backing layer

Q4. The improvement in transmission into and out of the patient provided by a matching layer:
a. Is greatest at the centre frequency of the pulse
b. Is the same for all frequencies present in the pulse
c. Is greater for the higher frequencies present in the pulse
d. Is greater for the lower frequencies present in the pulse
e. Is least for the centre frequency of the pulse

Q5. A linear-array transducer:
a. Consists of a single transducer element
b. Is made up of a pair of elements; one for transmission, one for reception
c. Involves mechanical sweeping of the beam to build up the 2D image
d. Involves electronically stepping a beam along the array to build up a 2D image
e. Typically contains over a hundred transducer elements

Q6. In a linear-array transducer:
a. A group of adjacent elements receives echoes from along a scan line
b. Half the elements are used for transmission and half for reception
c. One element is fired to produce one line, then the next element for the next line and so on
d. The number of elements used for receiving echoes from one scan line can change during the reception period

e. A group of elements is used for transmission and the identical group is used for reception

Q7. An array transducer:
a. Has one or more matching layers to help improve sensitivity
b. Has no matching layers as it is inappropriate to use these with linear arrays
c. Has a backing layer to help prevent ringing
d. Has no backing layer
e. Usually has a cylindrical lens for focusing in the elevation plane

Q8. In a linear array, electronic focusing in transmission can be achieved by firing the elements in the active transmit group:
a. At the same time
b. With inner elements first, then the outer ones
c. Sequentially one after the other
d. With outer elements first, then the inner ones with the central elements last
e. In a random manner

Q9. In a linear array, electronic focusing in reception is achieved by:
a. Delaying electrical echoes to account for different arrival times of the ultrasound echoes at different elements in the active receive group
b. Amplifying electrical echoes from the outer elements of the receive group less than those from the inner elements
c. Amplifying electrical echoes from the outer elements of the receive group more than those from the inner elements
d. Firing the outer elements in the transmit group first, then the inner ones with the central element last
e. Increasing the aperture of the receive group for echoes from larger depths

Q10. In a linear array, apodization is concerned with:
a. Beam steering
b. Increase in frame rate
c. Side lobe suppression
d. Increase in sensitivity
e. Improving lateral and contrast resolution

Q11. In an array transducer, dynamic focusing:
a. Is used for the receive beam
b. Is used for the transmit beam

c. Involves sweeping the beam from side to side

d. Involves continual adjustment of the element delays during reception

e. Involves continual adjustment of the element delays during transmission

Q12. In a phased array, beam steering is achieved by:

a. Firing all the elements at the same time

b. Firing only the elements in one-half of the array

c. Firing the elements sequentially one after the other

d. Firing the outer elements first, then the inner ones with the central element last

e. Firing the inner elements first, then the outer ones

Q13. Grating lobes:

a. Are weak replicas of the main beam occurring on both sides of the main beam

b. Can occur for any array transducer with regularly spaced elements

c. Only occur for linear-array transducers but not for phased-array transducers

d. Only occur for arrays in which the centre-to-centre spacing of the elements is less than half a wavelength

e. Occur when the echoes from adjacent elements are separated by a whole number of cycles

Q14. Beam steering with a linear array is used in:

a. Conventional scanning with scan lines perpendicular to the array face

b. Trapezoidal field of view scanning

c. Compound imaging

d. Generation of an oblique Doppler beam

e. Mixed B-mode/M-mode scanning with scan lines perpendicular to the array face

Q15. 1.5D arrays:

a. Have several rows of elements, one above the other

b. Consist of a square array of typically 128-by-128 elements

c. Improve focusing in the elevation plane

d. Improve focusing in the scan plane

e. Offer no improvement over the use of an acoustic lens

Q16. Parallel beam forming (multiple line acquisition) involves:

a. Several transmit beams and one receive beam for each scan line

b. One transmit and one receive beam for each scan line

c. An improvement in frame rate

d. A broad transmit beam within which are several simultaneous receive beams

e. Significant noise reduction in the image

Q17. Retrospective transmit focusing:

a. Uses several wide overlapping beams, each from a different transmission

b. Makes use of parallel beam forming in reception

c. Leads to uniformly good lateral resolution over a wide range of depths

d. Makes use of narrow transmit beams

e. Usually results in increased frame rate

Q18. The single plane wave imaging technique:

a. Produces an image using a single transmission

b. Has a low frame rate, limited to approximately 100 Hz

c. Makes use of diverging waves

d. Makes use of parallel beam forming in reception

e. Has poorer lateral resolution than can be achieved with a modern linear array

Q19. A 2D matrix array transducer:

a. Produces a cylindrically shaped 3D field of view

b. Sweeps the beam through a 3D volume using mechanical sweeping of the array

c. Usually consists of a square or rectangular array of several thousand transducer elements

d. Can produce images of 2D slices that are parallel to the transducer face

e. Overcomes the slice thickness limitation associated with linear arrays

Q20. Diagnostic probes in which a single element transducer is mechanically swept back and forth with a linear reciprocating motion:

a. Exhibit grating lobes

b. Cannot produce real-time B-mode images

c. Typically use frequencies as low as 3 MHz

d. Typically use frequencies over 10 MHz

e. Are used for endo-cavity applications

Short-Answer Questions

Q1. Describe the construction of a single element ultrasonic transducer, explaining the functions of the various layers.

Q2. Why is the thickness of the piezoelectric plate chosen to be half a wavelength?

Q3. Explain the process by which the beam from a linear-array transducer is moved along the scan plane to generate consecutive B-mode image lines.

Q4. How is electronic transmit focusing in the scan plane achieved with an array transducer?

Q5. In line-by-line scanning, why is the transmit beam from an array transducer wider than the receive beam?

Q6. Describe the advantages and disadvantages of multiple-transmit-zone focusing.

Q7. Explain the process used with a phased-array transducer to steer the transmit beam in different directions.

Q8. Explain the origin of grating lobes in a beam from an array transducer.

Q9. Why is the image quality from a phased-array transducer better in the centre of the sector than it is near the edges?

Q10. In line-by-line scanning, how are temporal resolution, imaged depth and image quality related?

Q11. Describe a method by which a transmission focus can be achieved at each point in the field of view, without actually transmitting a pulse focused to every point?

REFERENCES

Bercoff J. 2011. Ultrafast ultrasound imaging. In: *Ultrasound Imaging: Medical Applications*, O. Minin (Ed.), InTech, Available at: https://www.intechopen.com/books/ultrasound-imaging-medical-applications/ultrafast-ultrasound-imaging

Bradley C. 2008. Retrospective transmit beam formation. Siemens white paper. Available at: https://static.healthcare.siemens.com/siemens_hwem-hwem_ssxa_websites-context-root/wcm/idc/groups/public/@global/@imaging/@ultrasound/documents/download/mdaw/mtmy/~edisp/whitepaper_bradley-00064734.pdf

Caronti A, Caliano G, Carotenuto R et al. 2006. Capacitive micromachined ultrasonic transducer (CMUT) arrays for medical imaging. *Microelectronics Journal*, 37, 770–777.

Chen J, Panda R, Savord B. 2006. Realizing dramatic improvements in the efficiency, sensitivity and bandwidth of ultrasound transducers. Philips Medical Systems white paper. Available at: http://incenter.medical.philips.com/doclib/enc/fetch/2000/4504/577242/577260/593280/593431/Philips_PureWave_crystal_technology.pdf%3fnodeid%3d1659121%26vernum%3d-2

Dausch DE, Gilchrist KH, Carlson JB et al. 2014. In vivo real-time 3-D intracardiac echo using PMUT arrays. *IEEE Transactions on Ultrasonics, Ferroelectrics, and Frequency Control*, 61, 1754–1764.

Foster FS, Mehi J, Lukacs M et al. 2009. A new 15–50 MHz array-based micro-ultrasound scanner for preclinical imaging. *Ultrasound in Medicine and Biology*, 35, 1700–1708.

Jensen JA, Nikolov SI, Gammelmark KL et al. 2006. Synthetic aperture ultrasound imaging. *Ultrasonics*, 44, E5–E15.

Khuri-Yakub BT, Oralkan O. 2011. Capacitive micromachined ultrasonic transducers for medical imaging and therapy. *Journal of Micromechanics and Microengineering*, 21, 054004–054014.

Otake M, Tanaka H, Sako A et al. 2017. Development of 4G CMUT (CMUT linear SML44 probe). *MEDIX*, 67, 31–34.

Proulx TL, Tasker D, Bartlett-Roberto J. 2005. Advances in catheter-based ultrasound imaging intracardiac echocardiography and the ACUSON AcuNavTM ultrasound catheter. *IEEE Ultrasonic Symposium Proceedings*, 669–678.

Mclaughlin G. 2012. *Zone Sonography: What It Is and How It's Different*. Zonare Medical Systems. Available at: http://res.mindray.com/Documents/2016-12-14/1a81fa00-b1fc-4c45-8cca-229d4df58699/zone_sonography_what_it_is.pdf

Misaridis T, Jensen JA. 2005. Use of modulated excitation signals in medical ultrasound. Part III: High frame rate imaging. *IEEE Transactions on Ultrasonics, Ferroelectrics, and Frequency Control*, 52, 208–219.

Montaldo G, Tanter M, Bercoff J et al. 2009. Coherent plane-wave compounding for very high frame rate ultrasonography and transient elastography. *IEEE Transactions on Ultrasonics, Ferroelectrics, and Frequency Control*, 56, 489–506.

Nikolov SI, Kortbek J, Jensen JA. 2010. Practical Applications of Synthetic Aperture Imaging. In: *Proceedings of IEEE International Ultrasonics Symposium*, 350–358.

Qiu Y, Gigliotti JV, Wallace M et al. 2015. Piezoelectric micromachined ultrasound transducer (PMUT) arrays for integrated sensing, actuation and imaging. *Sensors (Basel)*, 15, 8020–8041.

Szabo TL, Lewin PA. 2007. Piezoelectric materials for imaging. *Journal of Ultrasound in Medicine*, 26, 283–288.

Tanaka H. 2017. Technology introduction to CMUT (Capacitive Micro-machined Ultrasound Transducer), Hitachi's next generation linear matrix transducer. *European Congress of Radiology (ECR 2017)*, 1–5 March 2017, Vienna.

Tanter M, Fink M. 2014. Ultrafast imaging in biomedical ultrasound. *IEEE Transactions on Ultrasonics, Ferroelectrics, and Frequency Control*, 61, 102–119.

Thiele K, Jago J, Entrekin R et al. 2013. Exploring nSIGHT Imaging – A totally new architecture for premium ultrasound. Philips Medical Systems white paper. Available at: https://sonoworld.com/Common/DownloadFile.aspx?ModuleDocumentsId=68

Ustener K. 2008. High information rate volumetric ultrasound imaging. Siemens white paper. Available at: https://static.healthcare.siemens.com/siemens_hwem-hwem_ssxa_websites-context-root/wcm/idc/groups/public/@global/@imaging/@ultrasound/documents/download/mdaw/mtc1/~edisp/whitepaper_ustuner-00064729.pdf

Zhou Q, Lam KH, Zheng H et al. 2014. Piezoelectric single crystals for ultrasonic transducers in biomedical applications. *Progress in Materials Science*, 66, 87–111.

4

B-mode instrumentation

KEVIN MARTIN

SIGNAL AMPLITUDE PROCESSING

The beam-forming techniques described in the previous chapter are used to acquire echo information from different parts of the imaged cross section by selection of the transducer array elements and manipulation of the relative timings of their transmit and receive signals. These yield echo sequences, which represent the B-mode image lines and define the spatial properties of the image. The brightness of the image at each point along the B-mode line is determined by the amplitude of the echo signals received at the transducer. However, the echo signals as received at the transducer must be processed in a number of ways before they can be used in the construction of the B-mode image. The processing methods used are described in this chapter.

Figure 4.1 illustrates, in block-diagram form, the essential elements of the complete B-mode system, and shows that the B-mode amplitude information is processed in various stages before the image is formed and stored in the image memory, from where the image is displayed. In Chapter 3, it was implied that the echo signals from the transducer elements went directly to the beam former. However, in practice the signals must undergo several important processing steps before beam formation. In some of the processes which follow beam formation, the echo signal still contains the original transmit frequency. This signal is referred to as the radio-frequency (RF) signal. Processing of such signals is referred to here as coherent image formation as echoes may be processed or combined using their amplitude and phase. Demodulation of the RF signal involves removal of the transmit frequency leaving only the amplitude or envelope of the echo signal. Processing after this is referred to as non-coherent. Some processing may be carried out after the image memory (post processing) to improve the displayed image or optimize its characteristics for a particular clinical application.

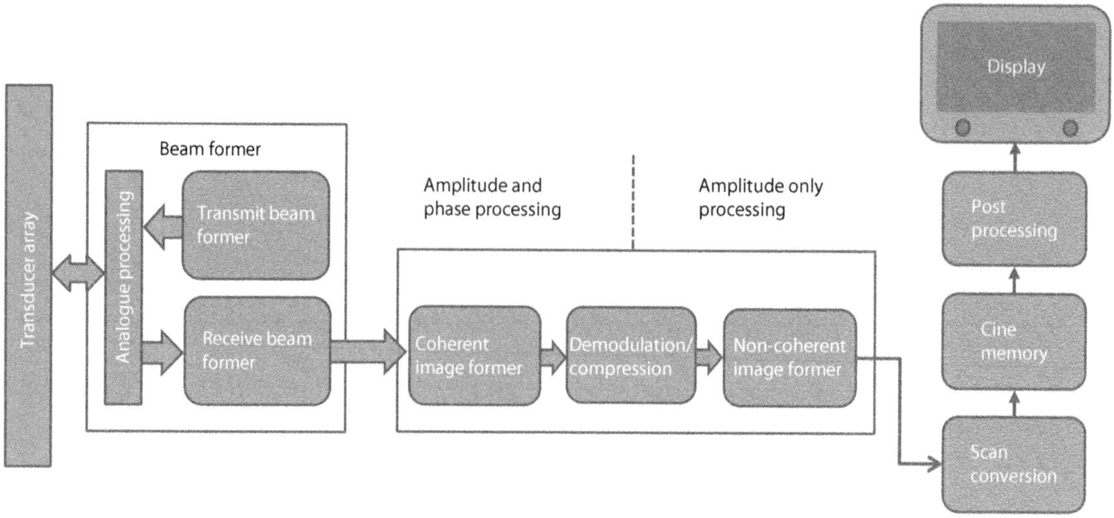

Figure 4.1 A block diagram of a B-mode imaging system. After beam formation, the lines of echoes must be processed in various ways before the B-mode image can be displayed. Some of these processes are coherent, i.e. they involve the amplitude and phase of the echoes. Others are non-coherent and involve only the amplitude of the echoes.

TRANSMIT BEAM FORMER

The pulse-echo sequence begins when electrical transmit signals are sent from the transmit beam former to the transducer elements as described in Chapter 3. The transmit beam former controls the timing and the amplitude and shape of the signals sent to each element. These signals are generated digitally but are then converted to analog signals, that is, they become continuously variable signals. Various user controls interact with the transmit beam former, such as output power, field of view and focusing. The output power control alters the amplitude of the electrical signals sent to the transducer elements and hence the amplitude of the pulse transmitted by the transducer. This control may allow the user to reduce the transducer output from its maximum level in steps of several decibels (e.g. 0, −3, −6, −9 dB) or it may be labelled as a percentage of the maximum power level available for the chosen application. Reducing the amplitude of the transmitted pulses reduces the amplitudes of all resulting echoes by the same number of decibels. Reducing the transmit power reduces the exposure of the patient to ultrasound and the risks of any adverse effects. Altering the output power

has a direct effect on the values of the displayed safety indices, mechanical index (MI) and thermal index (TI) (see Chapter 16), which change as the control is adjusted.

AMPLIFICATION

The echo signals generated at the transducer elements are generally too small in amplitude to be manipulated by the receive beam former and need to be amplified (made bigger). Figure 4.2a shows the conventional symbol for an electronic amplifier. This device in reality consists of numerous transistors and other electronic components, but can be treated as a single entity with an input terminal and an output terminal. The voltage signal to be amplified (V_{in}) is applied to the input terminal, and the amplified voltage signal (V_{out}) is available at the output terminal. The degree of amplification is expressed as the ratio V_{out}/V_{in} and is referred to as the voltage gain of the amplifier. Figure 4.2a illustrates the effect that a simple amplifier would have on a stepped voltage input signal.

There are two points to note. First, in the output signal, each step is larger than the corresponding step in the input signal by the same ratio. That is,

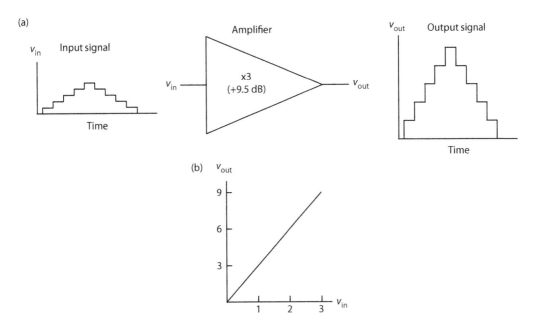

Figure 4.2 Linear amplification. **(a)** The amplifier gain (×3) is the same for all signal levels. The gain is also constant with time. **(b)** A graph of output voltage against input voltage is a straight line.

the voltage gain (in this case ×3) is the same for all voltage levels in the input signal. This is referred to as linear amplification, because a graph of V_{out} against V_{in}, as illustrated in Figure 4.2b, is a straight line. Second, the voltage gain is constant with time. Each of the downward steps in the second half of the signal is amplified to the same extent as the corresponding upward steps in the first half. An amplifier of this type is used to amplify all echo signals equally, irrespective of when they return to the transducer. The overall gain control, available to the user on most B-mode systems, applies this type of gain to the echo signals. The effect on the image is to make all echoes brighter or darker, whatever their depth in the image.

The effect is similar to changing the transmit power. In many circumstances, a reduction in transmit power can be compensated for by increasing the overall gain. However, where echoes of interest are weak due to a weakly scattering target or attenuated due to overlying tissue, their amplitudes may be reduced to below the system noise level. Increasing the overall gain cannot lift the signal above the noise level as the noise will be amplified with the signal. The operator should set the transmit power level to the minimum level which allows all relevant echoes to be displayed clearly after adjustment of the overall gain.

TIME-GAIN COMPENSATION

Attenuation

As described in Chapter 2, when a transmitted ultrasound pulse propagates through tissue, it is attenuated (made smaller). Echoes returning through tissue to the transducer are also attenuated. Hence, an echo from an interface at a large depth in tissue is much smaller than that from a similar interface close to the transducer. The attenuation coefficient of tissues (measured in dB cm^{-1}) is described in Chapter 2. For example, if a particular tissue attenuates an ultrasound pulse by 1.5 dB cm^{-1}, the amplitude of the pulse will be reduced by 15 dB when it reaches an interface 10 cm from the transducer. The echo from this interface will be attenuated by 15 dB also on its journey back to the transducer, so that compared to an echo from a similar interface close to the transducer, the echo will be smaller by 30 dB. In this tissue, echoes received from similar interfaces will be smaller by 3 dB for each centimetre of depth.

Time-gain control

In a B-mode image, the aim is to relate the display brightness to the strength of the reflection at each interface regardless of its depth. However, as we have just noted, echoes from deeper targets are much weaker than those from more superficial ones. Hence, it is necessary to compensate for this attenuation by amplifying echoes from deep targets more than those from superficial targets. As echoes from deep targets take longer to arrive after pulse transmission than those from superficial ones, this effect can be achieved by increasing the amplification of echo signals with time. The technique is referred to as time-gain compensation (TGC). It makes use of an amplifier whose gain may be controlled electronically, so that it can be changed with time.

As shown in Figure 4.3, the amplitude of received echo signals reduces exponentially with their depth of origin and hence their time of arrival at the transducer. This is the same as being reduced by the same number of decibels for each centimetre of depth. At the start of the pulse-echo sequence, as echoes are being received from the most superficial

interfaces, the echoes are relatively large and so the gain is set to a low value. It is then increased with time to apply a higher gain to echoes arriving from greater depths. For the previous example, when the pulse and the echo are each attenuated by 1.5 dB cm^{-1}, the gain must be increased by 3 dB cm^{-1}. Assuming a speed of sound in tissue of 1540 m s^{-1}, the go and return time of the pulse and echo is 13 μs for every centimetre of depth. So the gain must be increased by 3 dB for every 13 μs after transmission of the ultrasound pulse. After TGC, echoes from similar interfaces should have the same amplitude, regardless of their depth of origin.

The actual rate of attenuation of ultrasound with depth is determined by the ultrasound frequency and the type of tissue. The ultrasound system applies TGC to the received signals at a rate (in dB cm^{-1}) designed to compensate for attenuation in average tissue at the nominal transducer frequency. Adjustments to this base level, to compensate for changes in tissue type within the imaged cross section, can then be applied manually by the operator. The most common arrangement for manual TGC adjustment is a set of slide controls as illustrated in Figure 4.4. Each slide alters the gain of the TGC amplifier at specific times after transmission, i.e. for echoes returning

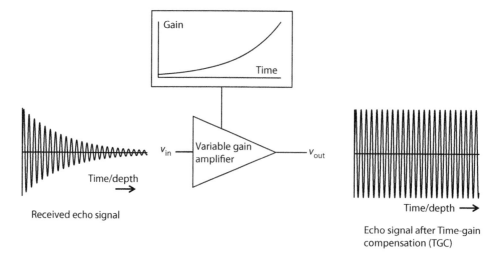

Figure 4.3 Time-gain compensation (TGC). The amplitude of received echo signals reduces exponentially with their depth of origin due to attenuation and hence with their time of arrival at the transducer. The gain applied by the TGC amplifier increases with time after transmission to compensate. After TGC, echoes from similar interfaces should be equal in amplitude regardless of depth.

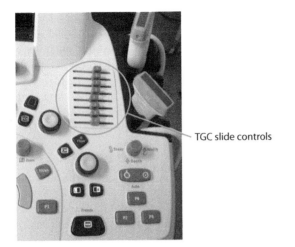

Figure 4.4 TGC is most commonly adjusted using a set of slide controls, each of which affects the gain at a different depth.

from a specific range of depths within the tissue. When all slides are in the central position, the average rate of TGC is applied, related to the frequency of the transducer. Moving the top slide to the right increases the gain applied to echoes from superficial tissues. The bottom slide adjusts the gain applied to the deepest echoes. In adjusting the

TGC, the operator seeks to eliminate any trend for the average image brightness to change with depth. There can be no question of making all echoes equal in brightness, of course, nor would this be desirable, since the differences in scattered echo strength between different tissues are crucial to the interpretation of the image.

Some manufacturers now incorporate fully automatic TGC systems, which analyse the overall image brightness to identify and correct for any trends in image brightness down or across the image. TGC controls may be absent from such systems.

ANALOGUE-TO-DIGITAL CONVERSION

The electrical signals produced at the transducer elements in response to received echoes are analogue signals. That is, their amplitudes vary continuously from the smallest to the largest value. The beam-forming techniques described in Chapter 3 are implemented using digital techniques. Hence, before beam forming, the echo signal must be converted from analogue to digital form. The process is carried out by an analogue-to-digital converter (ADC) as illustrated in Figure 4.5.

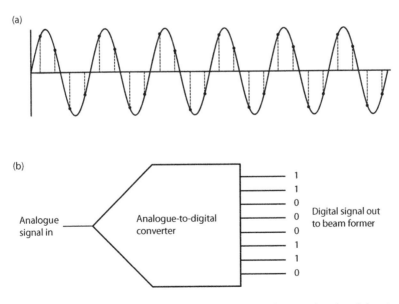

Figure 4.5 (a) Analogue to digital conversion involves measuring the amplitude of the signal (sampling) at regular intervals in time. The signal can then be stored as a set of digital values. (b) The analogue-to-digital converter (ADC) produces a binary number corresponding to the value of each sample. The sequence of samples of the echo signal is converted to a stream of such binary numbers.

At regular, frequent intervals in time, the amplitude of the analogue echo signal is measured, or sampled, producing a sequence of numbers corresponding to the amplitude values of the samples (Figure 4.5a). The digital output from the ADC for each sample is in the form of a binary number (Figure 4.5b). A stream of such binary numbers is produced corresponding to the sequence of signal samples. A binary number is formed from a set of digits, which can each have the value one or zero. Each digit of the binary number (1 or 0) is referred to as a bit (binary digit). The output from the ADC in Figure 4.5b is made up of 8 bits. There are 256 possible combinations of the 8 ones or zeros, so the digital output from this ADC can express 256 possible values of signal amplitude. Unlike the analogue signal, which was continuously variable, the digital signal can have only a finite number of values, which is determined by the number of bits. Each additional bit doubles the number of possible signal levels. The ADCs used in commercial B-mode systems are typically 12-bit, giving 4,096 possible values of signal level, so that the signal is a much more faithful recording of the analogue original. Also, since the smallest value that can be digitized is smaller, the range of values, i.e. the ratio of the largest to smallest value, is greater.

At the ultrasound frequencies used for imaging (3–15 MHz), the amplitude of the echo signal changes rapidly with time. To preserve detail in the stored image, the ADC must sample the echo signal at a high enough rate to capture these changes.

The sample rate must be at least twice the frequency that is being sampled (the Nyquist limit). As shown in Chapter 2, the ultrasound pulse waveform contains a range of frequencies centred on the nominal frequency. To ensure that the full range of frequencies in the pulse is captured, the sample rate is typically set to at least four times the nominal frequency of the transmit pulse, e.g. 40 MHz (i.e. 40 million samples per second) or more for a 10 MHz pulse.

Digital data can be stored in electronic memory without degrading. As it consists simply of a set of numbers, it is essentially immune to noise, interference and distortion, unlike analogue signals. Perhaps, the most important advantage of digitization for B-mode imaging is that it makes digital processing of echo information possible. Using built-in, dedicated computing devices, digital echo information can be processed by powerful mathematical techniques to improve the image quality. While some of these processes may be carried out in real time, others make use of information that has been stored temporarily in a digital memory. Some of the techniques described in Chapter 3, such as spatial compounding and retrospective transmit focusing, make use of stored digital data.

Figure 4.6 shows the early stages of analogue signal processing that must take place before the beam former. The analogue echo signal from each transducer element is amplified as described earlier. Time gain compensation is then applied before the signal is digitised by the ADC.

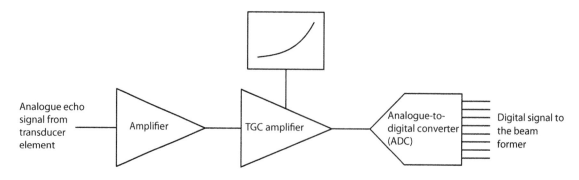

Figure 4.6 Several stages of analogue echo signal processing must take place before the beam former. The analogue echo signal from each transducer element is amplified. TGC is then applied before analogue to digital conversion. The digital signal is sent to the beam-former.

COHERENT AND NON-COHERENT IMAGE FORMATION

In receive mode, the beam former takes in digital information from each active element in the transducer array and combines it as described in Chapter 3 to form the B-mode image lines. The combination process is a coherent process, i.e. it takes into account the amplitude and phase of the various echo signals. The echo signals can interfere constructively or destructively when they are combined, according to their relative phase, as in the beam-formation process described in Chapter 2. The resulting image line data is in digital format but still contains information on the amplitude and phase of the echo signals. This allows further coherent combination of data from different B-mode lines to improve the B-mode image. In Figure 4.1, the coherent image former is where coherent processes such as retrospective transmit focusing are carried out as described in Chapter 3. Following the various coherent processes, the signal is demodulated, that is, the transmit frequency and its phase information are removed. Processing after this stage is described as non-coherent. Any combination of signals is additive and no phase cancellation can take place.

Following the beam-forming stage, various coherent processing techniques are commonly used to improve the quality of the image by suppressing the effects of acoustic or electronic noise.

CODED EXCITATION

One of the fundamental challenges for manufacturers and users of ultrasound is to optimize the choice of transmit frequency for different clinical applications. A high frequency allows the formation of a narrow beam and a short transmit pulse, leading to good spatial resolution (see Chapter 5). However, high frequencies are attenuated more rapidly, giving reduced penetration into tissue. Conventionally, high frequencies, e.g. 10–15 MHz, would be used only to image superficial structures, such as the thyroid or arteries of the neck. A low frequency, e.g. 3 MHz, gives good penetration but relatively poor resolution and might be used to image the heart or the liver of a large patient.

Attenuation of the transmitted pulse and returned echoes increases with depth, and the limit of penetration for a conventional pulse corresponds to the depth from which echoes are no longer larger in amplitude than the background system noise. Greater penetration can be achieved by increasing the amplitude of the transmit pulse. However, maximum pressure amplitudes generated in tissue by the transmit pulse are limited by regulation and other safety issues (see Chapter 16).

Coded excitation of the transmit pulse can be used to increase the signal-to-noise ratio for a given frequency and hence help to improve penetration without increasing the pulse amplitude. Pulse-coding methods employ transmit pulses which are much longer than a conventional three to four cycle imaging transmit pulse. The longer transmit pulse contains more energy and leads to an improvement in signal-to-noise ratio, but without further processing would result in very poor axial resolution. By embedding a digital code into the long transmit pulse, good axial resolution can be recovered and signal-to-noise ratio improved by identifying the transmitted code within the received echoes (Chiao and Hao 2005; Nowicki et al. 2006).

Figure 4.7a shows an example of a digital code (1,1,0,1) that might be used. To embed the code within the transmit pulse, the pulse is divided into four time intervals, in this case, each two wave periods long. The value of the digital code in each segment (1 or 0) is represented by the phase of the excitation waveform used. The starting phase in the '0' segment is opposite to that in the '1' segments. As the echoes of the coded pulse are received, they are stepped through a matched filter, which contains a copy of the transmitted code. In Figure 4.7b, the envelope of the detected echo is given a value of +1 where the '1' phase is detected and −1 for the '0' phase. This is a correlation filter, in which the received signal is convolved (as indicated by the circled cross symbol) with a time-reversed version of the transmitted pulse. At each time step (equal to the time segments in the transmit pulse), the code values of each segment are multiplied by the values they overlap in the filter and added together. The resulting signal has amplitude of +4 at the time step where the echo exactly overlies the code

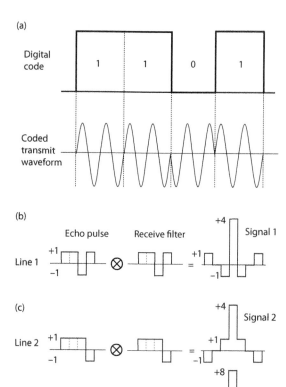

Figure 4.7 **(a)** In coded excitation, a digital code can be embedded into the transmit pulse by dividing it into a series of time zones and assigning a phase (normal or inverted) to each. The time zones corresponding to a digital '1' can be represented by the normal phase and those corresponding to a digital '0' by the inverted phase. **(b)** The digital code can be extracted from the returned echoes using a matched filter, which contains the original code. A large signal is received when the code and filter match, but with range side lobes before and after. **(c)** Range side lobes can be cancelled out using a pair of complementary codes known as a Golay pair.

in the matched filter. However, positions before and after this also result in non-zero values, which give rise to range side lobes on the signal. Longer codes and more complex sequences (e.g. Barker codes) may be used to achieve greater increases in signal-to-noise ratio and reduced range side lobe levels (Chiao and Hao 2005).

The range side lobes can be cancelled out by transmitting a second pulse along the same

B-mode line with a complementary code as shown in Figure 4.7c; the first pulse and the complementary pulse are described as a 'Golay pair' (Chiao and Hao 2005). When the output signals from the two lines are added together, the range side lobes cancel (phase +1 cancels with phase −1) and the resulting signal has an amplitude of +8 and a duration of one time segment. However, the use of multiple pulse-echo cycles on each line results in a reduction in frame rate and may lead to artefacts if the target tissue is not stationary.

An alternative approach to embedding a digital code in the transmit pulse is to transmit a chirp (Pedersen et al. 2003). This is a long pulse, in which the transmit frequency is swept from a low to a high value within the bandwidth of the transducer (Figure 4.8a). The elongated pulse again contains more energy, leading to improved signal-to-noise ratio. The received echoes are passed through a matched filter (Figure 4.8b) which is a time-reversed version of the transmitted chirp. This convolution process results in a shortened pulse which restores the axial resolution of the system (Figure 4.8c). Chirp excitation requires only a single pulse per line but requires a more complex and costly transmission pulse generator to generate the chirp waveform.

B-Flow

Coded excitation is also used by some manufacturers to allow moving blood to be visualised in a real-time B-mode image using a method commonly referred to as B-flow. The method is described further in Chapter 11. The echoes scattered from blood are very much weaker than those scattered from tissue and are difficult to detect using B-mode processing techniques. The improvement in signal-to-noise ratio achieved by coded excitation makes it possible to detect signals from blood, but they are so much smaller than those from tissue and organ boundaries that they cannot normally be displayed at the same time, even after dynamic range compression (see later in this chapter). In the B-flow technique, echoes scattered from blood and tissue are detected using coded excitation, but the echoes scattered from tissue are then suppressed using a technique called tissue equalization, so that both can be displayed in the

(a)

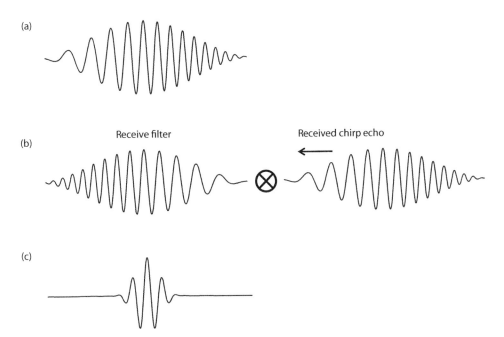

Receive filter   Received chirp echo

(b)

(c)

Figure 4.8 **(a)** The transmitted chirp is an elongated pulse in which the transmit frequency is swept from a low to high value within the bandwidth of the transducer. **(b)** The received echo signals are passed through a matched filter which is a time-reversed copy of the transmitted pulse. **(c)** The output of this convolution process is a shortened pulse whose spectrum is centred at the centre frequency of the chirp.

image at the same time (Chiao et al. 2000). Tissue equalization is achieved by applying a line-to-line high-pass filter. This suppresses echoes that do not change from line to line, such as those arising from stationary tissue. Echoes that do change from line to line, such as those from moving scatterers in blood, are emphasized. A simple way of achieving tissue equalization is to use two consecutive pulse-echo cycles for each image line. The echo signal from the first line is compressed (i.e. the waveform is decoded) and stored. The second line is compressed and then inverted before being added to the first (equivalent to subtraction). The stationary echoes from tissue are removed while those from moving blood are retained. The high-pass filter effect can be improved by using more pulse-echo cycles per line and altering the polarity (positive or inverted) and amplitude of the lines before they are added together.

The B-flow image gives an impression of moving blood within blood vessels. As it emphasizes changing image features, the brightness of the moving blood increases with blood velocity. B-flow

is not a quantitative technique and cannot be used to measure blood velocity; however, it does not suffer from some of the limitations of Doppler techniques such as angle dependence and aliasing (see Chapter 11).

TISSUE HARMONIC IMAGING

Harmonic imaging is described here in its application to B-mode imaging of tissue, where it is effective in reducing artefacts that cloud the image. Harmonic imaging is also used to differentiate echoes produced by ultrasound contrast media from those produced by tissue. Contrast applications of harmonic imaging are described in Chapter 13.

As described in Chapter 2, harmonic imaging of tissue is useful in suppressing weak echoes caused by artefacts (see Chapter 5), which cloud the image and can make it difficult for the operator to identify anatomical features with confidence. Such echoes, often referred to as clutter, are particularly noticeable in liquid-filled areas, such as the heart, blood vessels or the bladder. Clutter can arise due to the transmitted pulse being reflected from superficial features such as the ribs or organ boundaries. The reflected pulses may be reflected or scattered again, giving rise to echoes which return to the transducer and appear as clutter deeper in the image. The image may also be degraded by superficial fat layers. Fat has a different speed of sound to tissue and may cause defocusing of the beam and loss of image quality.

A low-amplitude ultrasound pulse travels through tissue by linear propagation, retaining its sinusoidal shape (see Figure 2.23). Echoes produced by reflection and scattering have a similar waveform to the original transmitted pulse. When a high-amplitude ultrasound pulse travels through tissue, non-linear propagation effects cause the pulse waveform to become distorted, developing sharp transitions in pressure that were not present in its original sine wave shape (see Figure 2.27). The sharp transitions in pressure are associated with some of the energy at the transmitted frequency f_0 (the fundamental frequency) being transferred into harmonic frequencies $2f_0$, $3f_0$, etc. (see Chapter 2). Non-linear distortion occurs most strongly in the parts of the transmit beam where the acoustic pressure is highest, i.e. on the beam axis and in the focal region. The distortion increases with distance from the transducer until it is reduced by attenuation in the tissue. In harmonic imaging, the image is formed using only the harmonic or non-linear element of the returned echoes. The component of the echoes at the fundamental frequency, the linear component, is removed. As most of the harmonic energy arises from on or near the beam axis, the harmonic beam is narrower than the beam formed from the fundamental frequency. The beam width in elevation is also reduced. Multiple zone transmit focusing as described in Chapter 3 can be used to extend the range of the high-amplitude region where harmonic energy is produced.

The image clutter and artefacts described are associated with low-amplitude, linear propagation of the transmitted pulse and appear in the fundamental part of the returned echoes. The reflections from superficial features, such as ribs, that give rise to clutter are low-amplitude duplicates of the transmit pulse and produce little harmonic energy. These are removed in the harmonic image, resulting in a clearer image. Little non-linear distortion occurs in the low-amplitude parts of the beam, such as within side lobes and grating lobes. Clutter due to reflections from these is also reduced. Defocusing of the transmit beam by superficial fat has less effect on the harmonic beam as the harmonics are created deeper within the tissue, beyond the fat layer. Hence the distortion of the transmit beam is avoided.

Harmonic imaging can be achieved using a transducer with a wide bandwidth (i.e. a wide frequency response), which can transmit a pulse at frequency f_0 and then receive echoes at the fundamental frequency f_0 and its second harmonic $2f_0$, as illustrated in Figure 4.9. In transmission (Figure 4.9a), the transmit frequency f_0 is chosen to ensure that the pulse spectrum sits in the lower half of the transducer's frequency-response curve. In reception (Figure 4.9b), the received echoes contain information around the transmit frequency f_0 and its second harmonic $2f_0$ due to non-linear distortion. The second harmonic component of the echoes is received in the upper part of the transducer bandwidth. Higher harmonic frequencies ($3f_0$, $4f_0$, etc.) are also generated by non-linear propagation in the tissue but are not detected as they are outside of the frequency response of the transducer. Low-amplitude parts of the beam produce echoes centred about f_0 via linear propagation and scattering processes. To create a harmonic image, these fundamental frequencies must be removed to leave only the harmonic frequencies. A number of different techniques have been developed to achieve this.

Harmonic imaging by filtration

The original approach to suppression of the fundamental part of the echo was to pass the received echo signal through a band-pass filter tuned to $2f_0$ to allow through the second harmonic frequencies while rejecting those centred about f_0. However, as can be seen from Figure 4.9b, the fundamental and second harmonic spectra of a typical ultrasound pulse overlap and cannot be separated completely. If the filter is designed to allow through all of the second harmonic echo spectrum, it will also allow through some of the fundamental component. As described in Chapter 2, the harmonic frequencies in the distorted pulse spectrum are much weaker than the fundamental frequency (typically about

(a)

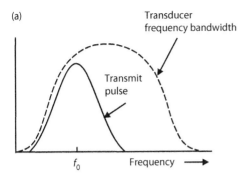

(b)

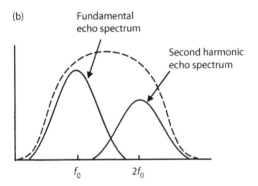

(c)

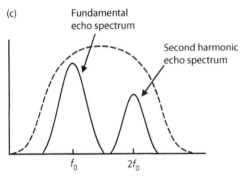

Figure 4.9 **(a)** In harmonic imaging, the pulse is transmitted at frequency f_0 using the lower half of the transducer bandwidth. **(b)** The received echoes contain components at the fundamental frequency f_0 due to low-amplitude linear propagation and at harmonic frequency $2f_0$ due to high-amplitude non-linear propagation. For a typical ultrasound pulse, the frequency bands at f_0 and $2f_0$ overlap making it difficult to filter out the fundamental component. **(c)** To allow the fundamental echoes to be filtered out, the spectrum of the transmit pulse must be made narrower so that the f_0 and $2f_0$ frequency bands do not overlap. The spectrum can only be narrowed by making the pulse longer and hence compromising axial resolution.

20 dB lower). So the fundamental frequencies that pass through the band-pass filter may swamp the harmonic signal, giving rise to clutter in the image. To allow the filter to achieve good suppression of the fundamental signal, the spectra of the transmit pulse and its second harmonic need to be well separated, with no overlap. To achieve this, the frequency spectrum of the transmitted pulse must be made narrower than for normal imaging as shown in Figure 4.9c. However, reduction of the width of the spectrum results in an increase in the length of the pulse (see Chapter 2), reducing the axial resolution of the system, as described in Chapter 5.

Pulse inversion harmonic imaging

A harmonic image can be produced also by a technique known as pulse-inversion (PI) imaging (see Figure 4.10). This technique requires two consecutive pulse-echo cycles for each beam position. In the first pulse-echo cycle, a high-amplitude pulse at the chosen fundamental frequency f_0 is transmitted and the line of echoes is received and stored digitally. A second pulse, which is an inverted version of the first, is then transmitted down the same B-mode line. The line of echoes generated by the second pulse is added to the stored first line. Low-amplitude echoes, from the edges of the beam or from multiple reflections, are generated by linear propagation and are undistorted replicas of the transmission pulse, containing only the fundamental frequency spectrum centred at f_0. Corresponding low-amplitude echoes in the first and second lines cancel out as each is an inverted version of the other. In the higher-amplitude parts of the beam, the transmitted pulse is distorted due to non-linear propagation and harmonic generation. The distortion is in the same direction for the normal and inverted pulses. The harmonic elements of the echoes from these regions do not cancel out, creating a harmonic image.

PI harmonic imaging has the effect of suppressing the odd-numbered harmonics, e.g. f_0, $3f_0$, $5f_0$ and enhancing the even harmonics, e.g. $2f_0$, $4f_0$ (Averkiou 2001). The effect can be understood by looking at the phase changes in the odd and even harmonics during inversion of the transmit pulse as illustrated in Figure 4.11. In this illustration, the transmitted pulse is represented by a few cycles of a sine wave but has become distorted by non-linear

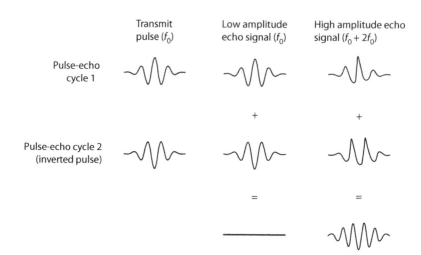

Figure 4.10 Pulse-inversion imaging. On pulse-echo cycle 1, a pulse is transmitted at f_0 and the line of echoes stored. On pulse-echo cycle 2, an inverted version of the pulse is transmitted along the same B-mode line. The line of echoes from pulse-echo cycle 2 is added to those from cycle 1. Echo signals at f_0 cancel out while those at second harmonic $2f_0$ are reinforced.

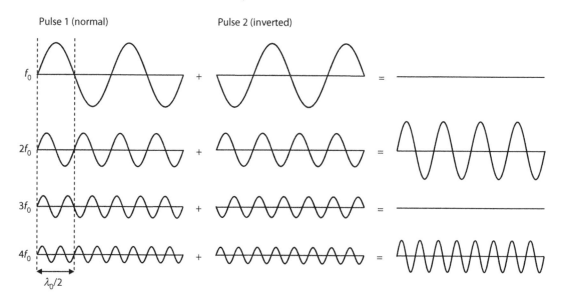

Figure 4.11 In PI harmonic imaging, the second transmit pulse is inverted. This is equivalent to a phase shift of half a wavelength at the fundamental frequency ($\lambda_0/2$). At even harmonics $2f_0$ and $4f_0$, the shift is equivalent to a whole number of wavelengths so the inverted versions of the even harmonics are in phase with the non-inverted versions and are reinforced when added. The odd harmonics (f_0 and $3f_0$) are in anti-phase when inverted and cancel.

propagation and is depicted in terms of its separate harmonic components (see Chapter 2). These are harmonic frequencies $2f_0$, $3f_0$ and $4f_0$ in addition to the fundamental component f_0. The phase relationship between the fundamental and harmonic

frequencies is fixed in the inverted and non-inverted states. Inverting the fundamental frequency f_0 is equivalent to a phase shift of half a wavelength $\lambda_0/2$. At even harmonic frequencies $2f_0$ and $4f_0$, the shift is equivalent to an even number of half wavelengths, so the inverted versions of $2f_0$ and $4f_0$ are in phase with their respective non-inverted versions and are reinforced when added. For the odd harmonics, the shift is equivalent to an odd number of half wavelengths and the inverted versions are in anti-phase with their non-inverted versions and cancelled when added (Whittingham 2005).

The advantage of the PI method is that it can remove the fundamental component of the echo even when the fundamental and second harmonic parts of the pulse spectrum overlap (as in Figure 4.9b). The pulse spectrum does not need to be reduced in width and hence axial resolution can be maintained. The disadvantage is that the need for two pulse-echo cycles per line reduces the frame rate and may cause artefacts if the imaged target moves between the two pulse-echo cycles.

Power modulation harmonic imaging

Power modulation (PM) is an alternative scheme for suppressing the components of echoes arising from low-amplitude linear propagation and scattering. It more correctly refers to modulation of the pulse amplitude rather than its power. It was noted earlier that the degree of distortion due to non-linear propagation increases with the amplitude of the transmitted pulse, i.e. its acoustic pressure.

The increased distortion of high-amplitude pulses leads to increased harmonic content. Power modulation suppresses the fundamental frequency by again transmitting two pulse-echo sequences for each image line. In this case, the amplitude of the transmit pulse is changed between lines, rather than the phase (Averkiou 2001).

As illustrated in Figure 4.12, in the first pulse-echo cycle, a high-amplitude pulse is transmitted, which becomes distorted as it propagates through tissue. Echoes due to reflection and scattering of the pulse are also distorted and have a high harmonic content. These are stored as they are received. The second transmit pulse is lower in amplitude, e.g. half of the amplitude of the first pulse and suffers less distortion as it propagates. Echoes from the second pulse contain energy mainly at the fundamental frequency, with little harmonic content. The line of echoes from the second pulse is amplified by two to compensate for the lower amplitude of the transmit pulse and then subtracted from the first line of echoes. This suppresses the fundamental frequency component of the echoes arising from linear propagation and scattering, giving a harmonic image. Power modulation retains both odd and even harmonic frequencies, although these are at a lower level than the second harmonic level in PI (Averkiou et al. 2008). PM also

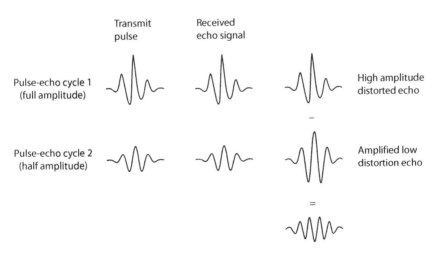

Figure 4.12 In power modulation harmonic imaging, a high-amplitude pulse is transmitted on the first pulse-echo cycle. Transmitting a pulse at half amplitude on the second pulse-echo cycle results in half amplitude echoes with low harmonic content. The half amplitude echoes are amplified by two and subtracted from the first line of echoes to remove the linear component of the first line, resulting in a harmonic beam.

retains some energy at the fundamental frequency. This fundamental element is not due to imperfect removal of the linear component but arises due to a difference in fundamental content of the high- and low-amplitude pulses. The high-amplitude pulse contains less energy at the fundamental frequency as some of this energy has been transferred to harmonic frequencies. Hence this fundamental component arises due to non-linear propagation and is part of the harmonic beam (Averkiou et al. 2008). As it is lower in frequency than the harmonics, the fundamental component suffers less attenuation in tissue and extends the range of the harmonic beam.

Pulse inversion power modulation harmonic imaging

Pulse inversion power modulation (PMPI) is a combination of the two techniques just described. The processing of the transmitted pulse and the returned echoes in these schemes may be described in terms of transmit and receive amplitude coefficients A_t and A_r. In the case of PI, the coefficients can be written as A_t (1, −1) and A_r (1, 1). The amplitudes A_t of the two transmit pulses are 1 and −1, showing that the amplitudes are the same but the second transmit pulse is inverted. The amplitudes and polarity of the received echoes are the same. The two lines of echoes are added. In power modulation, the coefficients for the scheme described are A_t (1, 0.5) and A_r (1, −2) showing that the second transmit pulse is at half amplitude and that the second line of echoes is amplified by two and inverted before addition, i.e. subtracted.

A possible PMPI scheme would be A_t (0.5, −1), A_r (2, 1) showing that the first transmit pulse is half amplitude and the second is inverted. The received echoes from the first pulse are amplified by two and added to those from the second pulse. This scheme has been shown to retain as much non-linear signal at f_0 as PM but increased harmonic energy at $2f_0$ and higher harmonics (Averkiou et al. 2008). More complex pulsing schemes using more transmit pulses with PI and amplitude modulation may be used commercially to suppress image clutter without restricting the imaged depth due to attenuation.

Harmonic imaging schemes which retain non-linear signal at f_0 allow greater penetration of the harmonic beam due to the lower attenuation at f_0 than at the harmonic frequencies. However, the lower frequency also results in poorer spatial resolution. In PI, the harmonic beam is formed almost entirely from the second harmonic, giving a narrow beam with good spatial resolution but reduced penetration due to the increased attenuation at $2f_0$. In commercial systems, manufacturers may change the harmonic imaging method used according to image settings chosen by the user. For example, if the user chooses a high-resolution setting (labelled RES in some systems), the transducer frequency may be increased and PI harmonic imaging used. Where more penetration is required, the transmit frequency may be reduced and a PM or mixed mode harmonic imaging technique used. For cardiology applications, where a high frame is needed, the filtration method may be used.

Suppression of movement artefact

A weakness of multiple pulse techniques for harmonic imaging is that tissue movement between pulse-echo cycles can cause misalignment of the echo signals, resulting in imperfect cancellation of the linear component. The loss of cancellation can result in a large temporary increase in the fundamental signal and increase in image brightness. For the case of limited but constant tissue motion, the artefact can be reduced by the use of a three-pulse PM scheme such as that shown in Figure 4.13. The scheme can be described using the amplitude coefficient notation as A_t (0.5, 1, 0.5), A_r (1, −1, 1). The amplitudes of the three transmitted pulses are 0.5, 1 and 0.5. Due to a small amount of tissue motion between pulse-echo cycles, the echo of pulse 2 from a particular scatterer is delayed by a small amount ∂t. The echo of pulse 3 is delayed by a further ∂t with respect to the pulse transmit time. Combining echoes 1 and 3 results in a full-amplitude pulse whose phase is halfway between those of the two pulses, i.e. the same as that of pulse 2. Inverting and adding pulse 2 then results in cancellation of the linear component of the echo signals. Where the tissue motion is not constant, for example where there is constant acceleration, longer pulse sequences can be used to compensate (Whittingham 2005).

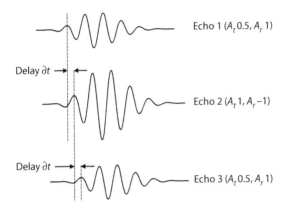

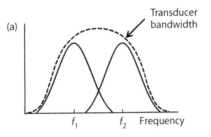

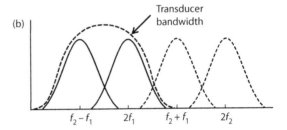

Figure 4.13 In two pulse schemes, small shifts ∂t in the time of arrival of echoes from consecutive pulse-echo cycles, due to axial motion of the scatterer, can result in poor cancellation of the linear echo signal. This motion artefact can be reduced by using a three-pulse scheme. Adding echoes from cycles 1 and 3 gives a larger echo signal with an average delay matching that of echo 2. Subtracting the second echo gives cancellation of the linear signal.

Dual-frequency tissue harmonic imaging

The tissue harmonic imaging methods described so far make use mainly of the harmonic component of the echo signal in forming the image, with some non-linear signal at the fundamental frequency. Limited use is made of the lower half of the transducer bandwidth, limiting the depth from which echoes can be detected. Dual-frequency tissue harmonic imaging, also referred to as differential tissue harmonic imaging, makes use of the full bandwidth of the transducer, including the lower half. The scheme is illustrated in Figure 4.14.

A transmit pulse containing two frequencies f_1 and f_2 is used, where f_2 is approximately twice f_1 (Figure 4.14a). The spectra of the pulses are designed to fit within the bandwidth of the transducer with limited overlap. Non-linear distortion of the transmitted pulse as it propagates through tissue results in the generation of harmonic frequencies $2f_1$ and $2f_2$. Non-linear propagation also causes mixing of the two frequencies, resulting in frequencies $f_2 - f_1$ and $f_2 + f_1$ in the spectrum of

Figure 4.14 Differential tissue harmonic imaging. (a) The transmit pulse is made up of two frequencies, f_1 and f_2, which fit within the transducer bandwidth. (b) Non-linear propagation within the tissue results in harmonic generation but also mixing of frequencies f_1 and f_2. The returned echoes contain harmonics $2f_1$ and $2f_2$ but also sum and difference frequencies $f_2 + f_1$ and $f_2 - f_1$. Frequencies $f_2 - f_1$ and $2f_1$ make full use of the bandwidth of the transducer. Frequencies $f_2 + f_1$ and $2f_2$ are outside the bandwidth of the transducer and are not detected.

the echo (Figure 4.14b). After subtraction of the original transmitted frequencies f_1 and f_2 using one of the methods described, the detected echo signal contains frequencies $f_2 - f_1$ and $2f_1$. Frequencies $f_2 + f_1$ and $2f_2$ are usually outside the bandwidth of the transducer and so not detected. The difference frequency $f_2 - f_1$ is approximately equal to f_1.

As for the single transmit frequency approaches described earlier, harmonic generation and frequency mixing occur most strongly in the high-amplitude parts of the ultrasound beam, resulting in a narrower beam with less clutter. However, the difference frequency $f_2 - f_1$ now makes use of the lower half of the transducer bandwidth, adding energy to the echo but also suffering less attenuation as it propagates back to the transducer. Differential tissue harmonic imaging increases the depth in tissue from which echoes can be detected (Chiou et al. 2007).

FREQUENCY COMPOUNDING

Compounding of images acquired with different beam angles was described in Chapter 3 as a means of clarifying curved image boundaries and reducing image speckle (see Chapter 5). Frequency compounding is an alternative form of image compounding, which can help to reduce the impact of speckle in B-mode images. The method requires the use of a transducer with a wide frequency response and a wideband transmit pulse, i.e. a short pulse (see Chapter 2). Reflection and scattering of the wideband transmit pulse produce echoes which also contain a wide range of frequencies (Figure 4.15a). The echo signal is passed through a set of narrow band-pass filters which separate the signal into a set of narrow band echo signals. These signals are separately demodulated and compressed (see later) before being summed back together (Figure 4.15b). Demodulating and compressing the signals separately before being adding back together mean that the summation process is non-coherent, i.e. the signals contain no information on the frequency and phase of the original transmit pulse. As the image speckle pattern is determined by the wavelength of the transmit pulse and the width of the beam, the different frequency bands give rise to different speckle patterns, which are averaged when combined. The result is a reduction in the amplitude and prominence of the speckle pattern. Adding the echo signals back together without demodulation would simply re-create the original echo signal and its speckle pattern.

Frequency compounding can be used in conjunction with some of the tissue harmonic imaging techniques described. For example, in differential tissue harmonic imaging (Figure 4.14), the two frequency components of the processed echoes $f_2 - f_1$ and $2f_1$ can be separated by band-pass filtration, then separately demodulated and compressed before being summed back together. This gives a small but worthwhile reduction in speckle. In power modulation tissue harmonic imaging, some non-linear fundamental signal is retained as well as the second harmonic. These can be treated in the same way to help reduce speckle.

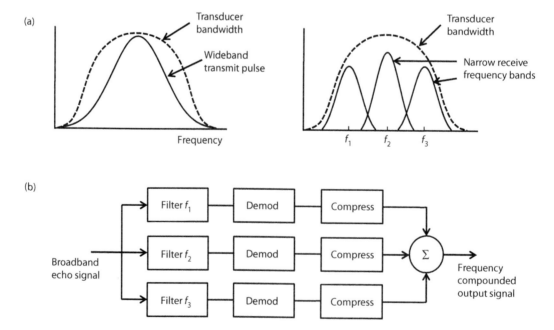

Figure 4.15 (a) In frequency compounding, a short pulse is used which contains a wide range of frequencies. The wideband received echo signal is divided into multiple narrow frequency bands by band-pass filters; in this case 3 frequency bands are used. (b) The 3 echo signals are separately demodulated and compressed before being summed to form the frequency compounded signal.

NON-COHERENT IMAGE FORMATION

Amplitude demodulation

The ultrasonic pulse transmitted by the B-mode system consists of several cycles of oscillation at the nominal transmit frequency, e.g. 5 MHz. Echoes from reflecting interfaces are of the same form. A typical echo signal due to a reflection at a single interface is illustrated in Figure 4.16a. It consists of oscillations above and below the zero baseline.

The echo can be described as a modulated sine wave. This means that the amplitude of the sine wave within the echo varies. In the case of a single echo, the amplitude varies from zero at the start of the pulse to a maximum and then back to zero at

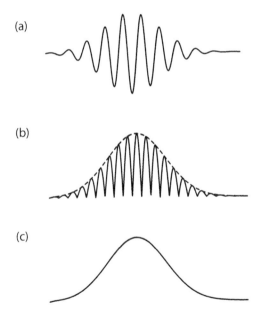

Figure 4.16 The echo signal is demodulated to remove the variations at the transmit frequency and recover the modulation signal, i.e. the echo amplitude information. It is the echo amplitude information that determines the brightness at each point in the B-mode image. (a) The echo signal from a single target is a pulse of several cycles at the nominal transmit frequency and can be described as a modulated sine wave. (b) The modulation can be extracted by first rectifying the pulse, i.e. making all half cycles positive. (c) Applying a low-pass filter removes the transmit frequency component, leaving the modulation signal.

the end of the pulse. This signal contains the transmit frequency and is referred to as the RF signal. It is the amplitude of this signal, i.e. its modulation, that determines the brightness of each point in the B-mode display. To form a B-mode image, it is hence necessary to extract this information. This is done through a process called 'demodulation'.

A simple method of demodulation is illustrated in Figure 4.16b and c. The RF signal is first rectified (Figure 4.16b), i.e. one-half of the waveform is inverted, so that all half cycles are aligned on the same side of the baseline. The rectified signal is then smoothed by passing it through a low-pass filter, which removes the high-frequency oscillations and retains the slowly varying envelope, the modulation (Figure 4.16c). In a B-mode image line, the RF echo signal amplitude varies continuously with time due to reflection and scattering within the tissue. The continuous RF signal can be demodulated as described for the pulse to extract the brightness information.

The demodulation technique described here is a simple technique. More complex techniques are used in commercial ultrasound imaging systems which make more use of the phase information in the modulated echo signal. These give a more faithful representation of the reflection and scattering events within the tissue.

The process of demodulation removes the transmit frequency and all information relating to its phase, retaining only the amplitude information. After demodulation, it is no longer possible to combine echo signals coherently such that they undergo constructive or destructive interference. Adding two brightness signals together will always result in an increase in brightness. All signal and image formation processes following demodulation are referred to as non-coherent processes. Several non-coherent image-forming processes are described later in this chapter.

Dynamic range of echoes

As discussed in Chapter 2, when an ultrasound pulse is incident on an interface or a scatterer, some of the incident intensity is usually reflected or scattered back to the transducer. For reflection at a large interface, as might be encountered at an organ boundary, the reflected intensity ranges from less than 1% of the incident intensity for a tissue-tissue interface to almost 100% for a tissue-air interface. The intensities of echoes received from small scatterers depend strongly on the size of the scatterer and the ultrasound wavelength, but are usually much smaller than echoes from large interfaces. Hence, the range of echo amplitudes detected from different targets is very large.

Figure 4.17 shows the relative voltages at the transducer produced by typical echoes from different targets. In this figure, the reference level used for expressing echo amplitudes in decibels is that from a tissue-air interface. Note that because the echoes are all weaker than that from the tissue-air interface, their decibel levels are negative. Also a voltage ratio of 10 between any two amplitudes corresponds to a difference of 20 dB (see Appendix A). The figure shows that if the echo from a tissue-air interface gives a transducer voltage of 1 V, that due to echoes from blood will be of the order of 10 μV. Even smaller signals will be produced by the transducer, but these are likely to be lost in background electrical noise and cannot be detected. The dynamic range of signals at the transducer is defined as the ratio of the largest echo amplitude which does not cause distortion of the signal, to the smallest that can be distinguished from noise. The dynamic range is expressed in decibels.

Weak echoes produced by scattering within tissue give information about organ parenchyma, while strong echoes from large interfaces give information about the size and shape of organs and other tissue features. To be useful diagnostically, a B-mode image should contain both types of echo. Figure 4.17 shows that the range of echo amplitudes displayed needs to be about 60 dB (−20 to −80 dB compared to a soft tissue–air interface) to include echoes from a typical tissue-tissue interface (e.g. liver-fat) as well as echoes due to scattering within tissue.

However, this range of echo amplitudes cannot be displayed without further processing. The ratio of the brightest level that the viewing screen can display to the darkest is typically about 20 dB. Therefore, if the gain of the B-mode system was adjusted to display an echo signal from a liver-fat interface at peak white level on the display, an echo from a liver-muscle interface would be displayed as the darkest available grey. All weaker echoes from tissue scattering would be displayed at black level and not visible. Alternatively, if the gain was increased to display weak echoes scattered from tissue (say at the −80 dB level in Figure 4.17) as dark grey, the echoes from within the liver would be displayed at peak white. Echoes with greater amplitudes (e.g. from blood-brain, liver-fat interfaces) would also be displayed at peak white and could not be distinguished.

Compression

To allow echoes from organ interfaces and organ parenchyma to be displayed simultaneously in a B-mode image, it is necessary to compress the 60 dB range of the echoes of interest into the 20 dB range of brightness levels available at the display. Compression is achieved using a non-linear amplifier as illustrated in Figure 4.18.

Unlike the linear amplifier described earlier for overall gain, this non-linear amplifier provides more gain for small signals than for large signals. Hence, weak echoes are boosted in relation to large echoes. A graph of the output voltage against input voltage for this amplifier is a curve (Figure 4.18a) rather than the straight line of the linear amplifier. In this example, an input voltage of 2 units gives an output voltage of 4 units (gain of 2) while an input voltage of 6 units gives an output voltage of 6 units

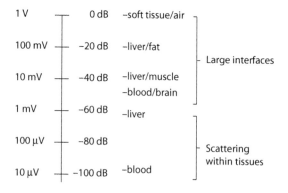

Figure 4.17 Echoes due to reflections at interfaces are much larger than those due to scattering from within tissues, leading to a large range of possible amplitudes of diagnostically relevant echoes. The B-mode image needs to display echoes from tissue interfaces at the same time as those due to scattering within tissue.

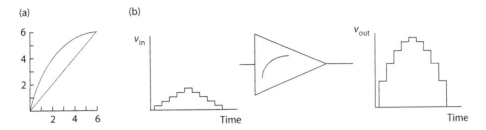

Figure 4.18 **(a)** To compress the dynamic range of echoes, an amplifier with a non-linear gain characteristic is used. **(b)** The amplifier applies more gain to small echoes than to large echoes so that both can be displayed at the same time.

(gain of 1). Normally, the amplifier used to compress the signal range has a logarithmic characteristic, so that the output voltage is related to the logarithm of the input voltage. Using this type of amplifier, weak echoes from scattering within the tissue can be boosted more than the large echoes from interfaces, so that both types of echo can be displayed at the same time.

In practice, the range of echo amplitudes which need to be compressed into the display range depends on the application. For example, when trying to identify local tissue changes in the liver, a wide dynamic range of echoes needs to be

displayed, so that these weak echoes are made to appear relatively bright in the image. For obstetric work, interpretation of the shapes in the image may be aided by reducing the displayed dynamic range, so that weak echoes are suppressed and areas of amniotic fluid are clearly identified.

Commercial B-mode systems provide a dynamic range control (sometimes labelled 'compression'). This control essentially allows adjustment of the curve in Figure 4.18a to alter the dynamic range of echo amplitudes displayed. Figure 4.19 shows the effects of altering the dynamic range control on an image of a liver and right kidney. In Figure 4.19a,

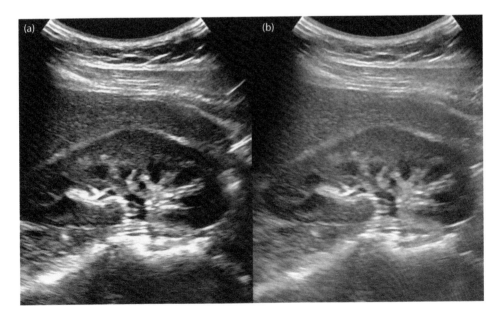

Figure 4.19 Increasing the dynamic range setting from **(a)** 40 dB to **(b)** 80 dB increases the gain for small echoes such as those due to scattering within the liver and kidney. Their brightness in relation to echoes from large interfaces is increased, making it easier to detect tissue irregularities.

the dynamic range is set to 40 dB while in Figure 4.19b it is set to 80 dB. In Figure 4.19a, the echoes from interfaces such as those near the top of the image are much brighter than those from within the liver and kidney. In Figure 4.19b, there is less contrast between echoes from within the liver and kidney and interface echoes. This makes it possible to display tissue echoes at a higher brightness level without saturating the interface echoes.

IMAGE FORMATION AND STORAGE

Scan conversion

The outcome of the various processing steps described so far in this chapter is a set of B-mode image lines corresponding to the particular scanning format, e.g. linear array, curvilinear or sector. Each line consists of a series of digital samples representing the amplitude of the echo signal, which progress in range from those received near the transducer face to those from the maximum imaged depth. The B-mode lines may be assembled and stored in this format in an area referred to in Figure 4.1 as the non-coherent image former. However, before they can be displayed as a B-mode image, they must be converted into a format which is compatible with a standard display monitor.

Display monitors divide up the screen into a rectangular matrix of picture elements (pixels), with each element displaying a level of brightness or colour corresponding to that small part of the image (Figure 4.20). The number of pixels in the matrix normally conforms to one of a number of display standards, such as 800×600 or 1024×768, where the two numbers refer respectively to the number of pixels in the horizontal and vertical directions. The B-mode lines are mapped onto the rectangular pixel matrix by a process called 'scan conversion' (Figure 4.21a). A commonly used scan conversion method is illustrated in Figure 4.21. In Figure 4.21b, the horizontal and vertical grey lines represent the rectangular grid of pixels, where the pixel is located at the intersection of the grey lines. The B-mode line is represented by the oblique black line and individual brightness samples along the line by the black dots. In the case illustrated, the B-mode line passes through the pixel location, but the samples fall before and after it. A brightness value for the pixel location is calculated by

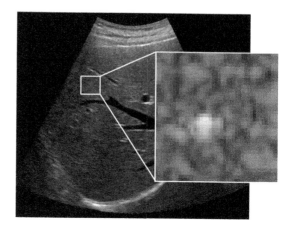

Figure 4.20 Display monitors divide up the screen into a rectangular matrix of picture elements (pixels), with each element displaying a level of brightness or colour corresponding to that small part of the image.

linear interpolation between the neighbouring sample values. In the example shown, the pixel is located 70% of the way between sample 1 and sample 2 (red arrow). So the brightness value assigned to the pixel will be 70% of the way between the values of sample 1 and sample 2.

For most pixels, a B-mode line does not pass directly through its location so interpolation is performed in two directions, referred to as bilinear interpolation. As shown in Figure 4.21c, the four nearest neighbour samples to the pixel are first interpolated in the range direction to the range of the pixel (red arrows) to give interpolated values on each side of the pixel location (red dots). Interpolation is then performed between the samples at the red dots to the location of the pixel (blue arrow).

Bilinear interpolation is a relatively simplistic method of calculating pixel brightness values from B-mode line samples. Alternative interpolation methods which more closely replicate the real profile of the ultrasound beam may be used but such methods require much more complex computations and hence increase the system cost.

Writing to and reading from the image memory

The scan conversion process is performed in real-time on the B-mode image lines during scanning

(a)

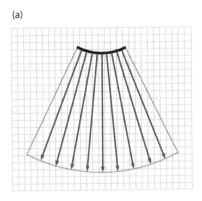

(b)

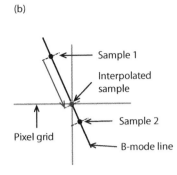

(c)

Figure 4.21 (a) Beam-forming and echo signal processing stages result in a set of B-mode image lines corresponding to the particular scanning format, e.g. curvilinear array. These must be mapped onto the rectangular matrix of picture elements (pixels) of the display monitor by scan conversion. (b) Where a B-mode line passes through a pixel location, the pixel sample value may be calculated by linear interpolation between nearest B-mode line samples. (c) Where B-mode lines do not pass through the pixel location, bilinear interpolation may be used to calculate the pixel value.

to produce a continuous series of image frames in pixel format. Multiple consecutive image frames can be stored in cine memory before further processing and display. The process of storing B-mode information in the image memory is referred to as writing to the memory.

To view the image, it must be read out to the display monitor in a sequence which is compatible with the display format. The display monitor screen is composed of a two-dimensional array of elements, each of which emits light when activated. These are addressed one at a time by the monitor in a sequence of horizontal lines or raster scan, starting at the top left of the screen and finishing at the bottom right. The stored image is read out from the image memory by interrogating each pixel in turn in synchronism with the display raster. The stored digital values may be converted to an analogue signal to control the brightness of the display or sent in digital format. The reading process does not degrade the information stored in the memory and can be repeated indefinitely, if required.

Write zoom

There are two ways to increase the magnification of part of an image. One is to use the conventional depth control, but this involves losing the deeper parts of the image from the display. Hence, features

which are deep within the tissue cannot be magnified by this means. The second method, known as write zoom, was introduced in Chapter 3 and allows the display of a selected region of interest (ROI) remote from the transducer. The user outlines the ROI by adjusting the position and dimensions of an on-screen display box, with the image set to a relatively large depth. When ROI display mode is activated, the B-mode system interrogates only those lines which pass through the ROI. For each of these lines, echoes originating from between the transducer and the near side of the ROI box are ignored. In each pulse-echo sequence, the system waits until echoes arrive from the depth marked by the near side of the box, and writes the subsequent echo amplitude information into the memory on a large scale. Information from within the ROI then fills the image memory.

Read zoom

Read zoom is another way of magnifying part of the image. During normal read-out, the display raster interrogates the whole area of the image memory. Where the imaged depth is larger than the imaged width, this can result in inefficient use of the screen area, and imaged features appear small on the display, as previously described. The read zoom function addresses and reads out a selected part of the stored image defined by the user with an on-screen ROI box similar to that used with the write zoom function. The selected area is expanded to fill the display screen.

An advantage of read zoom over write zoom is that read zoom can be applied to a previously stored image of the total area of interest and the zoomed area moved around to examine different parts of the image on an expanded scale. The disadvantage over write zoom is that when a large image magnification is used, the number of image memory pixels being displayed may become so small that individual pixels become obvious. A higher-quality image can be obtained by using write zoom, but the echo information for this needs to be acquired after the ROI box is set. Read zoom is rather like taking a wide-angle photograph of a scene and then examining parts of it with a magnifying glass. The details are enlarged, but the imperfections of the photographic process may be apparent. Write zoom is the equivalent of using the optical zoom or a telephoto lens on a camera to image a small part of the scene. The image quality and definition are maximized, but there is no information available from other parts of the scene.

IMAGE UPDATE MODES

Real-time display

In most medical imaging processes (e.g. X-ray, computed tomography and magnetic resonance imaging), there is a significant time delay between image acquisition and image display, so that diagnosis is made on information which may be many minutes old. Formation of a single ultrasound B-mode image takes typically of the order of 1/30 s, so that, if repeated continuously, 30 images can be formed each second. This rate is referred to as the frame rate and is measured in hertz. Also, the time delay between image acquisition and display is relatively short (a few tens of milliseconds). This leads to the description 'real-time' imaging, i.e. images of events within the patient are displayed virtually as they happen. The real-time nature of the process is an important aspect of ultrasound imaging and, coupled with the use of a handheld probe in contact with the patient, gives a highly interactive form of investigation. Real-time imaging allows

the study of the dynamic behaviour of internal anatomy, such as the heart, and can aid identification of normal anatomy (e.g. gut from peristalsis) and pathology (e.g. movement of a gallstone).

The display monitor typically displays a complete image in 1/25 s, giving 25 images per second. To avoid conflicts between the reading and writing processes, buffer memories are used to store echo data on a temporary basis, so that the two processes can proceed independently of each other.

Freeze mode

If the image acquisition process (writing) is stopped, so that data in the image memory remain unchanged, but are still read out repeatedly to the display, the image is said to be frozen. All commercial systems have a freeze button, which activates this mode. Freeze mode is used while measurements are made on features in the image and hardcopy records made. Normally, the transducer stops transmitting ultrasound pulses in this mode.

Cine memory

As previously described, during real-time scanning, multiple consecutive image frames may be stored in the cine memory. Most commercial B-mode systems can store up to 2,000 frames. In cine memory, each new image frame is stored in the next available image memory space, eventually cycling around to overwrite memory number 1 when number 2,000 is filled. The image can be frozen at the end of a dynamic event of interest or examination and the last 2,000 frames can be reviewed in real time, or individual frames selected and printed as hard copies. Sections of the stored data can be stored as movie clips as part of a patient report. For the heart, which has a regular, repetitive movement, the recorded frames can be displayed as a cine loop, repeating the cardiac cycle and allowing diagnosis after the examination has been completed.

POST-PROCESSING

Frame averaging

When consecutive B-mode images are formed and stored in the image memory, they contain small frame-to-frame variations due to random imposed electrical noise. If the stored frames are then read out directly to the display in the same sequence, the constantly changing random noise patterns in the image can be very distracting and mask weak displayed echoes. This random noise effect can be reduced by frame-to-frame averaging of images. Random noise events then tend to be averaged out, while constant ultrasonic image features are reinforced, improving the signal-to-noise ratio of the image. Frame averaging does not reduce speckle (see Chapter 5) as the speckle pattern is the same in each frame if there is no movement of the transducer.

In the process described earlier for reading out the image from the image memory to the display, the most recent B-mode frame was read out from the image memory, pixel by pixel in a raster format. This real-time image would contain obvious flickering noise patterns. To achieve frame averaging, the value of each pixel that makes up the raster pattern of the display is calculated as the average value of the corresponding pixels (e.g. row 10, pixel number 35) in each of the most recent few frames. The number of averaged frames is typically in the range of one to five, selected by the user.

Frame averaging is effective in suppressing random image noise, making it easier to study areas of anatomy that generate weak echoes, e.g. weakly scattering or deep tissues. However, the averaging effect slows the time response to changes in the image, resulting in smearing of the image of moving targets (e.g. heart valves). Long persistence (five frames) is useful for optimizing the image signal-to-noise ratio for relatively stationary images such as the liver, whereas short persistence (zero persistence being one frame) must be used to image rapidly moving targets, such as heart valves.

Grey-level transfer curves

In addition to controlling the dynamic range, as described earlier, most modern B-mode systems allow further modification of the gain curve that relates echo amplitude to displayed brightness level. Such processing would normally be applied to the stored image as it is read out from the image memory to the display and hence comes under the heading of post-processing. Figure 4.22 shows a set of commonly used grey-level curves (also referred to as gamma curves). In Figure 4.22a, there is

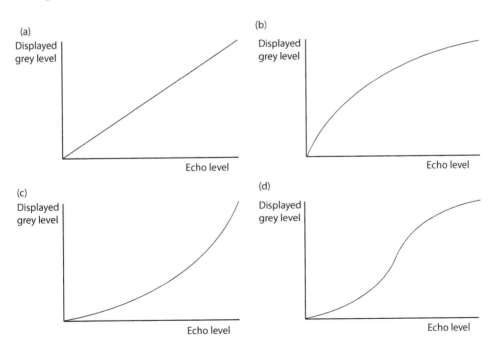

Figure 4.22 Greyscale post-processing curves. (a) linear, (b) small echo contrast enhancement, (c) high echo contrast enhancement and (d) mid-echo contrast enhancement.

a linear relationship between the stored image brightness level and the displayed image brightness, with contrast assigned equally to all levels. In Figure 4.22b, more contrast is assigned to low-level echoes such as those from within the liver to aid differentiation of normal from abnormal tissue regions. In Figure 4.22c, more grey levels are assigned to higher echo levels to aid diagnosis for organs whose tissues scatter more strongly and give a brighter image. Figure 4.22d shows a grey-level curve that may be used to enhance contrast for mid-level echoes. A selection of grey-level curves such as these is available to the user via the set-up menu of most commercial B-mode systems. Such grey-level processing curves may also be selected automatically by the system as part of the optimization of system parameters for particular clinical applications. For example, if the user selects a vascular application setting on the system, the grey-level curve chosen is likely to be different from that chosen to optimize for general abdominal or musculo-skeletal applications.

Edge enhancement and image smoothing

The B-mode images formed and stored in the image memory may contain a number of imperfections, such as indistinct or incomplete organ boundaries, due to variable orientations of the target interfaces to the direction of the ultrasound beam, or prominent speckle (see Chapter 5). Further image-processing methods would normally be applied to reduce such imperfections and improve the diagnostic quality of the images.

Where the targets of interest include blood vessels or anatomical interfaces, the boundary echoes can often be made clearer or more complete by applying edge enhancement. The echo patterns across features such as vessel walls contain abrupt changes in image brightness. These abrupt changes give rise to higher spatial frequencies in the image. To enhance the appearance of the walls of vessels and other anatomical features, a spatial high-pass filter can be applied to the two-dimensional image. This has the effect of making the boundary echoes more prominent. Normally, different levels of edge enhancement can be selected by the user to suit the particular clinical investigation.

The B-mode image normally contains various forms of image noise such as pixel sampling noise and speckle (see Chapter 5). These would normally be suppressed by applying a low-pass spatial filter to the image, i.e. one with the opposite characteristic to the edge enhancement filter, which suppresses changes in brightness across the image and has a smoothing effect. The limitation of both types of spatial filter is that they are applied uniformly to the whole image and so cannot be applied at the same time. Their characteristics are set by the user to suit the clinical application and do not change in response to changes in the image.

Adaptive image processing

Adaptive image processing is now widely used in commercial B-mode imaging systems to allow edge enhancement and image smoothing to be applied to different elements of the image at the same time. The process is normally applied frame by frame on the stored image so that it adapts in real time to the changing characteristics of the image.

Adaptive image processing is a multi-stage process which analyses the image content and applies processing according to local image features. Each scan converted image is decomposed into a set of frames which separate image features according to their scale. This is achieved by spatial filtration so that large-, medium- and small-scale image features appear in separate frames. Each frame is analysed to identify low contrast isotropic regions dominated by noise and speckle separately from high contrast features. High contrast features are analysed to identify anisotropic elements such as organ boundaries and estimate the direction of anisotropy, i.e. the orientation of the boundary in the image. Following analysis of each decomposed frame, region-by-region spatial filtering is applied. In low-contrast regions and high-contrast regions with no anisotropic features, some smoothing is applied to suppress noise and speckle. For anisotropic features such as organ boundaries, smoothing is performed along the direction of the boundary and edge enhancement across the boundary. Following the filtration stage, the separated image frames are added back together to form the final processed image.

Adaptive image processing improves the appearance of B-mode images by enhancing tissue contrast and border definition without degrading the image content. It adapts in real time to the constantly changing image (Meuwly et al. 2003).

DISPLAY, OUTPUT, STORAGE AND NETWORKING

A modern ultrasound system has a flat-screen display, has some capacity locally for storing video clips and images, and is connected to an imaging network, so that any hard copy is performed using a high-quality communal printer, and all data are archived in a large central datastore. The era of the stand-alone ultrasound system which is connected to its own printer and own video recorder is rapidly becoming a thing of the past. Many older ultrasound systems still in use will have older technology: video recorders, thermal printers, cathode ray tube displays, etc. This section concentrates on flat-screen network-based ultrasound systems. Details of older technology can be found in the first edition of this book (Hoskins et al. 2003).

Display

Recent years have seen a move away from traditional cathode ray tube displays, to flat-screen displays. Generally LCD (liquid crystal display) or LED (light-emitting diode) monitors are used. These are widely used as computer and TV displays and are able to produce high-quality images suitable for real-time video data from ultrasound systems.

Local storage

A modern ultrasound imaging system usually contains an on-board computer. Individual images and image sequences may be stored on the local hard disc, and retrieved and displayed as necessary. Older ultrasound systems tended to store image and video data in proprietary formats, which made it difficult to transfer images to an imaging network (see section 'Integration with an Imaging Network') or for the operator to incorporate digital images into reports and slide shows. Modern systems allow storage of images in several widely used formats, such as avi and mpeg for video, tiff and jpeg for individual images, and DICOM format for

networking. They also have computer-compatible recording facilities such as a USB port and a DVD writer.

Integration with an imaging network

The modern radiology department is based on imaging systems, including ultrasound scanners, which are connected to an imaging network. This system is generally referred to as a picture archiving and communication system (PACS). A PACS typically has workstations for review and reporting of image data, printers for producing hard copy of images and reports, and a large memory store for archiving patient data. Most reporting for ultrasound is performed in real time, with the operator saving images or video clips which illustrate the lesion (if present) or the measurement. A hard copy of images for inclusion in patient notes is usually done using a high-quality laser printer, though there may be no hard copy of images, with inclusion of only the report in the notes. Instead, if the referring clinician wants to inspect the images he or she can do so at a local terminal, connected to the hospital network, on the ward or in the clinic.

Compared to film-based radiology, PACS has considerable advantages including rapid access of images throughout the hospital, reduced requirement for hard copy and hard copy storage, reduction in lost images, and hence reduction in repeat examinations due to lost images.

As PACS involves technologies which evolve relatively rapidly, further up-to-date information is best found at relevant websites such as Wikipedia.

ACKNOWLEDGEMENT

The images in Figure 4.19 were obtained with the assistance of Hitachi Medical Systems, United Kingdom.

QUESTIONS

Multiple Choice Questions

Q1. The 'B' in B-mode stands for:
   a. Boring
   b. Beam

c. Bulk modulus

d. Brightness

e. Beam forming

Q2. The purpose of the TGC (time-gain control) is to:

a. Increase transmit power for ultrasound which travels to deeper depths

b. Decrease gain with depth to account for attenuation of echoes from deeper depths

c. Ensure the gain is constant for echoes from all depths

d. Ensure that the gain remains fixed for all times of the day

e. Increase gain with depth to account for attenuation of echoes from deeper depths

Q3. Which of the following are not parts of the B-mode system components?

a. Receive beam-former

b. Transmit power control

c. Amplitude processing

d. Doppler gain

e. PACS

Q4. Demodulation refers to:

a. Compression of the received ultrasound signal

b. Removal of the transmit frequency components from the received ultrasound signal

c. Adding consecutive frames to reduce noise

d. Compression of the transmit ultrasound signal

e. Removal of low-frequency components

Q5. If signal compression is not used, then the B-mode image shows:

a. Echoes from strong reflectors such as bone but none from tissue scattering

b. Only echoes from fat

c. Only echoes from weak reflectors such as blood

d. Nothing at all

e. Only echoes from bone

Q6. The echo amplitude from blood is how many decibels less than that from liver?

a. 0 dB

b. 100 dB

c. 1000 dB

d. 10 dB

e. 30 dB

Q7. Concerning coded excitation:

a. It is concerned with the collection of data at the third harmonic frequency

b. The transmit pulse is longer than usual and forms a code

c. It splits the receive echoes into different frequency bands

d. It results in increased penetration depth

e. It results in less speckle

Q8. Concerning high-amplitude pulse propagation through tissues:

a. The waveform changes with distance, developing sharp transitions in pressure

b. The frequency composition changes with increased amplitude at the second but not other harmonics

c. The harmonic energy is located mainly on the beam axis but not much near the transducer face

d. The frequency composition changes with increased amplitude at the second and higher harmonics

e. The harmonic energy is located mainly at the transducer face and not much at greater depths

Q9. Harmonic processing may involve:

a. Transmission at the fundamental frequency and image formation with the linear component of the received echoes

b. Transmission at the fundamental frequency and image formation using the second harmonic frequency

c. Transmission at the fundamental frequency and image formation using the third harmonic frequency

d. Transmission at the fundamental frequency and twice the fundamental frequency with image formation using both of these frequencies

e. Transmission at twice the fundamental frequency and reception at the same frequency

Q10. Harmonic processing is useful for:

a. Removing speckle in imaging of the fetus

b. Improving image quality in cardiology by removal of haze and clutter within the chambers

c. Reducing beam distortions due to refraction artefacts from the superficial fat layers

d. Improving the frame rate

e. Improving the field of view

Q11. Splitting receive echoes into different frequency bands then compounding the images is called:

a. Harmonic imaging

b. Spatial filtering

c. Frequency compounding

d. Compound imaging

e. Coded excitation

Q12. Compound imaging involves:

a. Using only the current image but adding data from different parts of the image

b. Adding frames together, each frame taken with the beam steered in a different direction

c. Collecting images with different transducers and adding them together

d. Collecting images at different frequencies and adding them together

e. Adding previous frames together, each frame taken using an unsteered beam

Q13. Scan conversion is used to:

a. Convert sector scan lines to a square array of pixels

b. Convert a linear array image to a sector image

c. Change the clinical application preset

d. Change the transducer frequency

e. Demodulate the echo signals

Q14. Interpolation of imaging data involves:

a. Adjusting the transmit power to ensure that echoes from deep tissues are seen

b. Removing the high-frequency components of the signal

c. Creating additional echoes between true echoes

d. Filling in pixel values in the B-mode image which are located between ultrasound beam lines

e. Averaging frames to reduce noise

Q15. Concerning spatial filtering:

a. This can be performed on each frame in turn

b. This can only be performed when there are several frames present

c. This is performed on the RF signal

d. Typical spatial filtering operations are smoothing and edge enhancement

e. This is performed on the final processed data just before they are displayed

Q16. Write zoom involves:

a. Magnifying an image written to the memory to study small detail

b. Increasing the transducer frequency

c. Writing the image more quickly to the image memory

d. Increasing the speed of sound assumed by the imaging system

e. Imaging a selected region of interest on a larger scale before writing to the image memory

Q17. Which of the following statements are true concerning post-processing?

a. It is designed to degrade image quality

b. Image filtering is an example of post-processing

c. It is designed to improve image quality

d. Beam formation is an example of post-processing

e. Frame averaging is an example of post-processing

Q18. Frame averaging (persistence):

a. Reduces random electronic noise in the image

b. Increases the frame rate

c. Reduces speckle

d. Should be used for rapidly moving targets

e. Is used to interpolate between pixels

Q19. B-mode speckle may be reduced by:

a. Harmonic imaging

b. Spatial filtering

c. Frequency compounding

d. Compound imaging

e. Coded excitation

Q20. Grey-level transfer curves (gamma curves):

a. Relate the stored pixel level to the displayed image brightness

b. Are used to control the gain at different depths

c. Are used to interpolate between pixels

d. Can be used to enhance low-level echoes

e. Are applied before storage in the image memory

Short-Answer Questions

Q1. How do the overall gain and the transmit power affect the B-mode image? How should the operator optimize their settings?

Q2. A 5 MHz transducer is used to image a region of tissue which attenuates at a rate of 0.7 dB cm^{-1} MHz^{-1}. Explain how the scanning system compensates for the attenuation and how the operator sets the TGC.

Q3. Explain what is meant by the dynamic range of echoes received at the transducer. Why is it necessary to be able to display a wide dynamic range of echoes?

Q4. Why is it necessary to compress the dynamic range of echoes received at the transducer and how is this achieved by the scanning system? How does the dynamic range setting affect the B-mode image?

Q5. Describe the trade-off that the operator must make in choosing the ultrasound transmit frequency. How can coded excitation help?

Q6. How is a tissue harmonic image formed?

Q7. What are the benefits of tissue harmonic imaging?

Q8. Explain what is meant by 'interpolation' and where it is used in the formation of ultrasound images.

Q9. Explain the meanings of the terms 'write zoom' and 'read zoom'. What is the disadvantage of write zoom?

Q10. What are the benefits and limitations of frame averaging (persistence)?

REFERENCES

Averkiou MA. 2001. Tissue harmonic ultrasonic imaging. *Comptes Rendus de l'Académie des Sciences Paris*, 2: Serie IV, 1139–1151.

Averkiou MA, Mannaris C, Bruce M, Powers J. 2008. Nonlinear pulsing schemes for the detection of ultrasound contrast agents. *155th Meeting of the Acoustical Society of America*, Paris, 2008, 915–920.

Chiao RY, Mo LY, Hall AL et al. 2000. B-mode blood flow (B-flow) imaging. *IEEE Ultrasonics Symposium*. 1469–1472.

Chiao RY, Hao X. 2005. Coded excitation for diagnostic ultrasound: A system developer's perspective. *IEEE Transactions on Ultrasonics, Ferroelectrics, and Frequency Control*, 52, 160–170.

Chiou SY, Forsberg F, Fox TB, Needleman L. 2007 Comparing differential tissue harmonic imaging with tissue harmonic and fundamental gray scale imaging of the liver. *Journal of Ultrasound Medicine*, 26, 1557–1563.

Hoskins PR, Thrush A, Martin K, Whittingham TA. 2003. *Diagnostic Ultrasound Physics and Equipment*. London: Greenwich Medical Media.

Meuwly JY, Thiran JP, Gudinchet F. 2003. Application of adaptive image processing technique to real-time spatial compound ultrasound imaging improves image quality. *Investigative Radiology*, 38, 257–262.

Nowicki A, Klimonda Z, Lewandowski M et al. 2006. Comparison of sound fields generated by different coded excitations – Experimental results. *Ultrasonics*, 44, 121–129.

Pedersen MH, Misarids TX, Jensen JA. 2003. Clinical evaluation of chirp-coded excitation in medical US. *Ultrasound in Medicine and Biology*, 29, 895–905.

Whittingham TA. 2005. Contrast-specific imaging techniques; technical perspective. In: Quaia E (Ed.), *Contrast Media in Ultrasonography. Basic Principles and Clinical Applications*. Berlin: Springer. pp. 43–70.

5

Properties, limitations and artefacts of B-mode images

KEVIN MARTIN

INTRODUCTION

The B-mode image-forming processes described so far have assumed an ideal imaging system interrogating an ideal tissue. As described in Chapter 2, real ultrasound beams have significant width and structure, which change with distance from the transducer, and ultrasound pulses have finite length. The speed of sound and the attenuation coefficient are not the same in all tissues. Real properties such as these give rise to imperfections in the image, which are essentially all artefacts of the imaging process. However, those that are related primarily to the imaging system (beam width, pulse length, etc.) are usually considered as system performance limitations, as they are affected by the design of the system. Those that arise due to properties of the target tissue (e.g. changes in attenuation and speed of sound) are considered as artefacts of propagation.

IMAGING SYSTEM PERFORMANCE LIMITATIONS

The performance of a particular B-mode system can be characterized in terms of image properties which fall into three groups, i.e. spatial, amplitude and temporal. At the simplest level, spatial properties determine the smallest separation of targets which can be resolved. The amplitude properties determine the smallest and largest changes in scattered or reflected echo amplitude which can be detected. The temporal properties determine the most rapid movement that can be displayed. However, the ability to differentiate between neighbouring targets or to display targets clearly, may depend on more than one of these property types.

Spatial properties

A B-mode image is a scaled two-dimensional (2D) map of echo-producing features within the target tissues and is designed to give a faithful representation of their size, shape and relative position. Ideally, a point target would be represented by a bright point in the image. However, this would require that the ultrasound beam have zero width and the pulse have zero length. As described in Chapter 2, real ultrasound beams have finite width due to diffraction effects and the ultrasound pulse must have finite duration

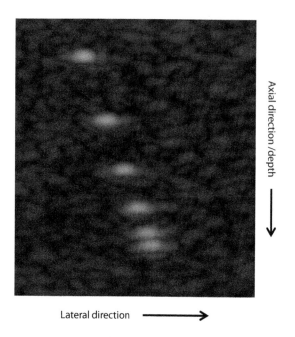

Figure 5.1 B-mode image of a set of test phantom filaments showing the spread of the images. The lower two images almost merge.

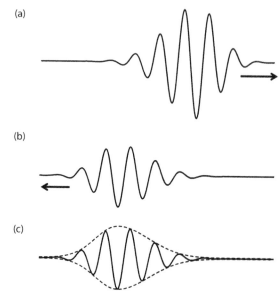

Figure 5.2 **(a)** A typical transmit pulse contains several cycles at the ultrasound transmit frequency. **(b)** The received echo has a similar shape to the transmit pulse but is much smaller. **(c)** The axial brightness profile in the image is a representation of the pulse envelope.

in order to define the transmit frequency. Hence, the image of a point target produced by a real imaging system is spread out so that it appears as a blurred dot or streak. This also means that the blurred images from two adjacent point targets may merge so that they cannot be distinguished. Figure 5.1 shows the image from a test phantom containing a set of fine filament targets. It can be seen that the filaments do not show as fine points in the image and that the images from the two lower filaments almost merge.

AXIAL RESOLUTION

When a point target is imaged by a real imaging system, the spread of the image in the axial direction is determined by the length of the ultrasound pulse. Figure 5.2a illustrates the shape of a typical transmit pulse. It contains several cycles at the ultrasound transmit frequency. The echo of this pulse returned from a fine filament target is much smaller in amplitude, but has a similar shape as shown in Figure 5.2b. When the echo signal is received by the ultrasound system, it is demodulated as described in Chapter 4 to produce the brightness signal. The brightness signal from the

echo and its profile in the axial direction follow the envelope of the echo signal as shown in Figure 5.2c.

Axial resolution is defined usually as the smallest separation of a pair of targets on the beam axis which can be displayed as two separable images. Figure 5.3a shows the brightness profiles of echoes from two filaments on the beam axis separated by more than the length L of the transmit pulse along the beam axis. The echoes are well separated as the round trip from the first to the second filament is twice the distance shown between them. The profiles do not overlap and the images are separate.

When two filaments are separated by $L/2$, as in Figure 5.3b, the two-way trip from filament 1 to filament 2 involves a total distance L, and the leading edge of the echo from filament 2 meets the trailing edge of the echo from filament 1. Hence, the axial brightness profiles of echoes from the two filaments just begin to merge. As the filaments are brought closer together (Figure 5.3c), the brightness profiles of the echoes cannot be distinguished in the image. At a separation of approximately $L/2$, the images are just separable. Thus, the axial resolution of a B-mode system is approximately half

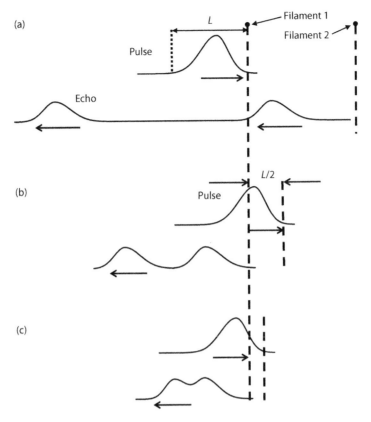

Figure 5.3 **(a)** A pulse of length L reflected from filament targets spaced along the beam axis by more than L, generates well separated echoes as the round trip distance between the targets is twice their spacing. **(b)** When the target spacing is $L/2$, the round trip distance between them is L and the echoes just begin to merge. **(c)** As the filaments are brought closer together, the echoes cannot be distinguished in the image.

the transmit pulse length and targets which are closer than this cannot be resolved separately.

LATERAL RESOLUTION

The images of the filaments in Figure 5.1 can be seen to have finite width as well as finite depth. The B-mode image is assembled from a set of adjacent lines of echo information as described in Chapter 3. The axial brightness profile of the echo from a filament as previously described will appear on the line which intersects the filament. If the ultrasound beam had zero width, it would appear on only a single line. However, because the beam has finite width, echoes also appear on adjacent image lines. Figure 5.4a shows a graph of echo amplitude plotted against distance across the beam, i.e. the lateral profile of the ultrasound beam. This is

essentially a graph of the transmit/receive response of the system to a single filament placed at different points across the beam. The response is strongest when the filament is on the beam axis and becomes weaker as it moves away from the central position. The position of the B-mode line in the image normally corresponds to the position of the beam axis, so a filament target in the edge of the beam will register as a weak echo in the position of the beam axis (Figure 5.4b). The ultrasound beams that form the image are usually close enough together for their profiles to overlap so that the echo from a single filament appears on several adjacent lines.

Figure 5.5 shows how the image of a point target appears on several adjacent lines. A filament is imaged with a stepped array system, which has a practical lateral beam width. Echoes from the

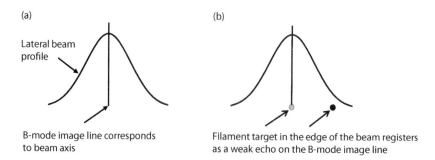

(a) Lateral beam profile

B-mode image line corresponds to beam axis

(b) Filament target in the edge of the beam registers as a weak echo on the B-mode image line

Figure 5.4 **(a)** The position of the B-mode image line corresponds to the axis of the beam but the lateral beam profile spreads to each side. **(b)** A filament target adjacent to the beam axis is overlapped by the edge of the beam. A weak echo is registered on the B-mode image line.

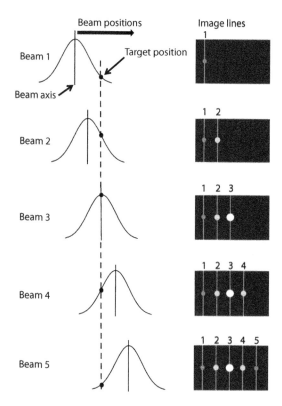

Beam positions
Image lines

Beam 1
Target position
Beam axis

Beam 2

Beam 3

Beam 4

Beam 5

Figure 5.5 Echoes are registered in the image at each beam position where the beam overlaps the target. The resulting lateral spread of the image depends on the beam width at the target depth.

filament target are returned at each beam position where the beam overlaps the target. However, the imaging system assumes that the beam has zero width, and so displays each echo on the image line corresponding to the current beam axis.

At beam position 1, the right-hand edge of the beam just intercepts the target and a weak echo is produced, but is displayed on image line 1 corresponding to the current position of the beam axis. At beam position 2, the target is still off axis, but in a more intense part of the beam. A larger echo is produced, which is displayed on image line 2. At beam position 3, the target is on axis and the largest echo is produced and displayed on image line 3. At beam positions 4 and 5, the target is in progressively less intense parts of the beam, giving weaker echoes, which are displayed on image lines 4 and 5. The resulting brightness profile across the image due to the point target is similar in width and shape to the beam profile at that depth.

The lateral resolution of an imaging system is usually defined as the smallest separation of a pair of identical point targets at the same range or depth in the image plane which can be displayed as two separable images. Figure 5.6a shows the lateral brightness profiles along a line in the image passing through two filament targets. Here, the filaments are separated by a distance which is greater than the beam width, and two distinct images are displayed. As the filaments are moved closer together, their brightness profiles meet when their separation equals the beam width (Figure 5.6b). They then effectively overlap until there is no discernible reduction in brightness between the two and their images are not separable (Figure 5.6d). At a critical separation, the images are just separable (Figure 5.6c). The separation at this point, which is about half the beam width, is a measure of the lateral resolution at that range. Thus, at best, lateral resolution is half the beam width.

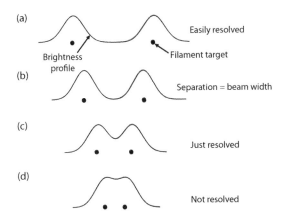

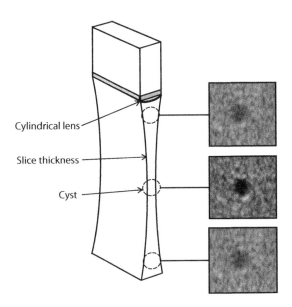

Figure 5.6 **(a)** The lateral brightness profiles from a pair of filament targets at the same depth spaced by more than the beam width are well separated and the targets are easily resolved in the image. **(b)** When the target separation equals the beam width, the lateral brightness profiles meet. **(c)** At a critical spacing of about half the beam width, the targets are just resolved. **(d)** At smaller separations, the brightness profiles overlap and the targets are not resolved.

Figure 5.7 The slice thickness is usually greater than the beam width in the scan plane and can lead to infilling of small cystic structures in the image, especially outside the focal region.

As described in Chapters 2 and 3, the beam width varies with distance from the transducer. It is narrow in the focal region, but wider at other depths. Hence lateral resolution may change with depth. Small lateral beam width and hence minimal image spread can be achieved by using multiple transmit focal zones and swept focus during reception. As described in Chapter 3, image processing techniques are now in commercial use which allow transmit focusing at all depths without resorting to multiple transmit beams focused to different depths, avoiding the associated time penalties.

SLICE THICKNESS

Lateral resolution as just described refers to pairs of targets within the scan plane. The ultrasound beam has significant width also at right angles to the scan plane, giving rise to the term 'slice thickness'. Slice thickness is determined by the width of the ultrasound beam at right angles to the scan plane (the elevation plane) and varies with range. Conventional array transducers have fixed focusing in this direction due to a cylindrical acoustic lens attached to the transducer face (Figure 5.7).

As the transducer aperture is limited in the elevation plane, focusing is relatively weak and so slice thickness is generally greater than the beam width that can be achieved in the scan plane, where wide apertures and electronic focusing are available.

The effect of slice thickness is most noticeable when imaging small liquid areas, such as cysts and longitudinal sections of blood vessels (see Figure 5.7). The fluid within a simple cyst is homogeneous and has no features which can scatter or reflect ultrasound. Hence, it should appear as a black, echo-free area on the image. The surrounding tissues, however, contain numerous small features and boundaries, which generate a continuum of echoes. A small cyst, imaged by an ideal imaging system with zero slice thickness, would appear as a clear black disc within the echoes from the surrounding tissues. However, when a small cyst is imaged by a real imaging system whose slice thickness is comparable to or larger than the diameter of the cyst, the slice may overlap adjacent tissues, generating echoes at the same range as the cyst. Such echoes are displayed within the cystic area in the image as if they were from targets, such as debris, within the cyst. The same artefact, often referred to as 'slice thickness artefact', is observed in a longitudinal image of a blood vessel whose diameter is comparable to, or smaller than, the slice thickness (Goldstein and Madrazo 1981).

For a conventional array transducer with a single row of elements and a fixed cylindrical lens, the beam width is fixed in the elevation direction, and there is little the operator can do to reduce this artefact, other than selecting a transducer with an elevation focus that matches the target depth. Slice thickness artefact can be reduced by the use of a multi-row transducer (see Chapter 3). Multi-row transducers are now available commercially and achieve reduced slice thickness by using electronic focusing and/or variable aperture in the elevation direction.

Image contrast

Spatial resolution is an important aspect of imaging system performance and is normally assessed in terms of the system response to a small, high-contrast target such as a nylon filament, i.e. a target which generates an echo which is much brighter than those from the surrounding medium. Contrast resolution describes the ability of the imaging system to differentiate between a larger area of tissue and its surroundings, where the overall difference in displayed brightness is small. The brightness of an echo from a large interface between tissues is determined by the reflection coefficient of the interface as described in Chapter 2. The brightness of echoes representing scatter from within tissues varies according to the tissue type and its state. The absolute value of brightness of such echoes in the display is affected by gain settings, frequency, etc. and is not normally of direct value in diagnosis. However, the use of relative brightness to differentiate tissues within the image is an important aspect of ultrasound diagnosis. The overall relative brightness of echoes within different organs (liver, spleen and kidney) is an aid to identification and can reflect pathological change. Of more importance to diagnosis, small local changes in echo brightness are often related to pathological change in that part of the tissue.

The smallest change in brightness which can be displayed clearly by the imaging system is limited by image noise, i.e. random fluctuations in brightness in different parts of the image. For all imaging systems, some noise arises from electronic processes in the image detection and processing circuits. In ultrasound images, there is also acoustic noise, the most important type of which is speckle.

SPECKLE

Speckle can be seen as a prominent granular texture in B-mode images, especially in areas of otherwise uniform parenchyma (Figure 5.8). It is tempting to imagine that the granular pattern represents genuine structure within the tissue. However, this is not the case. The pattern is random and arises from the coherent addition of echoes scattered from small-scale structures within the tissue. These structures are random in spacing and scattering strength, leading to random variations in image brightness. Speckle can severely restrict the detection of small genuine changes in echo brightness from neighbouring tissues and obscure fine anatomical detail.

As described earlier, echoes from interfaces or scatterers which are separated by more than approximately half a beam width (lateral and elevation) and half a pulse length are displayed as separate regions of brightness in the image. However, targets that are closer together than this cannot be resolved as separate features in the image. Another way of stating this is to say that targets which lie within a volume, called the sample volume, defined

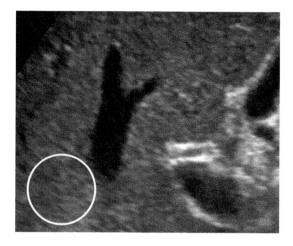

Figure 5.8 Echoes from small scatterers within the sample volume have random amplitude and relative phase. These add coherently to produce random fluctuations in the image brightness (as within the circle) called speckle.

by these three dimensions, cannot be resolved separately. The brightness displayed at any point in the image is the result of the coherent addition of echoes from scatterers within the sample volume.

By coherent addition, we mean that the size of the resulting echo is determined by the amplitudes and phases of the combining echoes. Figure 5.9 shows two examples of coherent addition of echoes. In Figure 5.9a, echo 1 is the result of scattering from a target off to the right and is returning to the transducer to the left. The overall shape of the echo shows the axial extent of the sample volume. If a second echo is generated by a scatterer slightly further to the right, such that it lags behind echo 1 by half a wavelength ($\lambda/2$), the peaks of echo 1 align with the troughs of echo 2 and they add destructively to produce a low amplitude echo. In Figure 5.9b, the second echo is from a scatterer that is slightly further again to the right than in the first case resulting in an echo that lags the first by a full wavelength. The peaks and troughs of the two echoes now coincide giving constructive interference and a large echo amplitude. The amplitudes of individual echoes are determined by the scattering strengths of the individual scatterers. The relative phase is determined by the distance of the scatterer from the transducer and whether it produces a normal or inverted echo, i.e. whether it has a positive or a negative reflection coefficient.

The simulation of two echoes combining from adjacent scatterers shown in Figure 5.9 can be extended to a line of scatterers along the beam axis. Figure 5.10a shows the scattering strength and sign (positive or negative) of the line of scatterers. Their spacings are random but much less than the length of the echo from a single scatterer shown in Figure 5.10b so that their echoes overlap, producing a continuous echo signal as shown in Figure 5.10c. In the continuous echo signal, the peak of each echo has been aligned with the position of the scatterer producing it. It can be seen that there is no correlation between the amplitude of the continuous echo signal and the scattering strengths and signs of the scatterers. The amplitude is large where the echoes happen to combine constructively and small where they tend to cancel out, but the variations are essentially random. This one-dimensional (1D) simulation describes only the interaction of echoes from scatterers along the beam axis. As the sample volume also has width in the lateral and elevational directions, scattered echoes from off-axis positions within the sample volume also contribute to the amplitude of the echo signal from each depth. However, the final echo pattern is still random in nature.

It might be noted that in the 1D simulation, the variations in the amplitude of the echo tend to be of similar extent to the length of the single pulse,

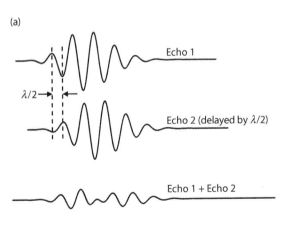
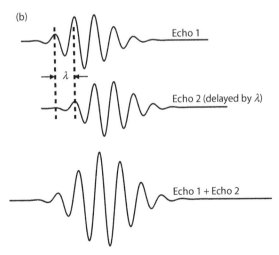

Figure 5.9 **(a)** Combination of echoes separated in time by half a wavelength ($\lambda/2$) results in destructive interference and low echo amplitude. **(b)** Combination of echoes separated in time by a wavelength (λ) results in constructive interference and high echo amplitude.

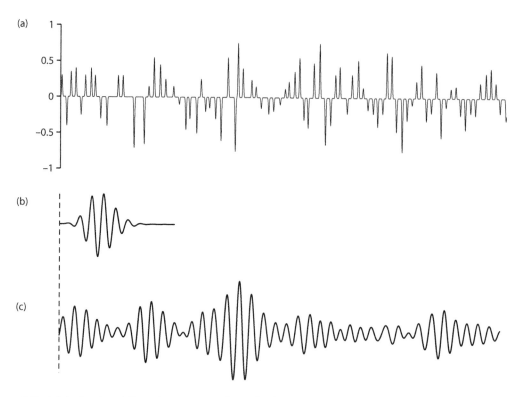

Figure 5.10 **(a)** A simulated line of scatterers along the beam axis has randomly varying spacing and scattering strength. However, the spacing is typically much less than the length of the echo from a single scatterer shown in **(b)**. **(c)** The line of scatterers gives rise to a continuous echo signal whose amplitude varies randomly due to the random way in which the echoes from each scatterer interact.

i.e. the axial extent of the sample volume. In a 2D B-mode image, the form of the speckle pattern tends to be related to the dimensions of the sample volume. For example, where the beam width in the lateral direction is greater than the pulse length, the pattern includes prominent streaks in the lateral direction whose lengths are similar to the beam width and axial extent similar to the pulse length. In the focal region of the beam, where the beam width is small, such streaks are much shorter. Figure 5.11 shows an example of speckle in a B-mode image. It can be seen that the speckle pattern is relatively fine in the region where the beam is focused. Beyond the focus, the beam diverges, leading to lateral streaks in the speckle pattern. The effect of beam width on the speckle pattern can be more exaggerated in phased-array images. The phased-array transducer has a small aperture and so is unable to achieve effective focusing in the

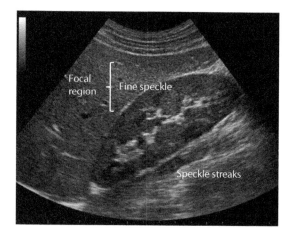

Figure 5.11 In this image of the liver and right kidney, the speckle pattern is relatively fine in the focal region. Beyond the focal region, the speckle pattern shows prominent streaks in the lateral direction.

deeper parts of the image. The lateral resolution in this part of the image may be poor, giving rise to prominent lateral streaks in the speckle pattern.

CONTRAST RESOLUTION

Contrast resolution refers to the ability of the imaging system to resolve areas of tissue whose displayed brightness is different from that of the surrounding tissue. As speckle introduces such large random variations in brightness, contrast resolution must be defined in terms of the statistics of the brightness variations. These may be described by the mean and standard deviation of brightness in the lesion or background tissue. The mean is the average brightness level over a defined area. The standard deviation (σ) is a measure of the typical degree of variation in brightness. A large standard deviation implies large random variations in brightness.

The contrast of a given lesion may be defined as the ratio of the mean difference in brightness (between the lesion and background) to the mean background brightness (Szabo 2014):

$$\text{Contrast} = \frac{\text{Mean } B_L - \text{Mean } B_B}{\text{Mean } B_B}$$

However, the detectability of the lesion, that is whether the lesion can be clearly observed in the image, will be affected by the level of acoustic noise present, i.e. speckle. A useful measure which takes this into account is the contrast-to-noise ratio (CNR). CNR is defined as the ratio of the mean difference in brightness to the standard deviation of the background brightness (σB_B):

$$\text{CNR} = \frac{\text{Mean } B_L - \text{Mean } B_B}{\sigma B_B}$$

If the mean difference in brightness is less than the standard deviation of the background brightness, then the lesion may not be detectable. Lesion detectability is also affected by the size of the lesion in relation to the typical dimensions of the speckle. The lesion will be easier to detect if it is much larger than the typical speckle size. Clearly, if the standard deviation of the speckle brightness is reduced, then the lesion will also be easier to detect.

Contrast resolution may also be made worse by other sources of acoustic noise. For example, in the case of a lesion of low echogenicity, e.g. a cyst, side lobes and grating lobes may result in the display of echoes within the cyst, reducing contrast with the surrounding tissue. The process is similar to that described earlier for lateral resolution. When the beam axis passes through the cyst, side lobes and grating lobes may overlap adjacent tissue, generating weak echoes which are displayed on the beam axis, i.e. within the cyst. Manufacturers design their transducers and beam-forming techniques to minimise the effects of such lobes on lesion contrast.

SPECKLE REDUCTION

Developments in ultrasound technology have helped to reduce the impact of speckle, including some developments aimed specifically at speckle reduction (Chang and Yoo 2014). From the previous description of speckle and the image in Figure 5.11, it can be seen that reducing the size of the sample volume leads to a finer (i.e. smaller scale) speckle pattern which is less distracting. As transducer technology and system sensitivity have improved, manufacturers have been able to use higher frequencies without loss of penetration depth. Higher frequencies lead to shorter pulses and reduced beam width, leading to a small-scale speckle pattern.

Compound scanning also leads to suppression of brightness variation due to speckle in ultrasound images. In compound scanning, as described in Chapter 3, the direction of the ultrasound beams is changed on each successive image frame, producing a set of overlapping images from up to nine different directions. The sample volume at each point in the image is oriented in a slightly different direction on each frame leading to a different speckle pattern from each. When the images are superimposed, the random speckle patterns are averaged out, giving an overall reduction in brightness variation due to speckle, but reinforcement of genuine features in the image. Note that it is effectively the envelopes of the echo signals from each image that are added together in this case rather than the radiofrequency signals (see Chapter 4). Addition of the radiofrequency signals from each, as previously described for speckle formation, would simply produce an alternative speckle pattern.

Frequency compounding, as described in Chapter 4, has also been used to reduce speckle. In this technique, a short transmit pulse is used, which means that it contains a wide range of frequencies. The wider range of frequencies in the returned echoes are then separated into three narrower frequency bands using band-pass filters, and three separate B-mode images are formed. The different frequencies used in each image give rise to different sample volumes and scattering behaviour, and hence different speckle patterns. Again, when the three images are superimposed, the speckle patterns are averaged to give reduced speckle. However, as the frequency range of the echoes is reduced, their lengths are increased, leading to some loss of axial resolution.

Movement

Imaging of rapidly moving structures, such as valve leaflets in the heart, requires that the image repetition rate (the frame rate) is high. To show the movement of a valve leaflet smoothly, the system needs to acquire a number of images (say 5–10) between the closed and open positions. As the leaflet may take only 20–50 ms to open, up to 10 images are needed in every 50 ms, a frame rate of 200 Hz.

Tissues such as the liver and other abdominal organs can be imaged successfully with much lower frame rates. Here, relative movement between the tissues and the ultrasound transducer is caused mainly by the patient's respiration and by slow operator-guided movement of the probe across the patient's skin, both of which can be suspended temporarily. Frame rates of less than 10 Hz can be tolerated, therefore, in the abdomen. In classical beam forming, as described in Chapter 3, the frame rate is limited by the two-way travel time of the transmit pulse and returning echo from the deepest part of the image.

Consider an image from a stepped linear array consisting of N ultrasonic lines interrogating the tissue to a maximum depth D (Figure 5.12a). To form each ultrasound line, the pulse must travel to

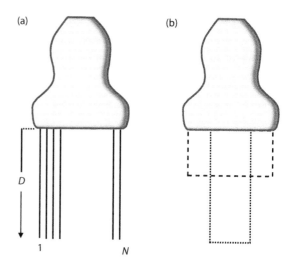

Figure 5.12 **(a)** The frame time, and hence the frame rate, is determined by the imaged depth D and the number of lines N. **(b)** The frame rate can be maximized by reducing the imaged depth or width to the minimum required.

depth D and echoes return from that depth before the pulse for the next line is transmitted.

The go and return time to depth D, i.e. the time per line $T_L = 2D/c$, where c is the speed of sound in the tissue.

The total time to form N lines is then $NT_L = 2DN/c$. This is the frame time T_F or time to form one complete B-mode image.

The frame rate $FR = 1/T_F$. That is,

$$FR = \frac{c}{2DN}\text{Hz}$$

For example, if the maximum imaged depth is 10 cm and the number of image lines is 256, then the maximum frame rate that can be achieved is: $1540/(2 \times 0.1 \times 256) = 30$ Hz.

The frame rate equation shows that increasing the depth D or the number of lines N reduces the frame rate, whereas decreasing D or N increases it. Hence, reducing the imaged depth to the minimum required to see the tissues of interest will help to avoid low frame rates. Choosing a narrower field of view, if this is available, may reduce the number of lines and hence increase the frame rate (Figure 5.12b).

Some image-forming techniques can introduce frame rate penalties. For example, coded excitation and some harmonic imaging techniques require more than one transmission along each beam position (see Chapter 4). Spatial compounding requires the superimposition of a number of frames at different beam angles to form the composite image, giving a reduction in overall frame rate.

Fortunately, newer beam-forming techniques, such as parallel beam forming, synthetic aperture sequential imaging and plane wave imaging (see Chapter 3) can achieve substantial increases in frame rate, negating the frame rate penalties just described and allowing further image processing steps to be carried out (Bercoff 2011).

ARTEFACTS

When forming a B-mode image, the imaging system makes a number of assumptions about ultrasound propagation in tissue. These include:

1. The speed of sound is constant.
2. The beam axis is straight.
3. The attenuation in tissue is constant.
4. The pulse travels only to targets that are on the beam axis and back to the transducer.

Significant variations from these conditions in the target tissues are likely to give rise to visible image artefacts. Most artefacts may be grouped into speed-of-sound artefacts, attenuation artefacts or reflection artefacts according to which of the conditions is violated.

Speed-of-sound artefacts

RANGE ERRORS

As described earlier, a B-mode image is a scaled map of echo-producing interfaces and scatterers within a slice through the patient in which each echo signal is displayed at a location related to the position of its origin within the slice. The location of each echo is determined from the position and orientation of the beam and the range of the target from the transducer.

The distance d to the target is derived from its go and return time t, i.e. the time elapsed between transmission of the pulse and reception of the echo from the target. In making this calculation, the system assumes that $t = 2d/c$, where the speed of sound c is constant at 1540 m s^{-1}, so that t changes only as a result of changes in distance d. However, if the speed of sound c in the medium between the transducer and the target is greater than 1540 m s^{-1}, the echo will arrive back at the transducer earlier than expected for a target of that range (i.e. t is reduced). The system assumes that c is still 1540 m s^{-1} and so displays the echo as if from a target nearer to the transducer. Conversely, where c is less than 1540 m s^{-1}, the echo arrives late and is displayed as if it originated from a more distant target. Such range errors may result in several forms of image artefact, depending on the pattern of changes in the speed of sound in the tissues between the transducer and the target. These include:

1. Misregistration of targets
2. Distortion of interfaces
3. Errors in size
4. Defocusing of the ultrasound beam

Misregistration of targets occurs as illustrated in Figure 5.13a, where the average speed of sound between the transducer and target is greater or less than 1540 m s^{-1}. A discrete target will be displayed too near or too far away from the location

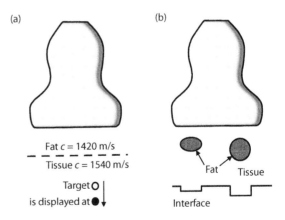

Figure 5.13 **(a)** The low speed of sound in a superficial fat layer results in the images of all targets beyond it being displaced away from the transducer. **(b)** Superficial regions of fat can result in visible distortion of smooth interfaces.

of the transducer face due to early or late arrival of the echo. In practice, this might be due to a thick layer of fat under the skin surface. The speed of sound in fat could be as low as 1420 m s^{-1}, approximately 8% less than the assumed speed of sound of 1540 m s^{-1}. For a uniform layer of fat, all targets beyond the fat would be displaced away from the transducer, but this error would not be noticeable, as it would be applied to all targets equally.

BOUNDARY DISTORTION

Range errors due to speed-of-sound variations may be more obvious in the presence of non-uniform regions of fat superficial to a smooth interface (Figure 5.13b). Here the parts of the interface which are imaged through a region of fat will be displaced to greater depths with respect to other parts and the resulting irregularities in the interface can be detected readily by eye.

SIZE ERRORS

Errors in displayed or measured size of a tissue mass may occur if the speed of sound in the region deviates significantly from 1540 m s^{-1}. For example, if the speed of sound in the mass is 5% less than 1540 m s^{-1}, the axial dimension of the displayed mass will be 5% too large. For most purposes, an error of 5% in displayed size is not noticeable. However, for measurement purposes, an error of 5% may need to be corrected.

PHASE ABERRATION

As described in Chapter 3, electronic focusing in transmit and receive is achieved by calculating the time of flight of the pulses and echoes from each element in the active aperture to the focal point. Electronic delays are then applied to ensure that the transmit pulses from each element arrive at the focus at the same time and the echoes received from the focus by each element are aligned in phase before they are added together. In transmit, the element delays are designed to produce a circular wavefront which converges to a focus (Figure 5.14a). The time-of-flight calculations are based on the assumption that the speed of sound in the intervening tissues is uniform at a value of 1540 m s^{-1}. Test object studies have shown that where the speed of sound in the medium is not

1540 m s^{-1}, the ultrasound beam becomes defocused (Dudley et al. 2002; Goldstein 2004).

In real tissues, the speed of sound does vary by a few percent from the assumed mean value of 1540 m s^{-1} (see Chapter 2), resulting in discrepancies between the calculated time of flight for each element and the real value. If there are variations in the speed of sound within the volume of tissue through which the beam passes, this gives rise to distortion of the wavefront and defocusing (Figure 5.14b). This effect is a significant problem when the beam passes through subcutaneous fat, resulting in significant loss of resolution for deeper tissues. The effect is often referred to as phase aberration and is a major limitation on the ultimate performance of ultrasound imaging systems, especially at higher frequencies where the effects of time errors on signal phase are greater (Flax and O'Donnell 1988; Trahey et al. 1991). A complete solution to phase aberration would effectively require the speed of sound in all regions of the image plane to be mapped so that corrections could be applied for every individual path between the transducer elements and all points in the image. Numerous solutions have been proposed for reducing the effects of phase aberration, but implementation has proved to be difficult.

Partial solutions to phase aberration correction are now available commercially based on the use of alternative values for the assumed speed of sound. In one approach, a strongly reflecting target within the tissue, referred to as a beacon, is identified. The system then forms several versions of the image using a different assumed speed of sound for each. The value which gives the most effective focus can be identified automatically by the system as that which gives the sharpest image of the beacon (McGlaughlin 2007). An alternative approach is to provide a speed-of-sound control which the operator can adjust to give the best image.

Many obese patients have a superficial layer of fat, which causes defocussing of the transmit and receive beams. One commercially available solution for such patients calculates the element transmission times and receive delays using a two-layer model to mimic the fat and tissue layers (Gauthier and Maxwell 2007). This assigns a speed of sound of 1450 m s^{-1} for the superficial fat layer and

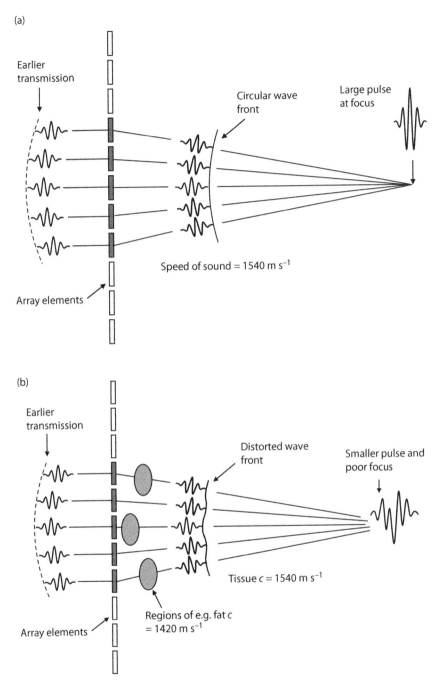

Figure 5.14 **(a)** A transmit focus is produced with an array transducer by delaying the transmit pulses from the central elements of the aperture. The delays are designed to produce a circular wave front in a medium with speed of sound of 1540 m s^{-1}, which converges to a focal point. **(b)** Where the pulses from the elements pass through different regions of tissue, in which the speed of sound is not 1540 m s^{-1}, e.g. fat, the resulting wave front is distorted resulting in a longer, lower amplitude pulse at a less well-defined focus. The effect is referred to as phase aberration.

1540 m s^{-1} for the tissue beneath. The thickness of the fat layer assumed in the calculation is not critical and so a fixed value is assumed. The operator can select this mode of operation for obese patients where needed.

REFRACTION

As described in Chapter 2, an ultrasound wave will be deflected by refraction when it is obliquely incident on an interface across which there is a change in the speed of sound. When an ultrasound wave propagates in the form of a beam, the direction of propagation is the beam axis, which may be deviated by a change in the speed of sound. In writing an ultrasonic line of echoes into the image memory, the B-mode system addresses a line of pixels across the memory assuming that the beam axis is straight, and displays all echoes at points along the assumed scan line at a range corresponding to their time of arrival. Hence, echoes received via a refracted beam will be displayed as if they originated on an undeviated axis (Figure 5.15) and will be displaced from their correct location in the image.

A refraction artefact is seen in some subjects when imaging the aorta in cross section through the superficial abdominal muscles. Here, the

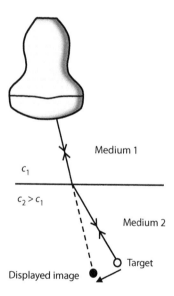

Figure 5.15 An echo received via a refracted beam is displayed in the image as if originating from a straight beam and hence is displaced from its correct position.

ultrasound beams may be refracted towards the centre of the abdomen by the oblique interface at the medial edges of the abdominal muscles. As the muscle structure is symmetrical, two side-by-side images of the aorta may be formed (Bull and Martin 2010).

EDGE SHADOWING ARTEFACT

Edge shadowing artefact is most commonly seen as narrow shadows extending distally from the lateral edges of a cyst. The clinical example in Figure 5.16a shows a pair of vertical dark streaks beneath the edges of the image of an epidermoid cyst. A number of explanations have been proposed for this effect, but as it is strongly associated with cysts in which the speed of sound is different from that in the surrounding tissue, it is most likely to be due to refraction (Ziskin et al. 1990; Steel et al. 2004). Edge shadowing has been observed in cases of higher and lower speed of sound within the cyst. It has also been reported in association with solid tissue lesions and bone within tissue. Attenuation in a wall surrounding a cystic structure may also contribute to an edge shadow (see next section on attenuation artefacts).

Figure 5.16b illustrates the case of adjacent ultrasound beams, here represented as rays, passing through the edge of a cyst in which the speed of sound is lower than that in the surrounding tissue. The beams are refracted as they pass in and out of the cyst. The edge of the cyst acts rather like an optical lens and causes the beams passing through it to converge. This leaves a region of divergence between the refracted beams and the undeviated beam that passes immediately on the outside of the cyst (and possibly within the peripheral beam). The region of divergence shows as a region of reduced pulse-echo sensitivity. As the imaging system assumes that all echoes arise from positions along the direction of the transmitted beam, the shadow is displayed in line with the transmitted beam.

Edge shadowing can also occur at the edge of a cyst or lesion in which the speed of sound is higher than in the surrounding tissue. In this case, refraction causes the beams to diverge as they pass through the cyst (Figure 5.16c), but the angle through which they are refracted is greatest at the edge of the cyst, where total reflection may also occur (see Chapter 2). Again, a region

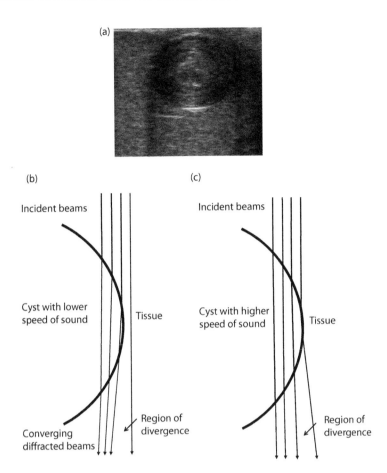

Figure 5.16 **(a)** Edge shadowing artefact is most commonly seen as narrow shadows extending distally from the lateral edges of a cystic region, in this case, an epidermoid cyst. **(b)** Edge shadowing artefact may result when ultrasound beams pass through the edge of a cyst, in which the speed of sound is slightly lower than that in the surrounding tissue. The beams converge due to refraction (like light rays through an optical lens). This leaves a region of divergence between the diffracted beams and the beam that passes immediately to the side of the cyst, resulting in a region of reduced pulse-echo sensitivity or shadow. **(c)** Edge shadowing can also occur at the edge of a cyst or lesion in which the speed of sound is higher than in the surrounding tissue. In this case, the beams diverge as they pass through the cyst, but the degree of divergence is greatest at the edge of the cyst, where total reflection may also occur. ([a] Reprinted from *Clinical Ultrasound*, Sidhu P., In: Allan PL, Baxter GM, Weston MJ [Eds.], Diseases of the testis and epididymis, London: Churchill Livingstone, Copyright 2011, with permission from Elsevier.)

of divergence occurs resulting in reduced pulse-echo sensitivity, which is displayed in line with the transmitted beam.

Attenuation artefacts

During each pulse-echo cycle of the B-mode imaging sequence, the outgoing pulse and returning echoes are attenuated as they propagate through tissue, so that echoes from deep targets are weaker than those from similar superficial targets. As described in Chapter 4, time-gain compensation (TGC) is applied to correct for such changes in echo amplitude with target depth. Most systems apply a constant rate of compensation (expressed in dB cm^{-1}), designed to correct for attenuation in a typical uniform tissue at the current transmit frequency. Also, the operator can usually make additional adjustments to the compensation via slide controls, which adjust the gain applied to specific depths in the image.

TGC artefacts may appear in the image when the applied compensation does not match the actual attenuation rate in the target tissues. A mismatch may occur due to inappropriate adjustment by the operator or to large deviations in actual attenuation from the constant values assumed. As the same TGC function is normally applied to each line in the B-mode image, inappropriate adjustment of TGC controls would result in bright or dark bands of echoes across the image of a uniform tissue (Figure 5.17a). Under some circumstances, these might be interpreted as abnormalities.

TGC artefacts due to substantial local deviations in tissue attenuation are of more interest as they usually have diagnostic value. Acoustic shadowing occurs distal to a region of increased attenuation, where all echoes in the shadow are reduced compared to those arising lateral to the shadow. This is because the TGC is set to compensate for the lower attenuation in the adjacent tissues, and so does not adequately boost echoes returning from beyond the region of higher attenuation. These undercompensated echoes are displayed at a reduced brightness. The most striking examples of acoustic shadowing are those due to highly reflecting and attenuating calcified lesions, such as gallstones or blood vessel plaque (Figure 5.17b). Shadowing may occur also

posterior to some more strongly attenuating tissue lesions, e.g. breast masses. Acoustic shadowing can be a useful diagnostic sign that indicates an attenuating lesion.

Post-cystic enhancement, which is the opposite of acoustic shadowing, occurs posterior to regions of low attenuation. Here, compensation is applied to echoes arising from behind the cyst assuming that these echoes have been attenuated by the same depth of tissue as those from regions lateral to the cyst. As the attenuation in the liquid of the cyst is very low compared to that in tissue, the echoes from tissues distal to the cyst are overcompensated and are displayed at a higher brightness level. Post-cystic enhancement (Figure 5.17c) can be a useful diagnostic indication of a liquid-filled lesion or one of low attenuation.

Reflection artefacts

SPECULAR REFLECTION

In Chapter 2, it was shown that when an ultrasound wave meets a large plane interface between two different media, the percentage of ultrasound intensity reflected is determined by the acoustic impedance values of the two media. A large change in acoustic impedance gives a strong reflection at the interface. The amplitude of the echo from an interface received back at the transducer is determined not only by the reflection coefficient at the interface and attenuation in the intervening medium, but also by

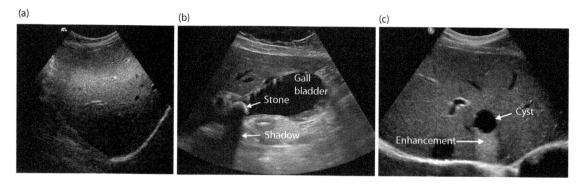

Figure 5.17 (a) Incorrect setting of the TGC controls can result in non-uniform image brightness. (b) The loss of echoes from beyond highly reflecting or attenuating objects such as this stone in the neck of the gall bladder results in acoustic shadowing. (c) Overcompensation of echoes from beyond low attenuation liquid-filled structures such as this liver cyst results in post-cystic enhancement. (Image courtesy of Hazel Edwards, East and North Herts NHS Trust, United Kingdom.)

the angle of incidence of the beam at the interface and the smoothness of the surface.

When reflection is from a large, smooth interface, i.e. an interface which is larger than the beam width, specular reflection occurs. That is, the reflected ultrasound propagates in one direction. When the angle of incidence is zero (normal incidence), the reflected echo from a large, smooth interface travels back along the same line as the incident beam to the transducer, where it is detected (Figure 5.18a). When the angle of incidence is not zero (Figure 5.18b), the beam is reflected to the opposite side of the normal at the angle of reflection. The angle of reflection is equal to the angle of incidence. Hence, when a beam is incident on a large, smooth interface at an angle of incidence of 10° or more, the reflected beam may miss the transducer so that no echo is received.

When the interface is rough or irregular on the scale of the ultrasound wavelength or smaller (Figure 5.18c), diffuse reflection occurs. Here, the incident pulse is reflected over a wide range of angles (scattered), so that echoes may be received back at the transducer even when the angle of incidence is quite large. Figure 5.18d shows a clinical example of specular reflection. The images show a longitudinal view of the Achilles tendon. In this orientation, the long fibres of the tendon act like specular reflectors. In the upper image, the fibres bend away from the line of the incident beams at the point where the tendon inserts into the calcaneus (circled) and the echo amplitude is reduced. In the lower image, the angle is adjusted so that the beams approach the tendon fibres at near normal incidence. The received echo amplitude is greater in the second case giving a brighter display of the tendon fibres. The reduced echo brightness in the upper image could be misinterpreted as a region of tendinopathy.

MIRROR-IMAGE ARTEFACT

Mirror-image artefact arises also due to specular reflection of the beam at a large smooth interface. It is most obvious when the reflection coefficient is large (e.g. at a tissue-air interface). If the reflected beam then encounters a scattering target, echoes from that target can be returned along a reciprocal path (i.e. via the reflecting interface) back to the transducer, where they are received (Figure 5.19a). As in the case of refraction artefact, the B-mode

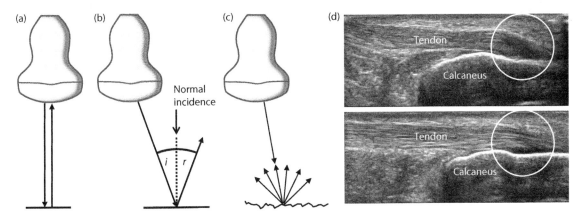

Figure 5.18 **(a)** Specular reflection occurs at a large, smooth interface. A strong echo can only be received at near normal incidence (beam at right angle to the interface). **(b)** When the angle of incidence (*i*) is not zero, echoes from the interface may not return to the transducer. The angle of reflection (*r*) is equal to the angle of incidence (*i*). **(c)** A large, rough interface gives a diffuse reflection, which may be received over a wider range of angles. Echoes are more likely to be returned to the transducer. **(d)** In the upper image of the Achilles tendon, the section where it inserts into the calcaneus (within the circle) appears dark, as the tendon fibres are angled away from the beam. In the lower image, the tendon fibres are closer to normal incidence and appear brighter in the image. The reduced brightness in the upper image could be mistaken for a region of tendinopathy. (Images courtesy of Hazel Edwards, East and North Herts NHS Trust, United Kingdom.)

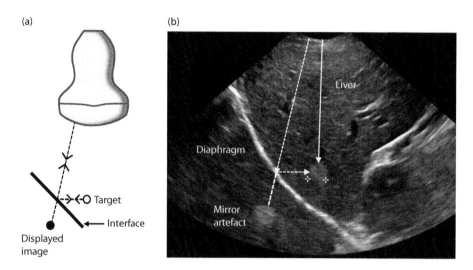

Figure 5.19 (a) Mirror image artefact can occur at a strongly reflecting interface. The transmit and receive beams are reflected by the interface. An image obtained via this route is displayed behind the interface as for reflection from a mirror. **(b)** In this liver image, the genuine lesion marked by crosses is imaged directly through the liver and appears in front of the diaphragm. A second image of the lesion appears behind the diaphragm due to mirror image artefact. (Image courtesy of Hazel Edwards, East and North Herts NHS Trust, United Kingdom.)

system assumes that all echoes arise from points along a straight beam. Hence, the reflected echoes are displayed in line with the original beam at a point beyond the strongly reflecting interface. The displayed effect is that of a mirror image of the scattering target behind the reflecting surface, as is observed when viewing a conventional optical mirror. A mirror-image artefact can often be observed posterior to the diaphragm. Figure 5.19b shows an example of a liver lesion near the diaphragm. A genuine image of the lesion (marked by crosses) is displayed in front of the diaphragm. This is produced by beams which intersect the lesion directly through the liver. However, a second image of the lesion is produced via beams that are reflected from the diaphragm. The image produced via this path is placed beyond the diaphragm in line with the transmitted beam. Other echoes from the liver produced by reflection from the diaphragm can be seen surrounding the artefactual image of the lesion.

REVERBERATIONS

Reverberations arise due to multiple reflections of pulses and echoes by strongly reflecting interfaces. They are most commonly associated with near normal incidence of the beam at superficial interfaces. As illustrated in Figure 5.20a, the original pulse transmitted by the transducer is reflected by an interface, giving rise to a genuine image of the interface at the correct depth. However, some of the energy in the returned echo may be reflected by the transducer face and return to the reflecting interface as if it was a weak transmitted pulse, returning as a second echo (reverberation). As the time taken for the second echo to arrive is twice that taken by the first echo, the B-mode system displays it at twice the depth of the first echo. This process can continue to give further weak echoes at multiples of the depth of the reflecting interface. The increasing weakness of these echoes is compensated partly by the TGC, which applies increased gain to echoes with longer go and return times. The returning first echo from the interface may also be reflected by more superficial interfaces before it reaches the transducer, giving rise to a continuum of weak echoes arriving back at the transducer. These are displayed beyond the image of the strongly reflecting interface.

Discrete reverberations can often be distinguished from genuine echoes by the way that they move in the image in response to gentle pressure on the skin surface exerted via the transducer. As

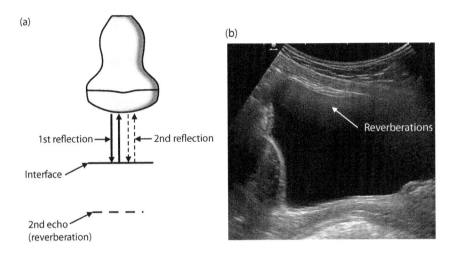

Figure 5.20 **(a)** Reverberations are generated when the echo from a strongly reflecting interface parallel to the transducer is partially re-reflected from the transducer face back to the interface. This generates a second echo which is displayed at twice the depth of the interface. **(b)** Reverberations are seen most commonly within liquid-filled areas such as the bladder, as shown here.

the operator presses the transducer slowly down into the tissue surface, genuine echoes in the image come up to meet the transducer at the same rate. However, as the path length for the first reverberation is twice that of the genuine echo, a 1 cm movement of the transducer results in a 2 cm movement of the reverberation in the image. Hence the reverberation will travel towards the transducer at twice the rate of the genuine echo.

Reverberations commonly appear in images of liquid-filled areas, such as the bladder, due to multiple reflections between the transducer and the anterior bladder wall and other intervening interfaces (Figure 5.20b). They are more obvious here as such regions are usually devoid of echoes. Reverberations may also be seen in other echo poor areas of tissue.

COMET TAIL ARTEFACT

In addition to multiple reflections between a large interface and the transducer, reverberations can occur within small highly reflecting objects or foreign bodies, leading to a long streak or series of echoes in line with the beam distal to the object. The artefact is known as comet tail artefact due to its resemblance to a comet tail (Ziskin et al. 1982). As illustrated in Figure 5.21a, the ultrasound pulse reverberates between the highly reflecting faces of the object, generating a long series of

echoes which are returned to the transducer and displayed as an extended tail distal to the object. Figure 5.21b shows an image of a distended gall bladder with small cholesterol crystals in the wall. These show short comet tails demonstrating their calcified nature. Figure 5.21c shows an image of an intrauterine contraceptive device (IUCD) within the uterine cavity. The pulse undergoes repeated reflections between the hard upper and lower surfaces, generating multiple images of the device.

A more intense comet tail artefact is more likely to be seen where the acoustic impedance mismatch between the implanted material and surrounding tissue is large, as in the case of the IUCD. Softer materials, e.g. wood, give rise to a less prominent comet tail (Ziskin et al. 1982). A large mismatch in acoustic impedance also occurs in the presence of gas within tissue structures. In ultrasound imaging of the chest, the presence of comet tails has been associated with the presence of pulmonary oedema (Lichtenstein et al. 1997). In an ultrasound image of the normal chest wall, a strong reflection is obtained from the interface between the tissues of the chest wall and air in the lung. However, in the presence of oedema, a thin layer of fluid may occur beneath the chest wall, sandwiched between the pleura and interlobular septa (Picano et al. 2006). The strong reflection at the lung surface enables the ultrasound pulse to reverberate across

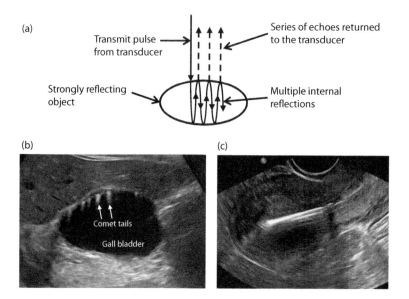

Figure 5.21 **(a)** Multiple internal reflections between the surfaces of a highly reflecting foreign body can give rise to a long series of closely spaced echoes described as comet tail artefact. **(b)** A comet tail artefact can be seen extending below small cholesterol crystals in the wall of the gall bladder. **(c)** An IUCD within the uterus shows multiple images as the pulse undergoes multiple reflections between the hard surfaces of the device. (Images courtesy of Hazel Edwards, East and North Herts NHS Trust, United Kingdom.)

the fluid region giving rise to multiple comet tails. Comet tail artefact has also been observed at other gas-tissue interfaces including the interface between the bowel wall and gas-filled lumen and in the presence of gas due to an abscess (Thickman et al. 1983).

ACKNOWLEDGEMENT

The image in Figure 5.20 was obtained with the assistance of Hitachi Medical Systems, United Kingdom.

QUESTIONS

Multiple Choice Questions

Q1. Which of the following describes spatial resolution?
  a. The difference between the true and measured distance
  b. The minimum detectable separation in time between two events
  c. The minimum detectable separation in space between images of two point objects
  d. The difference between the true and measured blood velocity
  e. The maximum depth to which echoes can be observed

Q2. Which of the following statements are correct?
  a. The axial resolution describes the spatial resolution along the beam axis
  b. The temporal resolution describes the spatial resolution along the beam axis
  c. The slice thickness determines the spatial resolution in the scan plane perpendicular to the beam
  d. The slice thickness determines the spatial resolution at 90° to the scan plane
  e. The lateral resolution describes the spatial resolution at 90° to the scan plane

Q3. Axial resolution is mainly determined by:
  a. Slice width
  b. Transmit frequency

c. Apodization

d. Pulse length

e. Width of the beam

Q4. Which of the following statements are true?

  a. Axial resolution is independent of transmit frequency

  b. Axial resolution gets larger (i.e. gets worse) as transmit frequency increases

  c. Axial resolution is optimum at 5.5 MHz

  d. Axial resolution is typically about half the pulse length

  e. Axial resolution gets smaller (i.e. improves) as transmit frequency increases

Q5. Lateral resolution in the scan plane:

  a. Is determined by the pulse length

  b. Is approximately half the beam width

  c. Becomes worse as frequency increases

  d. Can be improved by focusing

  e. Is the same at all depths

Q6. Slice thickness:

  a. Is about 1 mm

  b. Depends on the pulse length

  c. Affects images of small cysts

  d. Is less than the beam width in the scan plane

  e. Can be reduced using a multi-row transducer

Q7. Contrast resolution:

  a. Is the difference between axial and lateral resolution

  b. Is affected by the difference between lesion and background brightness

  c. Is affected by speckle

  d. May be affected by grating lobes

  e. Is adjusted in the display monitor settings

Q8. Concerning speckle:

  a. Speckle is an interference pattern arising from ultrasound scattered from within the tissues

  b. Speckle size is related to the size of structures in the tissues

  c. Speckle size decreases with increasing frequency

  d. Speckle size increases with increasing frequency

  e. Speckle is electronic noise generated from the amplifier

Q9. Artefacts:

  a. Are errors in imaging or measurement which occur when an assumption made during image formation is violated

  b. Can occur when the speed of sound is not constant in the field of view

  c. Can occur when the attenuation is not constant in the field of view

  d. Can occur when the ultrasound beam does not travel in a straight line

  e. Are very old ultrasound machines

Q10. If the speed of sound in tissues is less than 1540 m s^{-1}, then:

  a. Echoes from a target are placed deeper in the image than they should be

  b. Echoes from a target are placed less deep in the image than they should be

  c. Echoes from a target are placed at the correct depth in the image

  d. The attenuation is higher

  e. The attenuation is lower

Q11. Refraction of ultrasound:

  a. Is the deviation of a beam at an interface between tissues with different speeds of sound

  b. Can lead to sideways misplacement of echoes from targets

  c. Is the attenuation of a beam caused by absorption of sound

  d. Occurs all the time in ultrasound images

  e. Is responsible for the increase in image brightness seen below fluid-filled structures

Q12. The increased brightness often seen below fluid-filled structures is caused by:

  a. There being no difference in the attenuation between the fluid and surrounding tissue

  b. Increased attenuation in the fluid compared to the surrounding tissue

  c. The transmit power being set too high

  d. Decreased attenuation in the fluid compared to the surrounding tissue

  e. The B-mode gain being set too high

Q13. Specular reflection occurs:

  a. Typically within the parenchyma of tissues

b. Where the object size is much less than the wavelength
c. Where the object size is much greater than the wavelength
d. Where the object size is similar to the wavelength
e. Typically from organ boundaries, tendons, ligaments and vessel walls

Q14. Mirror image artefact:
a. Occurs when the transducer is the wrong way round
b. Occurs when the beam is partially reflected at an interface
c. Occurs when the transmit power is increased too much
d. Occurs when the B-mode gain is increased too much
e. Commonly involves the diaphragm

Q15. Reverberation artefacts:
a. May be due to echoes bouncing between two reflective tissue interfaces
b. Are commonly seen in solid areas such as tumours
c. Are commonly seen in liquid-filled areas such as the bladder
d. May be due to echoes bouncing between the transducer and a reflective interface
e. May be due to change in beam direction due to refraction

Short-Answer Questions

Q1. Define the term 'lateral resolution' and explain how its value is determined by the beam width.
Q2. Explain why the slice thickness of an array transducer is normally greater than the beam width in the scan plane.
Q3. Define the term 'axial resolution'. How is axial resolution related to the transmit pulse length?
Q4. Describe the origin of speckle in ultrasound images and its dependence on beam characteristics.
Q5. In a linear array scanner using classic beamforming techniques, how might increasing the imaged depth lead to a decrease in frame rate?

Q6. A fetal femur is aligned horizontally in the image plane of a phased-array sector scanner, with the centre of the femur in the centre of the field of view. There is a layer of fat with speed of sound 1420 m s^{-1} just beneath the skin surface; the speed of sound elsewhere is 1540 m s^{-1}. How will the fat layer affect the measured length of the femur?
Q7. Explain the process that leads to the formation of an acoustic shadow behind a calcified gallstone.
Q8. Describe the appearance in an ultrasound image of a large smooth plane interface lying parallel to and several centimetres below the skin surface, when imaged with a curvilinear array.
Q9. How do reverberations occur in an ultrasound image? How can the operator distinguish a reverberation from a genuine echo?
Q10. Describe the formation of a mirror image artefact. What anatomical feature commonly produces this effect?

REFERENCES

Bercoff J. Ultrafast ultrasound imaging. In: Minin O (Ed.), *Ultrasound Imaging – Medical Applications*. -1, InTech, 2011. Available at: https://www.intechopen.com/books/ultrasound-imaging-medical-applications/ultrafast-ultrasound-imaging

Bull V, Martin K. 2010. A theoretical and experimental study of the double aorta artefact in B-mode imaging. *Ultrasound*, 18, 8–13.

Chang JH, Yoo Y. 2014. Speckle reduction techniques in medical ultrasound imaging. *Biomedical Engineer Letter*, 4, 32–40.

Dudley NJ, Gibson NM, Fleckney MJ, Clark PD. 2002. The effect of speed of sound in ultrasound test objects on lateral resolution. *Ultrasound in Medicine and Biology*, 28, 1561–1564.

Flax SW, O'Donnell M. 1988. Phase-aberration correction using signals from point reflectors and diffuse scatterers: Basic principles. *IEEE Transactions on Ultrasonics, Ferroelectrics, and Frequency Control*, 35, 758–767.

Gauthier TP, Maxwell DR. 2007. Tissue aberration correction. Philips Medical Systems white paper. Eindhoven: Philips Medical Systems.

Goldstein A. 2004. Beam width measurements in low acoustic velocity phantoms. *Ultrasound in Medicine and Biology*, 30, 413–416.

Goldstein A, Madrazo BL. 1981. Slice-thickness artifacts in gray-scale ultrasound. *Journal of Clinical Ultrasound*, 9, 365–375.

Lichtenstein D, Meziere G, Biderman P et al. 1997. The comet tail artifact – An ultrasound sign of alveolar-interstitial syndrome. *American Journal of Respiratory and Critical Care*, 156, 1640–1646.

McGlaughlin GW. 2007. Practical aberration correction methods. *Ultrasound*, 15, 99–104.

Picano E, Frassi F, Agricola E et al. 2006. Ultrasound lung comets: A clinically useful sign of extravascular lung water. *Journal of the American Society of Echocardiography*, 19, 356–363.

Sidhu P. 2011. Diseases of the testis and epididymis. In: Allan PL, Baxter GM, Weston MJ (Eds.), *Clinical Ultrasound*. London: Churchill Livingstone.

Steel R, Poepping TL, Thompson RS, Macaskill C. 2004. Origins of the edge shadowing artifact in medical ultrasound imaging. *Ultrasound in Medicine and Biology*, 30, 1153–1162.

Szabo TL. 2014. *Diagnostic Ultrasound Imaging: Inside Out*. Oxford UK: Academic Press/ Elsevier.

Thickman D, Ziskin MC, Goldenberg NJ et al. 1983. Clinical manifestations of the comet tail artifact. *Journal of Ultrasound in Medicine*, 2, 225–230.

Trahey GE, Freiburger PD, Nock LF, Sullivan DC. 1991. In vivo measurements of ultrasonic beam distortion in the breast. *Ultrasonic Imaging*, 13, 71–90.

Ziskin MC, Follette PS, Blathras K, Abraham V. 1990. Effect of scan format on refraction artifacts. *Ultrasound in Medicine and Biology*, 16, 183–191.

Ziskin MC, Thickman DI, Goldenberg NJ et al. 1982. The comet tail artifact. *Journal of Ultrasound in Medicine*, 1, 1–7.

6

B-mode measurements

NICK DUDLEY

INTRODUCTION

Measurements have a significant role in many areas of ultrasound practice. Some of the earliest ultrasound measurements were made in the field of obstetrics, originally in A-mode. A small range of measurements was quickly established in regular practice, and this range has been considerably developed over the years. The most frequently performed measurements are nuchal translucency in detecting abnormalities; crown-rump length, head circumference (HC), abdominal circumference (AC) and femur length for dating and growth assessment. All of these may affect the management of pregnancy and therefore accuracy and reproducibility of measurements are important.

There is also a long history of measurement in echocardiography. Early measurements were made using the M-mode image and this is still used in modern clinical practice. Echocardiography images a dynamic process; the advantage of M-mode is that it contains information on both distance and time in a single image, so that changes in dimensions during the cardiac cycle may be measured. Cine-loop displays and imaging workstations now provide this temporal information in B-mode in a highly accessible form, facilitating measurement of changes in areas and volumes, e.g. for estimating left ventricular ejection fraction.

Most abdominal and small parts examinations are qualitative. Measurements are sometimes used, ranging from the gall bladder wall and common bile duct (a few millimetres) to the kidneys and liver (centimetres).

In vascular ultrasound, vessel diameters have been measured for many years in the diagnosis of, for example, aneurysms. Vascular dimensions are often used in conjunction with Doppler ultrasound in assessing blood flow; accuracy is then more important since small errors in vessel diameter may result in large errors in blood flow estimation, particularly in small vessels. The high spatial resolution of current systems allows measurement of small distances, e.g. intimal thickness.

In the following sections the implementation and application of measurement systems are described. Measurement uncertainty is often overlooked in the development and use of clinical measurements; possible sources of error and practical means for improving accuracy and reproducibility are discussed.

MEASUREMENT SYSTEMS

Measurement systems have the sophistication and flexibility provided by digital electronics and software control. In the early days of ultrasound, particularly before the advent of digital scan converters, only simple axial measurements were possible on scanners. These were made on the A-mode display by aligning the ultrasound beam with the targets to be measured, and identifying and placing markers on the appropriate signals on the display. The scanner took the time interval between these positions on the A-mode display and converted it to distance using the assumed speed of sound. Other measurements, e.g. in the horizontal plane, could be made only on hard copy of the B-mode image using a ruler and scale factors derived from axial distance measurements, or in the case of non-linear measurements using a planimeter or even a piece of string.

These early measurements were difficult and required great care. It is remarkable that early ultrasound practitioners were able to demonstrate the value of measurements. The fact that some of the normal data produced are close to currently accepted values and are still referred to is a testament to the care and dedication of these pioneers.

Calliper systems

Measurements made using ultrasound callipers range from simple linear distance measurements to more complex volume measurements. Scanners will often be able to perform calculations using measurements.

Measurements are made on the image stored in the scan converter and therefore depend on the accurate placement of image data in the memory, as described in Chapter 4. Each ultrasound scanner will use an assumed ultrasound velocity, usually 1540 m s^{-1}, to calculate distance along the beam axis, and a variety of algorithms, depending on the probe geometry, to calculate the final location of each echo in the image memory.

Distances within the scan converter image memory are then measured from pixel centre to pixel centre using an electronic calliper system. The measurement will generate a distance in pixels, from the centre of one calliper to the centre

of the other calliper, which is then converted to millimetres or centimetres using the pixel size. The scanner will calculate pixel size depending on probe geometry and selected settings.

The most common type of calliper control is a track-ball, although some smaller or older scanners may have less sophisticated devices. The track-ball is the most user-friendly option, as it allows callipers to be placed quickly. Some track-ball systems, however, are very sensitive and difficult to accurately control; for linear measurements, this is irritating but should not compromise accuracy, but for non-linear measurements significant errors may be introduced. Many track-ball systems are adjustable in software, so that their sensitivity can be altered to suit the user.

Linear distance, i.e. a measurement in a straight line between two points, was the first type of measurement made and is still the most commonly used.

Non-linear distance, circumference and area

Non-linear distances, including irregular circumferences, may be calculated from a tracing. The calliper is fixed at one end or point on the structure and the outline is traced using the track-ball. During the tracing, the system plots a line guided by the track-ball movement. The system then calculates the length of the trace, usually by calculating and summing the linear distances between points at intervals along the tracing.

The circumference of a structure is the distance around the perimeter and is widely used in obstetrics, e.g. fetal AC. There are several possible measurement methods, the choice depending on the regularity of the structure. Circumferences may be traced in the same way as other non-linear distances, fitted or plotted using a number of methods.

Most systems now include ellipse fitting as an option for circumference and area measurement (see Figure 6.1a). This is a quick and useful method for structures that are truly elliptical. Callipers are initially placed on the extremes of the long or short axis of the ellipse (any diameter for a circle). The length of the other axis is then adjusted using the track-ball, making the ellipse larger or smaller to fit the measured structure. Most scanners allow

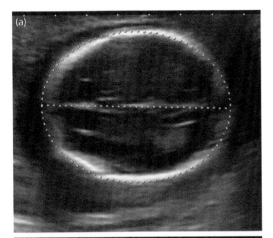

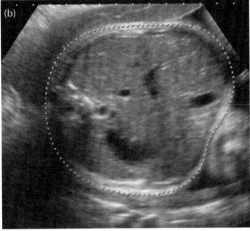

Figure 6.1 Examples of non-linear measurements. (a) Ellipse fitting; (b) tracing/plotting.

both size and position of the ellipse to be adjusted, so that it may be moved around the screen to match the structure. Once the size of the ellipse has been fixed, the system calculates the circumference by using an exact or approximate formula, e.g. Ramanujan (1914).

Another method was used in many early studies developing fetal growth charts and is recommended by the British Medical Ultrasound Society (Loughna et al. 2009). Here, two orthogonal diameters (d_1 and d_2) are measured, e.g. the biparietal and occipito-frontal diameters of the fetal head. The circumference is then given by the equation:

$$\text{Circumference} = \pi \times (d_1 + d_2)/2$$
$$= 1.57 \times (d_1 + d_2)$$

This is a good approximation unless the long and short axes are very different. It should be noted that some manufacturers offer a 'cross' method, where callipers may be placed at the ends of orthogonal diameters, but the circumference is often calculated using an exact or approximate formula for an ellipse, rather than the circumference equation.

Some manufacturers are offering a further alternative, where a series of points is plotted around a structure by successively moving and fixing the calliper (point-to-point method – see Figure 6.1b). The system then joins the points with straight lines or curves, depending on the manufacturer, and calculates the circumference. This is less time-consuming than tracing but more so than ellipse fitting, and is useful for measuring non-elliptical structures.

Cross-sectional area may also be calculated from the outline of a structure. Whichever method is used to generate the perimeter, the area is simply calculated by multiplying the number of pixels enclosed by the area of each pixel. This is not often used, although it has been suggested that it may be more accurate and reproducible than circumference (Rossavik and Deter 1984). Area measurements are more commonly used in combination to calculate a volume.

Volume

When estimating volume from two-dimensional images, most commonly, a calculation from three linear measurements is used, as shown in Figure 6.2. This method requires two images to be acquired at 90° to each other, e.g. sagittal and transverse images, and three orthogonal diameters

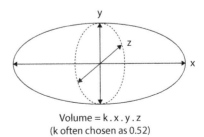

Volume = k . x . y . z
(k often chosen as 0.52)

Figure 6.2 Volume estimation from three orthogonal measurements.

to be measured. The volume is calculated by multiplying the diameters together and then multiplying by a constant value, k, appropriate for the shape. The shape is often assumed to approximate to a sphere, for which the required k value is 0.52. Where a sphere is not a good approximation, it may be possible to derive a more appropriate constant from measurements on a number of patients. This requires comparison with an accurate measurement of volume by another means, e.g. in measuring bladder volume the ultrasound measurements can be compared with the volume of urine collected when the bladder is completely emptied. Other structures measured in this way include the gestation sac and cardiac ventricles.

Some manufacturers offer a single image method, e.g. for bladder volume estimation, where only height and breadth of the structure are measured in a single plane. This method relies heavily on an assumption of the shape of the structure at 90° to the scan plane and should be used with extreme caution (Bih et al. 1998).

It is now possible to make volume measurements using three-dimensional (3D) ultrasound. Early examples include the prediction of birthweight (Lee et al. 1997), assessment of cardiac left ventricular volume and function (Kuhl et al. 1998) and measurement of neonatal cerebral ventricles (Kampmann et al. 1998).

Automatic measurement

Image-processing computers provide the opportunity for automating measurements. Although this is not a straightforward process, as the computer must perform some of the operations of the human eye-brain system in determining edges and boundaries, successful systems have been developed for cardiology and for obstetrics, which are both measurement-intensive operations.

Relatively simple automated measurements, such as the distance between adjacent surfaces (Wendelhag et al. 1997), have been available for some time. Considerable efforts have been made recently to automate the more complex measurements made in fetal ultrasound (Rueda et al. 2014) but the availability and use of commercial systems are limited (Carneiro et al. 2008; Yazdi et al. 2014). Rapid advances in machine learning may facilitate

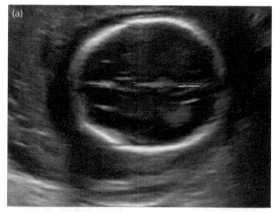

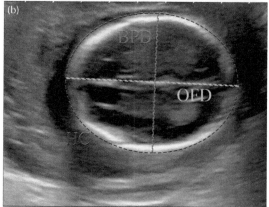

Figure 6.3 **(a)** Automatic recognition of central mid-line in the correct fetal head imaging plane. **(b)** Automated measurement of fetal head circumference (from bi-parietal and occipito-frontal diameters).

automatic recognition of the correct imaging plane as well as automation of measurements, as shown in Figure 6.3 (Zhang et al. 2017).

3D ultrasound offers the opportunity to automatically measure volume (Artang et al. 2009; Shibayama et al. 2013). The more widespread use of matrix array probes in echocardiography, and the high contrast between blood and myocardium, may lead to a rapid uptake of recently available automated measurement systems. Some caution is required as the limits of agreement with other methods, e.g. magnetic resonance imaging, are relatively large (Shibayama et al. 2013; Furiasse and Thomas 2015). However, advanced, knowledge-based, adaptive algorithms appear to offer time savings as well as results comparable with those of expert observers (Tsang et al. 2016) and are being applied in other fields, including obstetrics.

Calculations

Ultrasound scanners have the facility to perform and store measurements. It is, therefore, possible to programme the system to calculate other parameters from measurements made by the operator. Examples include bladder volume, as described earlier, fetal weight and cardiac volumes.

When a measurement function is selected, the operator must then make the measurements and store them under the appropriate name. The scanner may allow for several measurements of each parameter to be made and automatically averaged. Once the measurements have been stored, the result will be calculated using the programmed formula.

MEASUREMENT ERRORS

Measurements may be different from the true value for a number of reasons. The difference between the observed and true values is called the 'error of observation'. It is important in any measurement to consider two types of error, random and systematic. Many errors are a combination of these two types, although in different circumstances one or the other may be dominant. Both systematic and random errors affect accuracy. Reproducibility depends largely on random errors.

Most equipment manufacturers will specify measurement accuracy, but specifications often do not match clinical requirements. Axell et al. (2012) reported manufacturers' specifications in the range of 3%–5% for linear measurements, when the clinical requirement in obstetrics is much smaller.

Random errors

Random errors are accidental in nature and are often due to the observer, either directly or in the use of a system that lacks precision. For example, if the true diameter of a structure is 9.5 mm, and the calliper increment is 1 mm, the observer will never obtain a single true measurement. These errors are

Table 6.1 Ten measurements of the circumference of a test object, made by two observers

	Observer 1	Observer 2
	317	330
	317	325
	318	323
	321	315
	325	316
	321	318
	323	318
	319	322
	318	321
	322	314
Mean	320	320
CoV	0.9%	1.6%

often revealed by repeated measurements, where several values above and below the true value are likely to be obtained. Averaging these measurements will reduce the error and produce a result closer to the true value.

Random errors often follow a normal, or Gaussian, distribution which is characterised by its mean and standard deviation. The mean is the average of all the measurements, whereas the standard deviation is a measure of the spread of values about the mean. A widely used measure of random error is the coefficient of variation (CoV). This is the ratio of the standard deviation to the mean of a series of measurements, expressed as a percentage.

Table 6.1 shows two series of traced measurements made on a test object. Note that although the mean values are the same, the two observers have generated different random errors. The CoV reflects the fact that the second observer has made a wider range of measurements. It is also interesting to consider the potential consequences of making only a single measurement in a clinical situation, as there is a 13 mm difference between the first values obtained by the two observers. This shows the importance of making multiple measurements and averaging.

Systematic errors

Systematic errors are generally consistent in direction and relative size. Both observer- and instrument-related systematic errors are possible in ultrasound as illustrated by the following examples:

1. A sonographer consistently placing callipers just inside the edges of a structure will generate a systematically smaller result than colleagues placing callipers exactly on the edges.
2. If an ultrasound scanner is incorrectly calibrated to 1600 m s^{-1}, all axial measurements made in a medium with velocity 1540 m s^{-1} will be systematically larger by approximately 4%.

Systematic errors are not revealed by repeated measurements, but may be found by comparison between observers or by measurement of appropriate test objects, depending on the source of error. This is illustrated by Table 6.1, where the true circumference of the test object was 314 mm; there was a systematic error of 6 mm in both series of measurements.

Some degree of systematic error related to the imaging process is inevitable in ultrasound, but it is possible to reduce it to low levels by careful choice of scanner and methods.

Compound errors

Where measurements are combined in some way, e.g. as a ratio or in a volume estimation, the potential errors in the individual measurements are compounded, leading to a larger potential error in the result. The fractional error in a product or ratio is approximately equal to the sum of the fractional errors in each variable. For example, if the potential error in fetal AC and HC measurements is 3%, the potential error in the AC/HC ratio is approximately 6%.

SOURCES OF ERRORS IN ULTRASOUND SYSTEMS

Human error

There are many possible causes of human error, including inadequate or inappropriate training, inexperience, lack of locally agreed standards or failure to follow local procedures. These will result in either measurement of inappropriate images or incorrect calliper placement. A common source

of error is in the measurement of oblique, rather than longitudinal or transverse, sections, leading to overestimation. The frequency, magnitude and effect of human error in fetal measurement have been widely explored (Sarmandal et al. 1989; Chang et al. 1993; Dudley and Chapman 2002).

Failure to confirm or average measurements by repetition, and errors related to the measurement facilities, such as an over-sensitive track-ball, may also be considered as human error. A measurement system that is difficult to use will deter sonographers from making repeat measurements. This will increase the likelihood, and probably the size, of errors.

The use of evidence-based standards and protocols, training, audit and careful selection of equipment can reduce human errors.

Image pixel size and calliper precision

The ultrasound image is made up of pixels, as described in Chapter 4. The smallest distance that can be represented in the image is one pixel and so all distance measurements have an inherent uncertainty of ±1 pixel. For example, if an image is 512 pixels square and the image depth is set to 20 cm, each pixel represents one 512th of 20 cm, i.e. 0.39 mm.

A further limitation may be the calliper increment. All calliper systems have a limited calliper increment which may be larger than the pixel size, e.g. 1 mm. The true calliper increment is never smaller than the pixel size, as the calliper must move a minimum of 1 pixel at a time.

When measuring small structures, it is therefore important to magnify the real-time image using either depth/scale or write zoom controls. The pixel size in the scaled-up image will be smaller, reducing this uncertainty. These errors are random in nature.

Image resolution

Ultrasound images have a finite spatial resolution, as discussed in Chapter 5. The edges of structures may, therefore, appear blurred or enlarged, and callipers may be placed beyond the true dimensions. Errors due to lateral beam width may be minimised by optimizing focal settings, so that the resolution is best at the region of measurement.

Resolution may not always present a problem, as measurements are usually compared with a normal range or with previous measurements. If, however, measurements are made on a scanner with performance significantly different to that used to develop reference values, then they may be misleading. An area where this has been evaluated and documented is in the measurement of fetal femur length. Jago et al. (1994) demonstrated systematic differences between femur length measurements made with older and more modern scanners. Differences in beam width of about 2 mm generated femur length differences of about 1 mm; it is worth noting that differences of this magnitude can be generated on any scanner by an inappropriate focal depth setting.

Since lateral resolution is generally inferior to axial resolution, errors are more likely in nonaxial measurements. However, the distal margins of structures can be blurred as a result of poor axial resolution or due to reverberation in highly echogenic structures. For this reason some measurements, e.g. fetal nuchal translucency, are made between the proximal edges of structures (leading edge to leading edge).

Gain settings can have a significant effect on resolution, but as the technology develops this has become less of a problem, as gain is more easily managed on modern equipment. Resolution may be generally improved by the use of higher ultrasound frequencies where possible.

Velocity/distance calibration

Since measurements are made on the digitised image in the scan converter memory, it is important that echoes are accurately placed in the image. This depends on the velocity assumed for calculation of the axial origin of echoes and on the algorithms used to calculate the position of scan lines in the image. Any error in echo placement can lead to a measurement error.

If echoes have been correctly placed, accuracy of any measurement then depends largely on the accuracy of in-built conversion factors to translate the number of pixels between calliper positions into a real distance. A set of factors is required for each probe geometry and for scale or depth settings. With the increasing range of probes and the availability of continuously variable scale settings, it is important that the electronics or software used to make these conversions is carefully designed and tested.

The consequence of incorrect calibration is a systematic error in all measurements or, more subtly, differences between measurements made on different scale settings. Velocity errors inherent in the equipment can be avoided by rigorously testing calliper accuracy prior to purchase and at commissioning.

Ultrasound propagation

Ultrasound scanners are generally designed to assume an ultrasound velocity of 1540 m s^{-1}. This represents a mean velocity in human soft tissue. True soft tissue velocities vary about this mean by approximately 5% and this may lead to slight distortions in the image (as described in Chapter 5) and, therefore, to measurement errors.

The clinical impact of these distortions will depend on the anatomy under investigation. Measurements of a single organ may be incorrect if the velocity within the organ is not 1540 m s^{-1}, but so long as the velocity is the same in other patients there will be no error between patients. Measurements that cross structural boundaries may be adversely affected when compared with normal ranges or between patients, where the contribution of each structure to total distance varies. In most clinical situations, however, a single structure is measured. Where other structures are included, their size and velocity differences are often small, e.g. blood vessels included in an organ diameter measurement.

Advances in beam-forming techniques have enabled speed-of-sound corrections to be applied to real-time images. The average speed of sound in the imaged tissues is either assumed from prior knowledge or calculated from echo return times and the image is constructed using focusing delays based on this estimate rather than the usual 1540 m s^{-1}. A more accurate image is displayed and the absolute accuracy of measurements may be improved. Ultrasound scanners use the speed-of-sound correction solely for focusing and not for distance calibration so there will be no systematic

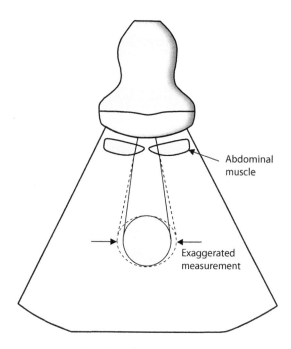

Figure 6.4 The effect of refraction by muscles close to the transducer face. Solid lines show the true ultrasound path (refracted by the abdominal muscles) and the true position of the structure. Dotted lines show the assumed ultrasound path and the refracted position of the structures in the image, leading to an exaggerated measurement.

measurement difference due to the correction, i.e. measurements should not be affected.

A further source of distortion is refraction, where the ultrasound beam changes direction as it crosses boundaries between tissues with different velocities. This has a greater potential clinical impact as distortions may be introduced into a structure by proximal tissues or by the structure itself. Refraction is described in more detail in Chapter 5.

An area where the possibility of measurement errors due to refraction should always be considered is obstetrics. Refraction around the maternal mid-line can lead to apparently stretched objects (Figure 6.4).

Errors in circumference and area

Circumference and area measurements are subject to all of the above errors. Traced measurements are particularly sensitive to human error, where calliper controls may be difficult to manipulate in generating an accurate outline. This leads to systematic and random errors, as every deviation is added to the circumference. In clinical practice, structural outlines are not always completely and reliably visualised and this may also lead to errors.

The magnitude of the systematic error in tracing depends on several factors. Difficulty in manipulating the track-ball accurately will increase the error, as will any lack of care on the part of the operator. The separation of points on the tracing used by the system to calculate the distance is also important. If the points are far apart, this will result in a shortening of the distance and if the points are too close together every small deviation will be included, increasing the measurement. Errors as high as 15% have been documented (Dudley and Griffith 1996). The optimum number of points depends on the sensitivity of the track-ball and the distance measured; this will be determined in the design of the equipment. For a fetal AC in the third trimester, one point per centimetre might be appropriate.

In the clinical situation, errors will usually be larger than those shown in Table 6.1, as measurements are rarely repeated 10 times and it is more difficult to obtain a good image from a patient than from a test object. Examples of clinical inter-observer CoVs for circumferences and areas of 3% and 5%, respectively, have been reported (Sarmandal et al. 1989; Owen et al. 1996).

Alternative methods are less prone to errors of operator dexterity. Ellipse fitting provides an accurate and reproducible measurement, provided that the outline of the structure is clearly visualised and is elliptical in shape. The point-to-point methods described earlier can provide accurate and reproducible results, with the advantage that the points may be fitted to any shape. This method is likely to give the most accurate and reproducible results in clinical practice, provided sufficient points are used.

Area measurements are somewhat more reliable where outline tracing is erratic, as small areas are both added to and subtracted from the true area by deviations outside and inside the outline. Mean errors are typically less than 1.5% with CoV of less than 0.5%.

Errors in volume

Since volume estimation methods always use a combination of measurements, they are subject to compound errors as described above. If the volume is the product of three diameters, each with a potential error of 5%, the potential error in the volume is approximately 15% (neglecting any assumptions about shape).

The human error may be large, as the method may rely on finding maximum diameters of an irregular structure or measuring multiple areas. In many cases, however, the largest errors may be introduced by assumptions used in calculations, especially when measurements used are from a single scan plane only. The shape of a structure will vary from patient to patient, reducing the validity of 'constants' used in the calculation.

One solution to the problem of variable shape is to develop different constants for the shapes encountered. This has been proposed for the measurement of bladder volume, where Bih et al. (1998) have derived constants for a number of shapes.

Summary of errors

In a controlled situation, e.g. test object, with careful measurement, errors can be summarised as shown in Table 6.2. In clinical practice, errors depend on additional factors such as the quality of image acquired, the shape of structures measured and any assumptions used when combining quantities in calculations. With careful imaging, measurement and choice of methods, it is possible to achieve errors similar to those shown in Table 6.2, although some clinical measurements may have

errors of two or three times this size. Calculated parameters, such as volume, may have errors of up to 100%.

INTERPRETATION OF MEASUREMENTS

Once measurements have been made, they must be interpreted. This requires an understanding of the source and magnitude of any possible errors, an aspect that is often ignored. Measurements that do not rely on operator dexterity or assumptions about shape tend to have the smallest errors. In the clinical situation, errors are influenced by other factors including the ease of obtaining and recognising the correct section for measurement. Where landmarks are well recognised and reproduced in clinical practice, measurements are widely used and are interpreted with confidence. The size of errors must be considered in the context of the normal biological variation of the measurement. A 5 mm measurement uncertainty is insignificant in a 50 mm normal range, but highly significant in a 5 mm normal range. Absolute accuracy is generally more important for small structures than for large structures, e.g. an error of 0.1 mm in a nuchal translucency measurement can make a significant difference to the estimated risk of Down's syndrome.

The variation in size of errors between equipment, together with inter-operator and inter-centre variation, may limit the value of clinical ultrasound measurements. Understanding and reducing this variation may allow ultrasound measurement to be used to its full potential. It is important to evaluate and minimise human error

Table 6.2 Dependency and errors in each type of measurement

Type of measurement	Error depends on	Approximate size of error
Linear	Pixel size	0.1–0.5 mm
	Calliper increment	0.1–1 mm
	Resolution	0.1–5 mm
Ellipse (circumference and area)	Ellipse fit and calculation	1%
Point to point (circumference and area)	Point-to-point calculation	1%
	Number of points	1–5%
Tracing (circumference)	Operator dexterity	2%
Tracing (area)	Operator dexterity	1%

in order to achieve the best value from the precision and accuracy of the equipment.

Use of normal reference data

Interpretation often involves a reference normal range or chart. This introduces a further potential source of human error in the reading of values or transcription of measurements. It is also important to consider how these data were derived. What technique was used? What was the uncertainty in the results? There may have been systematic errors and there are always random errors. Were these addressed?

Systematic errors in the derivation of normal data may be significant compared to the normal ranges, particularly when the equipment used was inferior to that available today. If these errors were not evaluated in the originating centre, this limits the value of the data. The uncertainty this generates is illustrated by the following example.

In generating normal ranges for fetal AC, Chitty et al. (1994) found the average perimeter tracing to be 3.5% greater than the circumference calculated from two orthogonal diameters. This is consistent with the findings of Tamura et al. (1986), who found an average difference of 3.1% between the methods. These measurements were made in the late 1980s and early 1990s. Dudley and Chapman (2002) found a difference of 1.5% between the two methods in a clinical setting, but even in a single centre this varied between sonographers, ranging from 0.5% to 2.1%. This difference has been reproduced in test-object measurements (Dudley and Griffith 1996) and represents the systematic error associated with operator dexterity and the sensitivity of the calliper control.

Normal ranges and charts may be evaluated against local practice and equipment by comparison with results from patient measurements. This is straightforward where large numbers of normal measurements are made, e.g. antenatal screening. A dating chart can be evaluated by plotting a number of measurements, say 20 at each gestational age, and assessing their distribution within the normal range; they should be centred on the mean and the majority should be within the normal range. Where a large amount of normal data is not available the evaluation requires more careful

judgement, considering where the measurements should fall with respect to the normal range based on knowledge of the likely composition of the patient population.

Measurement packages

Most ultrasound scanners now arrive equipped with 'measurement packages'. As well as providing a variety of measurement methods, these packages include programmed charts and automatic calculations to aid interpretation. Examples include fetal growth charts, fetal weight estimation and bladder volume calculation. Although charts and calculations will be based on published data the source may well be the choice of the manufacturer. Many manufacturers offer a selection of pre-installed charts and also provide the facility for users to enter their own preferred data.

Use of such packages removes a potential source of human error, as the sonographer is no longer required to make manual calculations or read from graphs. It is essential, however, that the equipment purchaser makes active decisions in the choice of charts and calculation algorithms, and ensures that these are thoroughly tested before being put into clinical use.

SUMMARY

Measurements are performed in many clinical ultrasound specialities. Calliper systems are generally flexible, offering a range of measurements and calculations.

Errors are present, to varying degrees, in most measurements and can arise from the instrumentation, the ultrasound propagation properties of tissue and the equipment operator. These errors can be minimised by adopting good practice at all stages of equipment selection and use.

The choice of equipment is important. Systematic and random errors can be minimised by thorough evaluation of measurement controls and packages prior to purchase, although reliable measurements and the highest-quality images may not always be available in the same instrument. At acceptance testing, systematic and random errors should be quantified, so that they may be reflected in the choice of normal reference data and reporting policies. Any programmed charts and calculations should be consistent with locally accepted practice.

Measurement technique is important. All users should understand locally accepted practice, which should be clearly defined in written procedures. Good practice should include the appropriate selection of probe type and frequency, the optimum use of image magnification, setting focal zones at the region of measurement, avoiding image distortion due to refraction whenever possible and the correct and careful placement of callipers. Methods should be appropriate for the selected normal reference data. Random errors can be greatly reduced by performing repeated measurements and using average results.

In interpretation of measurements, clinicians should always be aware of the size of any uncertainty in the result and the size of the normal biological variation. Where these are both small, results may be interpreted with confidence, otherwise a degree of caution is required.

In order to minimise errors, the following steps must be taken:

1. Buy a scanner with correct calibration and good reproducibility.
2. Train and assess staff to national standards where available. As a minimum, ensure local consistency by training and audit.
3. Use an appropriate method of measurement.
4. Overcome the limitations of pixel size and resolution by making the best use of scale/depth, magnification and focusing controls, so that the measured structure is large and optimally resolved within the field of view.
5. Repeat measurements to increase certainty. This either gives confidence where the same value is obtained or allows reduction of the random error by averaging.

QUESTIONS

Multiple Choice Questions

Q1. What is the conventionally assumed average speed of sound in human tissue?
a. 1600 m s^{-1}
b. 3×10^8 m s^{-1}
c. 1540 m s^{-1}
d. 1450 m s^{-1}
e. 330 m s^{-1}

Q2. Ultrasound scanners calculate distances based on:
a. Depth/scale setting
b. Resolution
c. Gain
d. Focal distance
e. Probe geometry

Q3. Regular circumferences are most often measured using:
a. String
b. Calculation from a single diameter
c. A ruler
d. Ellipse fitting
e. Visual estimation

Q4. Volume estimation is best done using:
a. A single measurement and an appropriate constant
b. The sum of areas
c. Two orthogonal measurements and an assumption about shape
d. Three orthogonal diameters and an appropriate constant
e. The sum of multiple measurements

Q5. Random errors may be:
a. Always larger than the true value
b. Distributed above and below the true value
c. Due to mis-calibration of the scanner
d. Due to random reflections
e. Due to human error

Q6. Systematic errors may be:
a. Distributed above and below the true value
b. Always larger than the true value
c. Consistent in direction and size
d. Due to random reflections
e. Due to human error

Q7. When measurements are combined in a formula, e.g. estimated fetal weight:
a. The final error is likely to be smaller than the individual errors
b. Speed of sound differences are averaged out
c. Errors do not really matter as it is an estimate

d. The final error is likely to be larger than any individual error

e. The result will be reproducible

Q8. Which of the following are potential sources of operator error?
   a. No agreed standard method
   b. Lack of training
   c. Incorrect calliper placement
   d. Refraction
   e. Scanner mis-calibration

Q9. If an image is displayed in a 512-pixel-square matrix, with a scale setting of 5 cm in order to make a small measurement (e.g. nuchal translucency, common bile duct, intimal thickness, valve leaflet thickness), what is the approximate pixel size?
   a. 0.01 mm
   b. 0.1 mm
   c. 5 mm
   d. 0.5 mm
   e. 100 mm

Q10. Which of the following is likely to affect the accuracy of reported measurements?
   a. Focusing
   b. Gain
   c. Scale
   d. Read zoom
   e. Repeating measurements

Short-Answer Questions

Q1. What information is used to determine linear distances in the axial and lateral directions?

Q2. What is circumference and what are the options for measurement?

Q3. Describe the two main types of error. List three possible causes for human error.

REFERENCES

Artang R, Migrino RQ, Harmann L, Bowers M, Woods TD. 2009. Left atrial volume measurement with automated border detection by 3-dimensional echocardiography: Comparison with Magnetic Resonance Imaging. *Cardiovascular Ultrasound*, 7, 16.

Axell R, Lynch C, Chudleigh T et al. 2012. Clinical implications of machine-probe combinations on obstetric ultrasound measurements used in pregnancy dating. *Ultrasound in Obstetrics and Gynecology*, 40, 194–199.

Bih LI, Ho CC, Tsai SJ, Lai YC, Chow W. 1998. Bladder shape impact on the accuracy of ultrasonic estimation of bladder volume. *Archives of Physical Medicine and Rehabilitation*, 79, 1553–1556.

Carneiro G, Georgescu B, Good S, Comaniciu D. 2008. Detection and measurement of fetal anatomies from ultrasound images using a constrained probabilistic boosting tree. *IEEE Transactions on Medical Imaging*, 27, 1342–1355.

Chang TC, Robson SC, Spencer JAD, Gallivan S. 1993. Ultrasonic fetal weight estimation: Analysis of inter- and intra-observer variability. *Journal of Clinical Ultrasound*, 21, 515–519.

Chitty LS, Altman DG, Henderson A, Campbell S. 1994. Charts of fetal size: 3. Abdominal measurements. *British Journal of Obstetrics and Gynaecology*, 101, 125–131.

Dudley NJ, Chapman E. 2002. The importance of quality management in fetal measurement. *Ultrasound in Obstetrics and Gynecology*, 19, 190–196.

Dudley NJ, Griffith K. 1996. The importance of rigorous testing of circumference measuring calipers. *Ultrasound in Medicine and Biology*, 22, 1117–1119.

Furiasse N, Thomas JD. 2015. Automated algorithmic software in echocardiography: Artificial intelligence? *Journal of the American College of Cardiology*, 66, 1467–1469.

Jago JR, Whittingham TA, Heslop R. 1994. The influence of scanner beam width on femur length measurements. *Ultrasound in Medicine and Biology*, 20, 699–703.

Kampmann W, Walka MM, Vogel M, Obladen M. 1998. 3-D sonographic volume measurement of the cerebral ventricular system: *In vitro* validation. *Ultrasound in Medicine and Biology*, 24, 1169–1174.

Kuhl HP, Franke A, Janssens U et al. 1998. Three-dimensional echocardiographic determination of left ventricular volumes and function by multiplane transesophageal transducer: Dynamic *in vitro* validation and *in vivo* comparison with angiography and thermodilution. *Journal of the American Society of Echocardiography*, 11, 1113–1124.

Lee W, Comstock CH, Kirk JS et al. 1997. Birthweight prediction by three-dimensional ultrasonographic volumes of fetal thigh and abdomen. *Journal of Ultrasound in Medicine*, 16, 799–805.

Loughna P, Chitty L, Evans T, Chudleigh T. 2009. Fetal size and dating: Charts recommended for clinical obstetric practice. *Ultrasound*, 17, 161–167.

Owen P, Donnet ML, Ogston SA et al. 1996. Standards for ultrasound fetal growth velocity. *British Journal of Obstetrics and Gynaecology*, 103, 60–69.

Ramanujan S. 1914. Modular equations and approximations to π. *Quarterly Journal of Pure and Applied Mathematics*, 45, 350–372.

Rossavik IK, Deter RL. 1984. The effect of abdominal profile shape changes on the estimation of fetal weight. *Journal of Clinical Ultrasound*, 12, 57–59.

Rueda S, Fathima S, Knight CL et al. 2014. Evaluation and comparison of current fetal ultrasound image segmentation methods for biometric measurements: A grand challenge. *IEEE Transactions on Medical Imaging*, 33, 797–813.

Sarmandal P, Bailey SM, Grant JM. 1989. A comparison of three methods of assessing interobserver variation applied to ultrasonic fetal measurement in the third trimester. *British Journal of Obstetrics and Gynaecology*, 96, 1261–1265.

Shibayama K, Watanabe H, Iguchi N et al. 2013. Evaluation of automated measurement of left ventricular volume by novel real-time 3-dimensional echocardiographic system: Validation with cardiac magnetic resonance imaging and 2-dimensional echocardiography. *Journal of Cardiology*, 61, 281–288.

Tamura RK, Sabbagha RE, Wen-Harn P, Vaisrub N. 1986. Ultrasonic fetal abdominal circumference: Comparison of direct versus calculated measurement. *American Journal of Obstetrics and Gynecology*, 67, 833–835.

Tsang W, Salgo IS, Medvedofsky D et al. 2016. Transthoracic 3D echocardiographic left heart chamber quantification using an adaptive analytics algorithm. *JACC Cardiovasc Imaging*, 9, 769–782.

Wendelhag I, Liang Q, Gustavsson T, Wikstrand J. 1997. A new automated computerized analyzing system simplifies readings and reduces the variability in ultrasound measurement of intima-media thickness. *Stroke*, 28, 2195–2200.

Yazdi B, Zanker P, Wagner P et al. 2014. Optimal caliper placement: Manual vs automated methods. *Ultrasound Obstet Gynecol*, 43, 170–175.

Zhang L, Dudley NJ, Lambrou T, Allinson N, Ye X. 2017. Automatic image quality assessment and measurement of fetal head in two-dimensional ultrasound image. *Journal of Medical Imaging*, 4, 024001.

7

Principles of Doppler ultrasound

PETER R HOSKINS

INTRODUCTION

Ultrasound systems may be used to detect the motion of blood in arteries, veins and in the heart. For several decades clinical practice has been concerned with two display modes; spectral Doppler for the display of velocity-time waveforms and colour flow as used for the two-dimensional (2D) display of flow. In each case these display modes are produced using the Doppler effect, and the general area is referred to as 'Doppler ultrasound' which is the theme of this chapter. Further details of spectral Doppler and colour flow are provided in Chapters 9 and 10.

For measurement of blood flow there are a number of other techniques available on commercial systems. These include high frame rate Doppler, measurement of flow in very small vessels (down to 50 μm) which then enables applications in the microcirculation, vector-flow techniques which provide information on direction of flow, and non-Doppler techniques for estimation of blood velocities. Further details of these techniques are found in Chapter 11.

The use of Doppler ultrasound for measurement of the motion of tissues goes all the way back to the beginning of this area in that the first Doppler ultrasound paper described measurement of the motion of heart tissues (Satomura 1957). The use of 2D Doppler ultrasound for measurement of cardiac tissue motion was described by McDicken et al. (1992) where this is called 'tissue Doppler imaging' (TDI). Though widely used in the research literature and available as an option on most high-end cardiac ultrasound systems, it is not widely used in clinical practice. Chapter 11 provides further details of TDI.

DOPPLER ULTRASOUND SYSTEMS

The Doppler effect

The Doppler effect is observed regularly in our daily lives. For example, it can be heard as the changing pitch of an ambulance siren as it passes by. The Doppler effect is the change in the observed frequency of the sound wave (f_r) compared to the emitted frequency (f_t) which occurs due to the relative motion between the observer and the source, as shown in Figure 7.1. In Figure 7.1a, both the source and the observer are stationary so the observed sound has the same frequency as the emitted sound. In Figure 7.1b, the source is moving towards the observer as it transmits the sound wave. This causes the wave fronts travelling towards the observer to be more closely packed, so that the observer witnesses a higher frequency wave than that emitted. If, however, the source is moving away from the observer, the wave fronts will be more spread out, and the frequency observed will be lower than that emitted (Figure 7.1c). The resulting change in the observed frequency from that transmitted is known as the Doppler shift, and the magnitude of the Doppler shift frequency

is proportional to the relative velocity between the source and the observer.

It does not matter whether the source or the observer is moving. If either one is moving away from the other, the observer will witness a lower frequency than that emitted. Conversely, if either the source or observer moves towards the other, the observer will witness a higher frequency than that emitted.

Ultrasound can be used to assess blood flow by measuring the change in frequency of the ultrasound scattered from the moving blood. Usually the transducer is held stationary and the blood moves with respect to the transducer, as shown in Figure 7.2. The ultrasound waves transmitted by the transducer strike the moving blood, so the frequency of ultrasound as experienced by the blood is dependent on whether the blood is stationary,

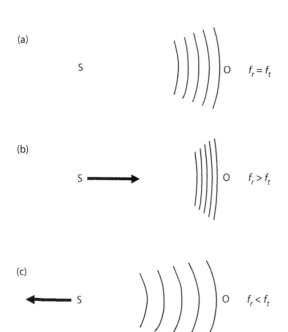

Figure 7.1 Doppler effect as a result of motion of the source (S) relative to a stationary observer (O). (a) There is no motion and the observer detects sound at a frequency equal to the transmitted frequency. (b) The source moves towards the observer and the observer detects sound at a higher frequency than that transmitted. (c) The source moves away from the observer and the observer detects a lower frequency than that transmitted.

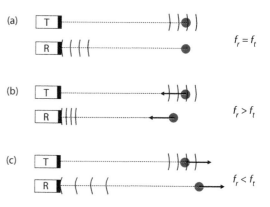

Figure 7.2 The Doppler effect occurring due to motion of blood. In each case the probe transmits ultrasound which strikes the blood. Separate elements are shown for transmission (T) and reception (R). Ultrasound is scattered by the region of blood, some of which returns to the transducer. (a) The probe and blood are not moving; the frequency of the ultrasound received by the transducer is equal to the transmitted frequency. (b) The blood is moving towards the probe; the blood encounters more wave fronts and the frequency of the ultrasound received by the transducer is greater than the transmitted frequency. (c) The blood is moving away from the probe; the blood encounters fewer wave fronts and the frequency of the ultrasound received by the transducer is less than the transmitted frequency.

moving towards the transducer or moving away from the transducer. The blood then scatters the ultrasound, some of which travels in the direction of the transducer and is detected. The scattered ultrasound is Doppler frequency shifted again as a result of the motion of the blood, which now acts as a moving source. Therefore, a Doppler shift has occurred twice between the ultrasound being transmitted and received back at the transducer (hence the presence of the '2' in Equation 7.1).

The detected Doppler shift frequency (f_d) is the difference between the transmitted frequency (f_t) and the received frequency (f_r). The Doppler shift frequency f_d depends on the frequency of the transmitted ultrasound, the speed of the ultrasound as it passes through the tissue (c) and the velocity of the blood (v) (Figure 7.3). This relationship can be expressed by the Doppler equation:

$$f_d = f_r - f_t = \frac{2f_t v \cos\theta}{c} \qquad (7.1)$$

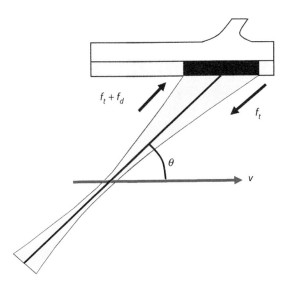

Figure 7.3 A beam of ultrasound is produced by a linear array in which the active elements are shown in black. Ultrasound is transmitted whose centre frequency is f_t. This strikes blood moving at velocity v. The angle between the beam and the direction of motion is θ. The received ultrasound frequency has been Doppler-shifted by an amount f_{d}, so that the detected frequency is $f_t + f_d$.

The detected Doppler shift also depends on the cosine of the angle θ between the path of the ultrasound beam and the direction of the blood flow. This angle is known as the angle of insonation. The angle of insonation can change as a result of variations in the orientation of the vessel or the probe. Many vessels in the upper and lower limbs run roughly parallel to the skin, although if the vessel is tortuous the angle of insonation will alter. In the abdomen there is considerable variation in the orientation of vessels which will result in many different angles of insonation. The operator is also able to alter the angle of insonation by adjustment of the orientation of the probe on the surface of the skin. The relationship between an angle and the value of its cosine is shown in Figure 7.4. It is often desirable that the operator can adjust the angle of insonation to obtain the highest Doppler frequency shift possible. The highest Doppler frequency shift from any specific vessel occurs when the vessel and the beam are aligned; that is, when the angle of insonation is zero and the cosine function has the value of 1.0. The least desirable situation occurs when the angle approaches 90° as then the Doppler frequency shift will be minimum. In clinical practice it is often not possible to align the beam and the vessel. As a rule, provided the angle is less than about 60°, good-quality spectral waveforms can be obtained.

If the angle of insonation of the ultrasound beam is known it is possible to use the Doppler shift frequency to estimate the velocity of the blood

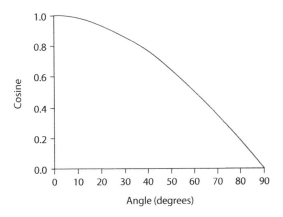

Figure 7.4 The cosine function. This has a maximum value of 1.0 when the angle is 0°, and a value of 0 when the angle is 90°.

using the Doppler equation. This requires rearrangement of Equation 7.1 to give:

$$\nu = \frac{cf_d}{2f_t \cos\theta} \quad (7.2)$$

In diseased arteries the lumen will narrow and the blood velocity will increase. This provides the means by which the lumen diameter may be estimated using Doppler ultrasound. The blood velocity is estimated within the narrowed region, and converted into a percentage stenosis using standard tables.

Doppler displays

The main display modes used in a modern Doppler system are as follows:

- *Spectral Doppler*: All the velocity information detected from a single location within the blood vessel is displayed in the form of a frequency shift-time plot (Figure 7.5). Vertical distance from the baseline corresponds to Doppler shift, while the greyscale indicates the amplitude of the detected ultrasound with that particular frequency.

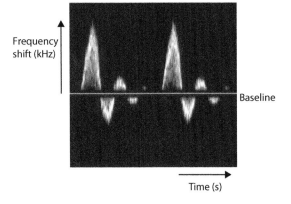

Figure 7.5 Spectral display. This is a display of the Doppler frequency shift versus time. The Doppler waveform from the femoral artery is shown. Vertical distance from the baseline corresponds to Doppler shift, while the greyscale indicates the amplitude of the detected ultrasound with that particular frequency.

- *2D colour flow imaging*: The Doppler signal is displayed in the form of a 2D colour image superimposed on the B-scan image (Figure 7.6). Colour represents the Doppler shift for each pixel, averaged over the area of the pixel.

The advantage of the 2D colour display is that it allows the user to observe the presence of blood flow within a large area of the tissue. The spectral Doppler display allows closer inspection of the changes in velocity over time within a single small area. Both colour flow and spectral Doppler systems use the Doppler effect to obtain blood-flow information. In this respect they are both Doppler systems. The two modalities complement each other, providing the user with a wide range of useful information.

Continuous-wave and pulsed-wave Doppler

There is not the restriction in Doppler systems as there is for B-mode devices that the ultrasound must be transmitted in the form of pulses. Some Doppler ultrasound systems, known as continuous-wave (CW) systems, transmit ultrasound continuously. Other Doppler systems, known as pulsed-wave (PW) systems, transmit short pulses of ultrasound. The main advantage of PW Doppler is that Doppler signals can be acquired from a known depth. The main disadvantage is that there is an upper limit to the Doppler frequency shift which can be detected, making the estimation of high velocities more challenging.

In a CW Doppler system there must be separate transmission and reception of ultrasound (Figure 7.7a). In the pencil probe this is achieved using two elements, one which transmits continuously and one which receives continuously. The region from which Doppler signals are obtained is determined by the overlap of the transmit and receive ultrasound beams. In a PW system it is possible to use the same elements for both the transmit and receive (Figure 7.7b). In the pencil probe only one element is needed, serving both transmit and receive functions. The region from which Doppler signals are obtained is determined by the depth of the gate and the length

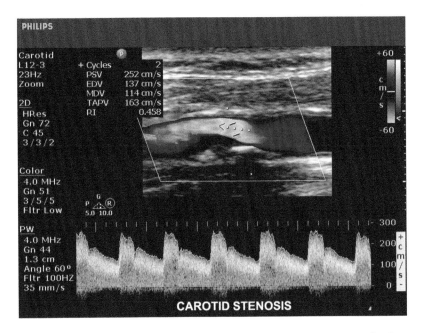

Figure 7.6 Colour Doppler display. This is a display of the mean (or average) Doppler frequency, at each point in a 2D slice of the tissue from blood, superimposed on the B-mode display. The example also shows simultaneous display of spectral Doppler. (Image kindly provided by Philips Medical Systems, Eindhoven, the Netherlands.)

of the gate, which can both be controlled by the operator.

The received ultrasound signal is processed by the Doppler signal processor to extract the Doppler frequency shifts, which are then displayed in the form of spectral Doppler or colour Doppler.

ULTRASOUND SIGNAL RECEIVED BY THE TRANSDUCER

Before considering in detail the Doppler signal processor, the nature of the ultrasound signal received at the transducer must be considered, as this will determine the subsequent signal processing that is performed.

When ultrasound is emitted from the transducer it will pass through regions of tissue and regions of blood contained in veins and arteries. The blood will almost certainly be moving, but some of the tissue may also be moving. For example, the arteries move during the cardiac cycle, the heart moves as a result of contraction, and tissues in contact with the arteries and with the heart will

also move. The received ultrasound signal consists of the following four types of signal:

- Echoes from stationary tissue
- Echoes from moving tissue
- Echoes from stationary blood
- Echoes from moving blood

The task for a Doppler system is to isolate and display the Doppler signals from blood, and remove those from stationary tissue and from moving tissue. Table 7.1 shows the amplitude of signals from blood and tissue, and it is clear that the signal from blood is extremely small compared to tissue, typically 40 dB smaller. The maximum blood velocity occurs when disease is present, and is about 6 m s^{-1}.

The maximum tissue velocity occurs during the systolic phase of the heart where the myocardium attains a velocity of up to 10 cm s^{-1}. In general, Doppler signals from blood are of low amplitude and high-frequency shift, whereas those from tissue are of high amplitude and low-frequency shift.

(a) CW system

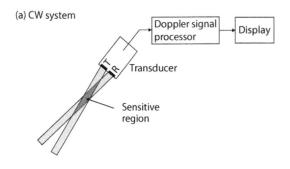

(b) PW system

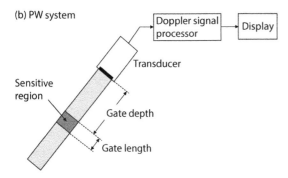

Figure 7.7 Schematic of CW and PW Doppler system consisting of the transducer, the Doppler signal processor and the display. **(a)** In a CW Doppler system there must be separate transmission and reception of ultrasound. In the pencil probe this is achieved using two elements, one which transmits continuously and one which receives continuously. The region from which Doppler signals are obtained is determined by the overlap of the transmit and receive ultrasound beams. **(b)** In a PW system it is possible to use the same element or elements for both transmit and receive. In the pencil probe only one element is needed, serving both the transmit and receive functions. The region from which Doppler signals are obtained is determined by the depth and length of the gate, which are both controlled by the operator.

Table 7.1 Typical velocities and signal intensities

	Velocity ranges	Signal intensity
Blood	0–600 cm s^{-1}	Low
Tissue	0–10 cm s^{-1}	40 dB higher than blood

These differences provide the means by which true blood flow signals may be separated from those produced by the surrounding tissue.

CONTINUOUS-WAVE DOPPLER SIGNAL PROCESSOR

Dedicated signal-processing algorithms are used to produce the Doppler signals from the received ultrasound signal. There are three steps in the process (Figure 7.8): 'demodulation' is the separation of the Doppler frequencies from the underlying transmitted signal; 'high-pass filtering' is the removal of the tissue signal; 'frequency estimation' is where the Doppler frequencies and amplitudes are calculated. There are important differences between the CW and the PW Doppler signal processor which are described in the next section.

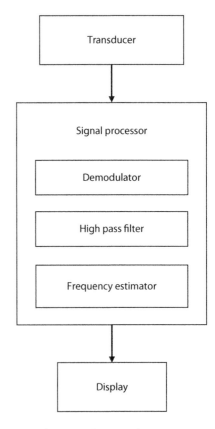

Figure 7.8 The Doppler signal processor can be considered to be made of three parts: the demodulator, the high-pass filter and the frequency estimator.

Demodulation

The Doppler frequencies produced by moving blood are a tiny fraction of the transmitted ultrasound frequency. Equation 7.1 shows that, if the transmitted frequency is 4 MHz, a motion of 1 m s^{-1} will produce a Doppler shift of 5.2 kHz, which is less than 0.1% of the transmitted frequency. Extraction of the Doppler frequency shift information from the ultrasound signal received from tissue and blood is called 'demodulation'. This process is generally invisible to the user as there are no controls which the user can adjust to alter this process. The process of demodulation consists of comparison of the received ultrasound signal (Figure 7.9b) with a 'reference' signal (Figure 7.9a) which has the same frequency as the transmit echo. This process is illustrated in Figure 7.9, which

shows the ultrasound signal obtained from a CW Doppler system being used with a small moving target. The received ultrasound signal has a slightly different frequency compared to the reference signal. The first step of the demodulation process consists of multiplying together the reference signal with the received signal. This process is called 'mixing'. This produces the signal shown in Figure 7.9c, which consists of a low-frequency component (the Doppler frequency) arising from the moving target, and a high-frequency component arising from the transmit frequency. The high-frequency component can be removed by simple filtering, revealing the Doppler frequency (Figure 7.9d). The end result of demodulation for the example of Figure 7.9 is the removal of the underlying high-frequency transmit signal, revealing the blood flow Doppler signal.

The previous example described demodulation when there was a single moving target present. In reality the sensitive region of the ultrasound system is unlikely to be placed within a region in which all the blood is moving at the same velocity. It is more likely that there will be a range of blood velocities present within the sensitive region, with low velocities present near the vessel wall and higher velocities present near the vessel centre. In addition the sensitive region for a CW system will usually encompass moving tissue. When there are signals from both blood and tissue present, and multiple velocities, the process of demodulation is best understood with reference to the detected ultrasound frequency (Figure 7.10). The received signal from a region in which blood is flowing consists of a high-amplitude clutter signal from stationary and slowly moving tissue, and Doppler-shifted components from the moving blood (Figure 7.10a). Blood moving towards the transducer gives rise to a positive shift, while blood moving away from the transducer gives rise to a negative Doppler shift. Demodulation removes the high frequencies arising from the transmit frequency, leaving both the Doppler shift signal from blood flow and the clutter signal from tissue motion (Figure 7.10b). At this stage in the process it is not possible to differentiate the tissue signal from the blood flow signal. This requires another step (high-pass filtering) which is described in the next section.

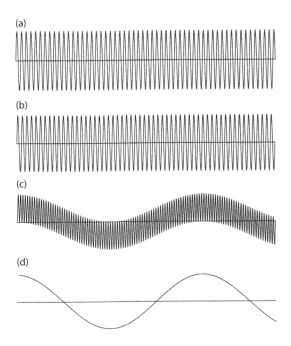

Figure 7.9 Doppler demodulation for CW Doppler. (a) Reference signal of frequency f_t. (b) Detected signal which has been Doppler-shifted and whose frequency is $f_r = f_t + f_d$. (c) The reference signal is multiplied by the detected signal. (d) The high-frequency oscillations are removed by low-pass filtering. The end result is the Doppler frequency shift signal f_d (the example shown contains no clutter).

(a)

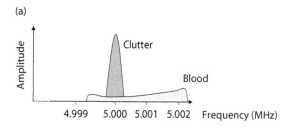

(b)

Demodulation

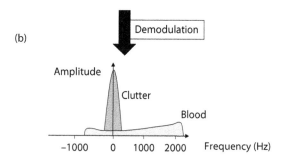

(c)

High pass filter

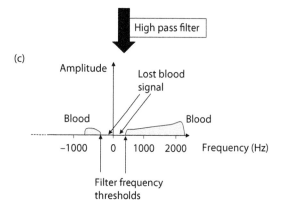

Figure 7.10 Demodulation and high-pass filtering. **(a)** The received signal, from a region in which blood is flowing, consists of a high-amplitude clutter signal from stationary and slowly moving tissue, and Doppler-shifted components from the moving blood. Blood moving towards the transducer gives rise to a positive shift, while blood moving away from the transducer gives rise to a negative Doppler shift. **(b)** Demodulation removes the high-frequency signals arising from the transmit frequency, leaving the Doppler shift signal from both blood flow and the clutter signal from tissue motion. **(c)** The high-pass filter acts to remove the clutter signal. A consequence of this is that the very lowest blood velocities are also removed.

Another way of viewing the process of demodulation is that it converts a high-frequency signal (MHz, or millions of cycles per second) into a low-frequency signal (kHz, or thousands of cycles per second). It is easier for the ultrasound system to digitise and process signals which are of lower frequency. This is important as computational efficiency is crucial to ultrasound system design. High-speed processing is expensive, and increases the cost of the system to the user.

High-pass filtering

We saw that the signal arising from demodulation can contain Doppler frequency shifts arising from both blood flow and tissue motion. If no further processing were performed then the estimated Doppler signals from blood would be overwhelmed by the large-amplitude signal from tissue. In order to correctly estimate the Doppler shift frequencies from blood, the Doppler signals from tissue must be removed. This step is referred to as the 'wall-thump filter' or 'cut-off filter' in spectral Doppler.

The tissue signals are called 'clutter'. This term originates from radar and refers to stationary objects such as trees and buildings which interfere with the detection of moving targets such as aeroplanes. Removal of clutter can be performed using a number of methods. The basis for discrimination between the blood signal and the tissue signal is that signals from tissue tend to be low frequency and high amplitude, and those from blood are high frequency and low amplitude. The simplest approach, and that adopted by many Doppler systems, relies on a frequency filter. Removal of components below a certain threshold frequency will remove the Doppler shift components arising from the tissue (Figure 7.10c). An unfortunate consequence of this process is that the Doppler frequency shifts from slowly moving blood will also be lost. In general the filter frequency is set by the user to just suppress tissue signals. In cardiology applications a high-filter setting of 300 Hz or more is necessary since myocardium and valves are travelling at high speeds (up to 10 cm s^{-1}), and the detection of high blood velocities is of interest. In obstetrics, low settings of 50–80 Hz are more typical as it is important to be able to detect the low velocities at end diastole. After filtering, the Doppler frequency shift data are in a form suitable for the third step in the process, the frequency estimator.

Frequency estimation

When information concerning blood velocity is required then some form of frequency estimator must be used. As CW systems provide no information on the depth from which the blood flow signal has returned, it is not possible to use a colour flow display. CW systems display the time-velocity waveform, either in the form of spectral Doppler, or in the form of a single trace.

A spectrum analyser calculates the amplitude of all of the frequencies present within the Doppler signal, typically using a method called the 'fast Fourier transform' (FFT), in which a complete spectrum is produced every 5–40 ms. In the spectral display (Figure 7.5) the brightness is related to the power or amplitude of the Doppler signal component at that particular Doppler frequency. The high speed with which spectra are produced means that detailed and rapidly updated information about the whole range of velocities present within the sample volume can be obtained.

Before the advent of real-time spectrum analysis, it was common to display only a single quantity related in some way to the blood velocity. A relatively simple electronics device known as the 'zero-crossing detector' produces an output proportional to the root mean square of the mean Doppler frequency. However, the zero-crossing detector is sensitive to noise, and the output depends strongly on the velocity profile within the vessel. Its use in a modern Doppler system is no longer justified.

ORIGIN AND PROCESSING OF THE DOPPLER SIGNAL FOR PW SYSTEMS

The Doppler signal

There is a crucial difference between CW and PW systems in that for PW systems a received ultrasound signal is not available continuously, as it is for CW systems. This is illustrated in Figure 7.11, which shows a point target moving away from the transducer, and the received ultrasound signal for

(a) Location of target for 4 consecutive pulses

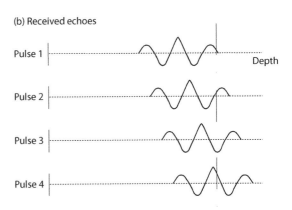
(b) Received echoes

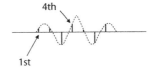
(c) Echo amplitude from consecutive pulses

Figure 7.11 Received echoes in PW Doppler. (a) A single target moves away from the transducer. Consecutive ultrasound pulses are emitted which strike the target at increasing depth. (b) The received echoes are displayed at increasing depth. A vertical dashed line is shown corresponding to a fixed depth. The amplitude of the echo changes as it crosses the dashed line for each consecutive pulse. The change in amplitude of the received echoes with depth contains the Doppler shift, which is extracted by demodulation. (c) Amplitude of consecutive echoes from the depth indicated by the dashed line in (b) shown as a function of time.

consecutive ultrasound pulses. For each consecutive pulse the target moves farther away from the transducer, and consequently the echo is received at a later time from the start of the transmit pulse. As the echo pattern moves farther away from the transducer with each consecutive pulse, so the amplitude of the signal at a specific depth changes, as shown in Figure 7.11. In PW Doppler systems the location of the Doppler gate defines the range of depths from which the Doppler signal arises. The Doppler signal is contained within the series of consecutive received echoes from the gate. The Doppler signal is then revealed by the subsequent signal processing.

Pulsed-wave Doppler signal processor

In PW Doppler systems the basic steps of demodulation, high-pass filtering and frequency estimation are essentially the same as for CW Doppler. Figure 7.9 has shown how demodulation works for CW systems. Figure 7.12 shows the demodulator in action for PW Doppler. Although the reference signal (Figure 7.12a) is produced continuously,

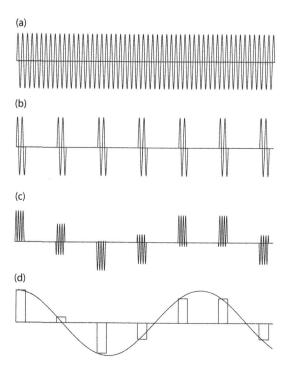

Figure 7.12 Doppler demodulation for PW Doppler. **(a)** Reference signal of frequency f_t. **(b)** The received ultrasound signal consists of consecutive echoes which have passed through the electronic gate. **(c)** The reference signal is multiplied by the detected signal. **(d)** The high-frequency oscillations are removed by low-pass filtering. The end result is the Doppler frequency shift signal f_d (the example shown contains no clutter).

the received ultrasound signal (Figure 7.12b) from the Doppler gate is only produced once for every ultrasound pulse. The process of 'mixing' (Figure 7.12c) and low-pass filtering (Figure 7.12d) produces the Doppler signal. This signal has the same overall shape as the demodulator output from the CW Doppler system. This is the reason that the displayed Doppler signals from CW and PW Doppler appear to be similar, often indistinguishable.

This process of Doppler frequency detection is usually referred to by physicists as the 'phase-domain' method; PW Doppler systems which use it are referred to as 'phase-domain systems'. The term 'phase' refers to the degree of correspondence between the received echo and the reference echo in the demodulation process. This can be explained by further consideration of the demodulation process. Figure 7.13 shows the received echoes from consecutive pulses with respect to the reference frequency. Figure 7.13 also shows the amplitude of the signal after the reference and received signals have been multiplied, and the amplitude of the detected Doppler signal. The peak positive amplitude of the detected Doppler signal arises when the reference and received signals are aligned or 'in phase'. As the reference and received frequencies gradually become misaligned the detected Doppler signal amplitude gradually decreases. When the reference and received signals are completely misaligned or 'out of phase' the demodulated amplitude reaches its peak negative value.

For spectral Doppler, estimation of the Doppler shift frequencies for the purpose of display is performed using the same FFT method as used for CW Doppler.

In colour flow imaging, the Doppler frequency estimator must calculate the mean (or average) Doppler shift frequency for each pixel of the 2D image. In order for colour images to be displayed at a suitable frame rate for real-time viewing, typically above 10 frames per second, the estimator must calculate this frequency value very quickly, usually in 1–2 ms. One possibility is to perform FFT, then calculate the mean frequency and colour code this value. This is not used in practice. The preferred option is to use a device called an 'auto-correlator'. This calculates the mean frequency

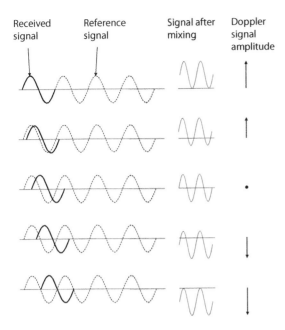

Received signal   Reference signal   Signal after mixing   Doppler signal amplitude

Figure 7.13 PW Doppler is a 'phase-domain process'. The received ultrasound signal is shown in comparison to the reference signal for five consecutive ultrasound pulses. Also shown are the signal after the reference and received signal have been multiplied, and the final detected Doppler signal after low-pass filtering. The first ultrasound signal is in phase with the reference signal and the detected Doppler amplitude achieves its peak positive value. For consecutive echoes the received and reference signals become misaligned until, in the fifth pulse of this example, they are 'out of phase' and the detected Doppler amplitude reaches its peak negative value.

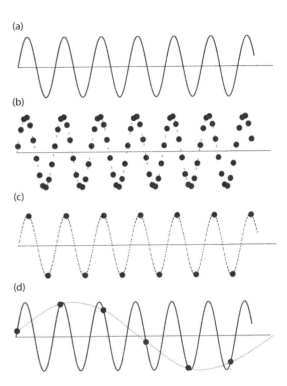

Figure 7.14 Aliasing. **(a)** Doppler signal from CW system with a single-frequency f_d. **(b)** PW Doppler, PRF $> 2f_d$: there are many samples for each cycle of the Doppler signal, and as a consequence the Doppler frequency is correctly detected. **(c)** PW Doppler; PRF $= 2f_d$: there are two samples per cycle and the Doppler frequency is estimated correctly. **(d)** PW Doppler, PRF $< 2f_d$: there are fewer than two samples per cycle, and the detected frequency (dashed line) is less than the true Doppler frequency (solid line).

directly, and is computationally more efficient (and hence faster) than the FFT approach.

Aliasing

The main difference between CW and PW Doppler, in terms of the display of blood velocities, is in the estimation of high velocity. Provided that there are a sufficient number of ultrasound pulses per wavelength it is possible to estimate the Doppler frequency shift accurately from the demodulated PW signal. If the pulse repetition frequency (PRF) is too low, the Doppler frequency shift cannot be estimated properly. This

phenomenon is called 'aliasing', and is a feature of pulsed Doppler systems, but not CW Doppler systems. This phenomenon can be explained with reference to Figure 7.14. The true Doppler signal is shown in Figure 7.14a. If the PRF is high (Figure 7.14b) then there are a sufficient number of samples to enable the Doppler system to detect the frequency correctly. As the PRF drops there comes a point at which the Doppler signal is only just sampled sufficiently. This is the case in Figure 7.14c where there are two samples per cycle. This is called the 'Nyquist limit'. This is equivalent to saying that the maximum Doppler frequency shift which can be detected is half of the PRF, i.e. $\text{PRF} = 2f_{d(max)}$. If the PRF drops further (Figure 7.14d) the Doppler system can no longer calculate the correct frequency.

Do pulsed-wave Doppler systems measure the Doppler effect?

There has been some debate in the literature about whether PW Doppler systems actually use the Doppler effect in their operation, and about whether PW Doppler systems have the right to be called 'Doppler' systems. This is not an issue which impacts upon the clinical use of Doppler systems and these readers may wish to skip the remainder of this section.

The Doppler effect can be considered to be a dilatation or expansion of the ultrasound wave. CW Doppler operates by directly measuring this dilatation or expansion. For a PW Doppler system there will also be dilatation or expansion of each individual echo arising as a result of the Doppler shift. However, PW Doppler systems do not measure this dilatation or expansion, hence strictly speaking are not true Doppler shift detectors. In other words, CW Doppler operates by the Doppler effect whereas the Doppler effect is an artefact for PW Doppler. PW Doppler measures the rate of change of phase between echoes from which the target velocity can be estimated using the same equation (Equation 7.1) as is used in CW Doppler (Evans et al. 2011). In quantitative terms, these considerations are unimportant for the clinical use of CW or PW Doppler. The interested reader is referred to Section 4.2.2.5 in the textbook by Evans and McDicken (2000) and Example 4.2 in the textbook by Jensen (1996) for further discussion.

TIME-DOMAIN SYSTEMS

Another method for calculation of the velocity of a moving object such as blood is to divide the distance the object travels by the time taken. This is the principle used in the time-domain Doppler system, which uses the PW approach. Figure 7.15 shows the position of a moving target and the corresponding echo for two consecutive

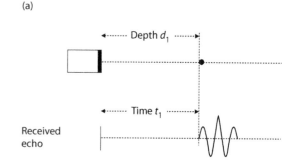

(a)

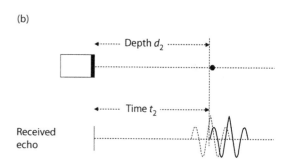

(b)

Figure 7.15 Velocity estimated from time-domain method. A single target moves away from the transducer. The target position and echo are shown for two consecutive pulses: **(a)** first pulse (time t_1) and **(b)** second pulse (time t_2). The change in depth is calculated by estimating the depth the target has moved, and dividing this by the time interval between consecutive pulses.

ultrasound pulses. For the second ultrasound pulse the target and hence the echo are located further from the transducer. The estimation of the target velocity is performed by the following steps.

Estimate distance travelled by target

As described in Chapter 1, estimation of the depth from which echoes are received is performed automatically by the machine from the time between transmission and reception of the echo, assuming that the speed of sound c is 1540 m s^{-1}. For the first and second pulses the depth of the moving target is estimated as:

$$d_1 = \frac{ct_1}{2}, \quad d_2 = \frac{ct_2}{2}$$

The factor of 2 above is to allow for the total distance travelled by the ultrasound, which is two times the depth of the target. The distance d_m moved between the two consecutive pulses is then estimated from the difference in estimated depths:

$$d_m = d_2 - d_1$$

$$d_m = \frac{c(t_2 - t_1)}{2} \quad (7.3)$$

The time-domain system estimates the difference in the time the echo is received between consecutive echoes using a process called 'cross-correlation', which is described in more advanced texts.

Calculate time taken for target to move

The time taken for the target to move is simply the time between consecutive echoes, which is called the 'pulse repetition interval' (PRI). For Figure 7.15 the PRI is the time between transmission of the first and second pulses, which is:

$$PRI = t_2 - t_1 \quad (7.4)$$

The PRI is equal to the inverse of the PRF:

$$PRI = \frac{1}{PRF}$$

For example, if the PRF is 1000 Hz, the PRI is 1/1000 s, which is 0.001 s or 1 ms.

Calculate velocity of target

The velocity of the target is calculated from the distance travelled and the time taken:

$$\nu = \frac{d_m}{PRI} \quad (7.5)$$

Using Equations 7.3 and 7.4 in combination with Equation 7.5 gives Equation 7.6, in which the blood velocity is expressed in terms of the measured time difference between consecutive echoes:

$$\nu = \frac{(t_2 - t_1)c\,PRF}{2} \quad (7.6)$$

Commercial time-domain Doppler systems

Phase-domain systems use demodulated data, whereas time-domain systems require calculations to be performed on the radio-frequency (RF) data. The increase in computational power with the use of RF rather than demodulated data makes time-domain systems more expensive than phase-domain systems. This increase in cost has been the main reason for the predominance of phase-domain Doppler systems, though a commercial time-domain system was produced for imaging of blood flow in the 1990s, and some

Table 7.2 Common features of Doppler systems

Feature	Description
Aliasing	The highest Doppler frequency shift that can be measured is equal to PRF/2
Angle dependence	Estimated Doppler frequency is dependent on the cosine of the angle between the beam and the direction of motion
Doppler speckle	Variations in the received Doppler signal give rise to a 'speckle' pattern seen in spectral and colour systems
Clutter breakthrough	Tissue motion giving rise to Doppler frequencies above the wall thump or clutter filter may be displayed on spectral Doppler or colour flow systems
Loss of low Doppler frequencies	Blood velocities which give rise to low Doppler frequencies (as a result of low velocity or angle near to 90°) will not be displayed if the value of the Doppler frequency is below the level of the wall thump or clutter filter

tissue-Doppler systems are based on time-domain processing.

OTHER FEATURES

The basic steps of both spectral Doppler and colour flow systems are similar, as illustrated in Figure 7.8, so there will be a number of similar features. Table 7.2 lists common features which are described in detail in the chapters on spectral and colour systems.

QUESTIONS

Multiple Choice Questions

Q1. The Doppler effect is the change in _____ due to the relative motion of source and observer.
   a. Speed of sound
   b. Velocity of sound
   c. Frequency
   d. Amplitude
   e. Wavelength

Q2. If a source emits sound of frequency f, an observer moving to the source will experience:
   a. Sound of a higher frequency than f
   b. Sound that is the same frequency as f
   c. Sound of a lower frequency than f
   d. Sound with two frequencies, one at f and one at twice f
   e. No sound

Q3. If a source emits sound of frequency f, an observer stationary with respect to the source will experience:
   a. Sound of a higher frequency than f
   b. Sound that is the same frequency as f
   c. Sound of a lower frequency than f
   d. Sound with two frequencies, one at f and one at twice f
   e. No sound

Q4. Ultrasound is generated at a frequency f. If blood moves away from the transducer the frequency of the received echoes will:
   a. Be lower than f
   b. Be higher than f
   c. Be the same as f

   d. Consist of two frequencies at f and $f/2$
   e. There will be no received echoes

Q5. In Doppler ultrasound the Doppler shift is:
   a. The difference in wavelength between the transmitted ultrasound and received ultrasound
   b. The difference in frequency between the transmitted ultrasound and received ultrasound
   c. The difference in amplitude between the transmitted ultrasound and received ultrasound
   d. The difference in speed of sound between the transmitted ultrasound and received ultrasound
   e. The difference in angle theta between the transmitted ultrasound and received ultrasound

Q6. In Doppler ultrasound the Doppler shift is:
   a. Proportional to the speed of sound
   b. Inversely proportional to the velocity of the target
   c. Proportional to the square of the velocity of the target
   d. Proportional to the velocity of the target
   e. Proportional to the cosine of the angle between beam and target

Q7. According to the Doppler equation the Doppler shift is zero when:
   a. The target velocity is zero
   b. The beam and direction of motion are aligned
   c. The angle between the beam and direction of motion is 45°
   d. The angle between the beam and direction of motion is 90°
   e. The target velocity is 1 metre per second

Q8. Typical Doppler signals from moving blood have frequencies:
   a. In the range 0–15 Hz
   b. In the range 0–15 kHz
   c. In the range 0–15 MHz
   d. In the range 0–15 GHz
   e. In the audible range

Q9. The spectral Doppler display is a display of the:
   a. Maximum Doppler frequency shift in a 2D region

b. Doppler intensity against time

c. Mean Doppler frequency shift within a 2D region

d. Doppler frequency shift against time

e. Doppler intensity against frequency

Q10. The colour Doppler display is a display of the:

a. Maximum Doppler frequency shift in a 2D region

b. Doppler intensity against time

c. Mean Doppler frequency shift within a 2D region

d. Doppler frequency shift against time

e. Doppler intensity against frequency

Q11. The clutter signal:

a. Is made from echoes from tissues which are stationary in the Doppler sample volume

b. Is made from echoes from blood

c. Provides useful information for Doppler ultrasound applications in blood flow

d. Can be removed by low-pass filtering

e. Can be removed by high-pass filtering

Q12. Spectrum analysis:

a. Is performed using the Fourier transform or some variant of this

b. Estimates only the mean Doppler frequency

c. Is performed using an autocorrelator or some variant of this

d. Estimates only the maximum Doppler frequency

e. Estimates the Doppler frequency components present within the Doppler signal

Q13. The cut-off filter:

a. Is designed to preserve the clutter signal

b. Is designed to remove high-amplitude signals from tissue

c. Removes Doppler signals from low blood velocity

d. Suppresses true blood flow if set too high

e. Maintains true blood flow if set too high

Q14. Pulse-wave Doppler:

a. Involves the transmission of ultrasound continuously

b. Allows the operator to control the depth in the tissue from which Doppler signals arise

c. Involves the transmission of ultrasound in pulses

d. Can only be done using single-element transducers

e. Can only be done using array transducers

Q15. Concerning aliasing:

a. There is an upper limit to the Doppler frequency shift which can be estimated

b. Occurs when there are more than two samples per wavelength of the Doppler signal

c. It is a feature of Doppler frequency shift estimation using continuous-wave Doppler

d. Occurs when there are less than two samples per wavelength of the Doppler signal

e. It is a feature of Doppler frequency shift estimation using pulsed-wave Doppler

Q16. Attempts to overcome aliasing include:

a. Using a higher transmit frequency

b. Using a lower transmit frequency

c. Turning up the velocity scale

d. Reducing the angle

e. Increasing the angle

Short-Answer Questions

Q1. What is the Doppler effect for ultrasound?

Q2. What are the two main types of display for Doppler ultrasound data?

Q3. In Doppler ultrasound estimation of velocity what does 'angle dependence' refer to?

Q4. In spectral Doppler what does the 'Doppler signal processor' do and what are its three component parts?

Q5. Describe two differences between CW and PW Doppler ultrasound.

Q6. What is demodulation in Doppler ultrasound?

Q7. What is the purpose of the wall-thump or clutter-filter in Doppler ultrasound?

Q8. In Doppler ultrasound what is aliasing?

REFERENCES

Evans DH, Jensen JA, Nielsen MB. 2011. Ultrasonic colour Doppler imaging. *Interface Focus*, 1, 490–502.

Evans DH, McDicken WN. 2000. *Doppler Ultrasound: Physics, Instrumentation, and Signal Processing*. Chichester: Wiley.

Jensen JA. 1996. *Estimation of Blood Velocities Using Ultrasound*. Cambridge: Cambridge University Press.

McDicken WN, Sutherland GR, Moran CM, Gordon LN. 1992. Colour Doppler velocity imaging of the myocardium. *Ultrasound in Medicine and Biology*, 18, 651–654.

Satomura S. 1957. Ultrasonic Doppler method for the inspection of cardiac function. *Journal of the Acoustics Society of America*, 29, 1181–1185.

8

Blood flow

ABIGAIL THRUSH

INTRODUCTION

The development of colour flow imaging has led to an increase in the investigation of blood flow to aid diagnosis of both vascular and non-vascular disorders. An understanding of the physical properties of blood flow is essential when interpreting colour flow images and Doppler spectra. For example, the presence of reverse flow within a vessel as seen on a colour image may be due to the presence of disease or may be normal flow within the vessel. Changes in the velocity of the blood or the shape of the Doppler spectrum can help in locating and quantifying disease. A poor understanding of how blood flows in normal and diseased vessels may lead to misdiagnosis or to loss of useful clinical information. Blood flow is a complex pulsatile flow of a non-homogeneous fluid in elastic tubes; however, some understanding can be obtained by considering the simple model of steady flow in a rigid tube (Caro et al. 1978; Nichols and O'Rourke 1999).

STRUCTURE OF THE VESSEL WALLS

The structure of the arterial wall can change due to disease and these changes may be observed using ultrasound. Artery walls consist of a three-layer structure. The inner layer, the intima, is a thin layer of endothelium overlying an elastic membrane. The middle layer, the media, consists of smooth muscle and elastic tissue. The outer layer, the adventitia, is predominantly composed of connective tissue with collagen and elastic tissue. The intima-media layer can be visualised with ultrasound in the carotid arteries and is normally of the order of 0.5–0.9 mm thick, when measured using ultrasound. Arterial disease will lead to changes in the vessel wall thickness that may eventually lead to a reduction of flow or act as a source of emboli. Vein walls have a similar structure to arteries but with a thinner media layer. Blood vessels not only act as a conduit to transport the blood around the body but are complex structures that respond to nervous and chemical stimulation to regulate the flow of blood.

LAMINAR, DISTURBED AND TURBULENT FLOW

The flow in normal arteries at rest is laminar. This means that the blood moves in layers, with one layer sliding over the other. These layers are able to move at different velocities, with the blood cells remaining within their layers. However, if there is a significant increase in velocity, such as in the presence of a stenosis, laminar flow may break down and turbulent flow occurs. In turbulent flow, the blood moves randomly in all directions at variable speeds, but with an overall forward flow velocity. In the presence of turbulence, more energy is needed, i.e. a greater pressure drop is required if the flow rate is to be maintained. Turbulent flow may be seen distal to severe stenosis, as shown in the spectral Doppler waveform in Figure 8.1. This demonstrates an increase in spectral broadening due to the range of velocities present within the turbulent flow, and this may be used to indicate the presence of disease. The transition from laminar flow to turbulent flow is displayed diagrammatically in Figure 8.2. The term 'disturbed flow' refers to regions of circulating flow (vortices or eddies) and flow reversal. Vortices are commonly seen in the distal part of an atherosclerotic plaque. The vortex may be static or it may be shed and travel downstream where it will dissipate after travelling a short distance. The appearance of disturbed flow does not by itself indicate the presence of disease and may be seen in a normal bifurcation. For example, flow reversal in the carotid bifurcation (as seen in Figures 8.3 and 8.8 later in this chapter) is considered to be a normal finding.

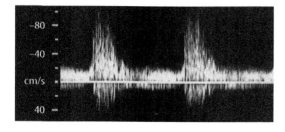

Figure 8.1 Spectral Doppler display demonstrating turbulent flow.

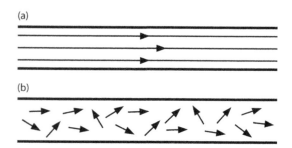

Figure 8.2 Schematic diagram showing **(a)** laminar flow and **(b)** turbulent flow.

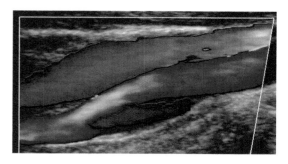

Figure 8.3 Colour image showing flow reversal (blue at bottom of image) in the carotid bulb of a normal internal carotid artery. The vein (blue at top of image) can be seen overlying the carotid artery.

VELOCITY PROFILES

The blood within a vessel may not all be moving with the same velocity at any particular point in time. In fact, there is usually a variation in the blood velocity across the vessel, typically with faster flow seen in the centre of the vessel and slowly moving blood near the vessel wall. This variation of blood velocity across the vessel is called the velocity profile. The shape of the velocity profile will affect the appearance of both the colour flow image and the Doppler spectrum.

Considering a simple system of steady (i.e. non-pulsatile) flow of a homogeneous fluid entering a long rigid tube from a reservoir, the flow will develop from a blunt flow profile, with all the fluid moving at the same velocity, to parabolic flow (Caro et al. 1978), as shown in Figure 8.4. This change of flow profile occurs due to the viscous

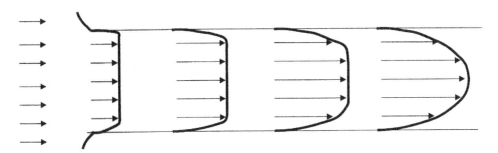

Figure 8.4 The change in velocity profile, with distance along a rigid tube, from a blunt to a parabolic profile. (Adapted from **Caro CG** et al., *The Mechanics of the Circulation*, 1978, Oxford: Oxford University Press, by permission of **Oxford University Press**.)

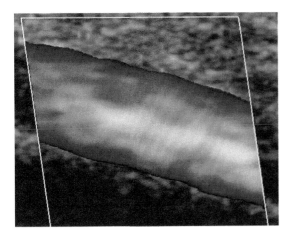

Figure 8.5 Colour flow image showing high velocities (yellow) in the centre of a normal superficial femoral artery with lower velocities (blue) near the vessel wall.

drag exerted by the walls, causing the fluid at the wall to remain stationary. This generates a velocity gradient across the diameter of the vessel. The distance over which the flow profile develops from blunt to parabolic flow depends on the diameter of the tube and velocity of the fluid but is usually several times the tube diameter. Similar differences in velocity profile can be seen in different vessels within the body. For example, the flow profile in the ascending aorta is typically blunt but the flow seen in the normal mid-superficial femoral artery tends towards being parabolic. However, the shape of the velocity profile is further complicated by the fact that blood flow is pulsatile. The flow profile across

a vessel may be visible on a colour flow image, as shown in Figure 8.5, with the slower moving blood seen near the vessel walls.

Velocity profiles in normal vessels

As the flow in arteries is in fact pulsatile, the velocity profile across an artery varies over time. The direction and velocity of the flow are dependent on the pressure drop along the length of the vessel. The pressure pulse, produced by the heart, travels down the arterial tree and is modified by the pressure wave that has been reflected back from the distal vessels. In the presence of high distal resistance to the flow, such as seen in the normal resting leg, this will lead to a reversal of flow during part of the diastolic phase of the cardiac cycle. This can be seen on the Doppler spectrum shown in Figure 8.6. This reversal of flow will affect the velocity profile seen in the vessel. Figure 8.7a and b shows the predicted velocity profile at different points in the cardiac cycle for a normal common femoral artery and a common carotid artery, respectively (Evans and McDicken 2000). This shows that reverse flow is not seen in a normal common carotid artery but flow reversal, during diastole, is demonstrated in the common femoral artery. Figure 8.7a also shows that it is possible for flow in the centre of the vessel to be travelling in a different direction from flow nearer to the vessel wall. These changes in flow direction during the cardiac cycle can be observed, using colour flow imaging, as a change from red to blue or blue to red depending on the direction of flow relative to the transducer.

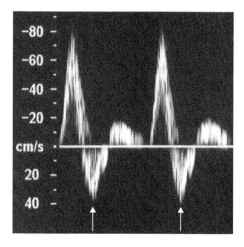

Figure 8.6 Velocity waveform from a normal superficial femoral artery (arrows indicate reverse flow).

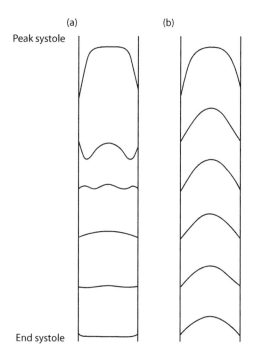

Figure 8.7 Velocity profiles from **(a)** a common femoral artery and **(b)** a common carotid artery, calculated from the mean velocity waveforms. (Adapted from Evans DH, McDicken WN. 2000. *Doppler Ultrasound: Physics, Instrumentation, and Signal Processing*. Chichester: Wiley, with permission of Wiley-Blackwell.)

The shape of the velocity profile will also affect the degree of spectral broadening seen on the Doppler spectrum. If a parabolic profile is visualised with a large sample volume, both the low velocities near the vessel wall and the higher velocities in the centre of the vessel will be detected. This leads to a higher degree of spectral broadening in the Doppler spectrum than would be seen for a small sample volume placed in the centre of the vessel (see Chapter 9).

Velocity profiles at branches and curves

The arterial tree branches many times along its length and this branching has an influence on the velocity profile. The velocity profiles found in the normal carotid bifurcation have been extensively studied (Reneman et al. 1985) and an example of a velocity profile in a carotid bifurcation is given in the schematic diagram in Figure 8.8. This shows an asymmetric flow profile in the proximal internal

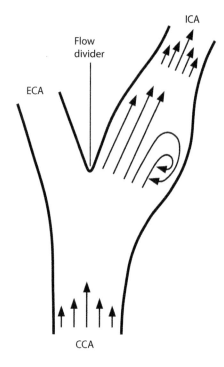

Figure 8.8 Diagram of an example of velocity patterns observed in the normal carotid bifurcation.

(a)

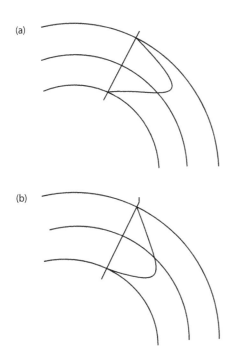

(b)

Figure 8.9 Distortion of **(a)** parabolic flow and **(b)** blunt flow due to curvature of the tube. (Adapted from Caro CG et al., *The Mechanics of the Circulation*, 1978, Oxford: Oxford University Press, by permission of Oxford University Press.)

carotid artery with the high-velocity flow occurring towards the flow divider and the reverse flow occurring near the wall away from the origin of the external carotid artery. This profile results from pulsatile flow through a vessel of varying dimensions and will depend on the geometry of the bifurcation. An area of flow reversal is commonly seen on colour flow images of the normal carotid bifurcation (Figure 8.3).

Curvature of a tube will also influence the shape of the velocity profile. The peak velocity within the tube will be skewed off-centre but whether this is to the inner or outer wall of the curve will depend on whether the underlying profile is blunt or parabolic. Figure 8.9 shows the effects of tube curvature on blunt and parabolic velocity profiles.

Velocity profiles at stenosis

The velocity profile in the vessel can also be altered by the presence of arterial disease. If a vessel becomes narrowed, the velocity of the blood

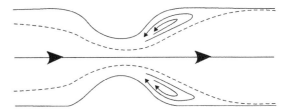

Figure 8.10 Flow through a constriction followed by a rapid expansion downstream showing the region of flow reversal. (From Caro CG et al., *The Mechanics of the Circulation*, 1978, Oxford: Oxford University Press, by permission of Oxford University Press.)

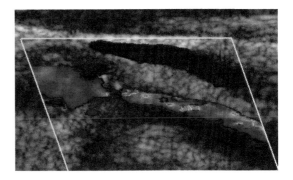

Figure 8.11 Colour flow image showing the velocity increase as the blood flows through a stenosis, from left to right, with an area of flow reversal (shown as blue) beyond the narrowing. The vein lying over the artery is also seen (blue).

will increase as the blood passes through the stenosed section of the vessel. Beyond the narrowing, the vessel lumen will expand again and this may lead to flow reversal (Caro et al. 1978) as shown in Figure 8.10. The combination of the velocity increase within the narrowing and an area of flow reversal beyond is often observed, with colour flow ultrasound, at the site of a significant stenosis (Figure 8.11). It is important to remember that the complex nature of both normal and abnormal blood flow is such that the flow is not necessarily parallel to the vessel walls.

VELOCITY CHANGES WITHIN STENOSIS

The changes in the velocity of the blood that occur across a narrowing are used in spectral Doppler

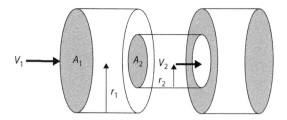

Figure 8.12 For constant flow, Q, through a tube, the velocity of the fluid increases from V_1 to V_2 as the cross-sectional area decreases from A_1 to A_2 (r is radius).

investigations both to identify the presence of a stenosis and to quantify the degree of narrowing. The relationship between the steady flow, Q, in a rigid tube of cross-sectional area A and the velocity of the fluid, V, is described by:

$$Q = V \times A$$

If the tube has no outlets or branches through which fluid can be lost, then the flow along the tube will remain constant. Therefore, the mean velocity at any point along the tube depends on the cross-sectional area of the tube. Figure 8.12 shows a tube of changing cross-sectional area (A_1, A_2) and, as the flow (Q) is constant, then:

$$Q = V_1 \times A_1 = V_2 \times A_2$$

This can be rearranged to show that the change in the velocities is related to the change in the cross-sectional area as follows:

$$\frac{V_2}{V_1} = \frac{A_1}{A_2}$$

This relationship actually describes steady flow in a rigid tube and so cannot be directly applied to pulsatile blood flow in elastic arteries. However, it does give an indication as to how the velocity may change across a stenosis.

The graph in Figure 8.13 shows how the flow and velocity within an idealised stenosis vary with the degree of diameter reduction caused by the stenosis, based on the predictions from a simplified theoretical model (Spencer and Reid 1979).

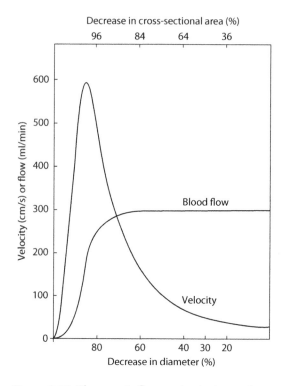

Figure 8.13 Changes in flow and velocity as the degree of stenosis alters, predicted by a simple theoretical model of a smooth, symmetrical stenosis. (Adapted from Spencer MP, Reid JM. 1979. *Stroke*, 10, 326–330, with permission of LWW Publications.)

This suggests that where the diameter reduction is less than 70%–80%, the flow remains relatively unchanged; however, as the diameter reduces further, the stenosis begins to limit the flow (known as a haemodynamically significant stenosis). The graph also shows that the velocity within the stenosis increases with diameter reduction and that these changes in velocity occur at much smaller diameter reductions than required to produce a flow reduction. For this reason, velocity changes are considered to be a more sensitive method of detecting vessel lumen reductions than measurements of flow. Eventually, there comes a point at which the flow drops to such an extent that the velocity begins to decrease and 'trickle flow' is seen within the vessel. The relationships between vessel diameter reductions and velocity described relate to simple steady-flow models, and the velocity criteria used clinically to quantify the degree

(a)

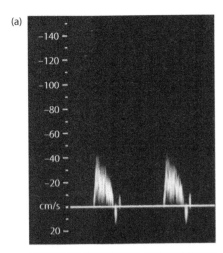

(b)

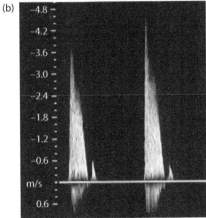

Figure 8.14 Doppler signal obtained **(a)** proximal to and **(b)** within a significant stenosis demonstrating a velocity increase from 40 cm s^{-1} to 2.2 m s^{-1}.

of narrowing are produced by comparing Doppler velocity measurements with arteriogram results. Figure 8.14 shows the velocity increase detected at the site of a superficial femoral artery stenosis.

RESISTANCE TO FLOW

The concept of resistance is used to describe how much force is needed to drive the blood through a particular vascular bed. Blood flow is due to the pressure drop between different points along the vessel and also depends on the resistance to the flow. This relationship was described by Poiseuille as:

$$\text{Pressure drop} = \text{Flow} \times \text{Resistance}$$

where the resistance to flow is given by:

$$\text{Resistance} = \frac{\text{Velocity} \times \text{Length} \times 8}{\pi \times \text{Radius}^4}$$

Therefore, in the presence of a given pressure drop, increased resistance would result in a reduction in flow. Again this equation describes non-pulsatile flow in a rigid tube and so cannot be directly related to arterial blood flow but indicates that flow not only depends on the pressure drop but is affected by the resistance to flow. It also shows how the resistance is highly dependent on the vessel diameter (r^4 term), with the resistance rapidly increasing as the diameter reduces. In the normal circulation, the greatest proportion of the resistance to flow is thought to occur at the level of the arterioles. The resistance to flow varies from one organ to another. For example, the normal brain, kidney and placenta are low-resistance structures, compared to muscles at rest. Changes in the resistance to flow may occur during disease, either of the arteries or of the arterioles. For example, abnormal development of the placenta may lead to the placenta being a high-resistance structure, making it difficult for blood to pass through the placenta, with the consequence that the fetus may obtain insufficient nutrition and oxygen and be small for its gestational age. Disease causing narrowing of the arteries, such as the superficial femoral artery, may lead to an increase in resistance to flow through the artery resulting in a reduction in blood flow, causing the patient to experience pain either when walking or at rest.

The waveform shape observed at different points in the body is dependent on the outflow resistance of the vessel, i.e. the vascular bed being supplied by the observed vessel. For example, the internal carotid artery supplies the brain, which offers little resistance to flow, and the Doppler waveform has continuous forward flow during diastole, as shown in Figure 8.15. The renal arteries also supply a low-resistance vascular bed and, therefore, a similar waveform shape is observed. However, the peripheral circulation in the normal resting leg offers a higher resistance to flow and the typical waveform seen in the superficial femoral artery, supplying

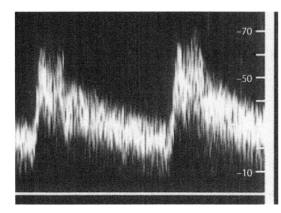

Figure 8.15 Spectral Doppler obtained from a normal internal carotid artery.

the leg, demonstrates a high-resistance waveform, with reverse flow seen during diastole (Figure 8.6).

PHYSIOLOGICAL AND PATHOLOGICAL CHANGES THAT AFFECT THE ARTERIAL FLOW

Tissue perfusion is regulated by changes in the diameter of arterioles, thus altering peripheral resistance. For example, the increased demand by the leg muscles during exercise will produce a reduction in the peripheral resistance, by dilatation of the arterioles, leading to an increase in blood flow. This also leads to a change in the shape of the waveform, as seen in Figure 8.16a, whereby the flow in diastole is now entirely in the forward direction (compared to Figure 8.6).

The presence of arterial disease can significantly alter the resistance to flow, with the reduction in vessel diameter having a major effect on the change in resistance. The Doppler waveform shape will be affected by an increase in the resistance distal to the site from which the waveform was obtained. This observed change in waveform shape may help indicate the presence of an occlusion distal to the vessel being observed. For example, the waveform in Figure 8.17 shows a Doppler spectrum obtained from a common carotid artery proximal to an internal carotid artery occlusion. This waveform demonstrates a short acceleration time (from the beginning of systole to peak systole) but with an absence of any diastolic flow

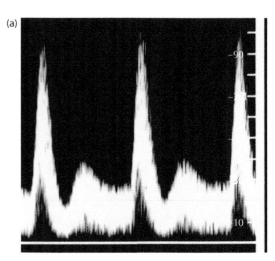

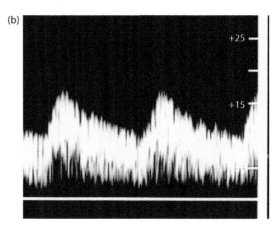

Figure 8.16 **(a)** Doppler waveform obtained from an artery in the foot, in a normal leg following exercise, demonstrating multi-phasic hyperaemic flow. **(b)** Waveform demonstrating low-volume monophasic flow seen in the foot distal to an occlusion of the artery. Higher-velocity flow is seen in the hyperaemic flow in **(a)** with a peak systolic velocity of 95 cm s^{-1} compared to the low peak systolic velocity of 15 cm s^{-1} seen distal to disease in **(b)**.

which is normally seen in the cerebral circulation (Figure 8.15).

Beyond an occlusion the spectrum may also appear abnormal. In the presence of severe disease in the superficial femoral artery, the arterioles may become maximally dilated to reduce the peripheral resistance to maximise the limited blood flow in an attempt to maintain tissue perfusion. A typical

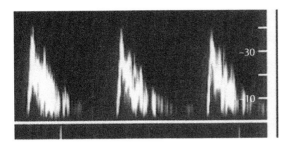

Figure 8.17 Doppler waveform obtained from a common carotid artery proximal to an occluded internal carotid artery demonstrating a high-resistance waveform.

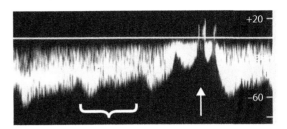

Figure 8.18 Doppler waveform demonstrating the effect of respiration (indicated by the arrow) and the cardiac cycle (indicated by the bracket) on the blood flow in a subclavian vein.

spectrum obtained distal to a superficial femoral artery occlusion is shown in Figure 8.16b with the characteristic damped waveform shape with longer systolic acceleration time and increased diastolic flow. The velocity of the flow seen in Figure 8.16b, obtained from a vessel distal to an occlusion, is lower than that seen in Figure 8.16a, which was obtained from a normal vessel with hyperaemic flow due to exercise. The presence of collateral flow, i.e. the blood finding an alternative route to bypass a stenosis or occlusion, will also influence the waveform shape both proximal and distal to the disease. Good collateral flow may alter the effect on the waveform shape expected for a given severity of disease, as the collateral pathway may affect the resistance to flow.

VENOUS FLOW

Veins transport blood back to the heart. To enable them to perform this function they have thin but strong bicuspid valves to prevent retrograde flow. There are typically a larger number of valves in the more distal veins. Venous flow back to the heart is enhanced by the influence of pressure changes generated by the cardiac cycle, respiration (Figure 8.18) and changes in posture as well as the action of the calf muscle pump. The flow and pressure in the central venous system are affected by changes in the volume of the right atrium which occur during the cardiac cycle. This pulsatile effect can be seen on Doppler spectra obtained from the proximal veins of the arm and neck due to their proximity to the chest. However, flow patterns in the lower limb veins and peripheral arm veins are not

significantly affected by the cardiac cycle due to the compliance of the veins, leading to a damping of the pressure changes. The presence of valves and changes in intra-abdominal pressure during respiration also mask the effect of the cardiac cycle on venous flow in the distal veins.

Changes in the volume of the thorax, due to movement of the diaphragm and ribs, also assist venous return. During inspiration, the thorax expands, leading to an increase in the volume of the veins within the chest, resulting in an increase in flow into the chest. During expiration, a decrease in flow is seen. The reverse situation is seen in the abdomen because the diaphragm descends during inspiration, causing the abdominal pressure to increase, encouraging flow into the thorax. When the diaphragm rises, the pressure drops, encouraging flow from the upper leg veins into the abdomen. Respiration effects can be observed on the spectral Doppler display as phasic changes in flow in proximal deep peripheral veins, such as the common femoral vein. Augmentation of flow using breathing manoeuvres is often used in the investigation of venous disorders.

Changes in posture can lead to large changes in hydrostatic pressure in the venous system. When a person is lying supine, there is a relatively small pressure difference between the venous pressure at the ankle and at the right atrium. However, when standing there is a column of blood between the right atrium and veins at the ankle and this produces a significant pressure gradient which has to be overcome in order for blood to be returned to the heart. This can be achieved by the calf muscle pump mechanism assisted by

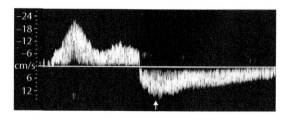

Figure 8.19 Sonogram demonstrating reversal of flow in the vein, reflux (shown by arrow), due to incompetent valves in the vein.

the presence of the venous valves. The muscle compartments in the calf contain the deep veins and venous sinuses that act as blood reservoirs. When the deep muscles of the calf contract, thus causing compression of the veins, blood flow is forced out of the leg and prevented from returning by the venous valves. This also creates a pressure gradient between the superficial and deep veins in the calf causing blood to drain from the superficial to the deep venous system. If there is significant failure of the venous valves in either the superficial or deep venous system, reflux will occur, leading to a less-effective muscle pump and a higher pressure than normal in the veins following calf muscle contraction and relaxation. This may eventually lead to the development of venous ulcers.

Colour flow imaging and spectral Doppler can be used to identify venous incompetence seen as periods of retrograde flow, away from the heart, following compression of the calf (Figure 8.19). Venous outflow obstruction leads to a loss of the normal spontaneous phasic flow generated by respiration detected by Doppler ultrasound.

QUESTIONS

Multiple Choice Questions

Q1. Concerning different flow states in blood flow, which of the following is true?
  a. Flow in normal arteries is mostly turbulent
  b. Turbulent flow is associated with movement of blood in layers which slide over each other
  c. In turbulent flow a greater pressure drop is required (compared to laminar flow) to maintain flow
  d. Flow reversal in the carotid bifurcation indicates disease
  e. Flow in normal arteries is mostly laminar

Q2. The velocity profile in a vessel is the variation of:
  a. Flow rate with position across the vessel
  b. Velocity with position across the vessel
  c. Viscosity with position across the vessel
  d. Position across the vessel with velocity
  e. Velocity with position along the vessel

Q3. Which of the following is true regarding velocity profiles?
  a. In parabolic flow the fastest velocity is seen in the centre of the tube
  b. A parabolic velocity profile will be skewed off centre when travelling around the bend in a tube
  c. In blunt flow, all the blood is travelling with a similar velocity
  d. During steady flow the velocity profile in a long straight vessel is parabolic
  e. During pulsatile flow the velocity profile in a straight vessel is parabolic

Q4. Concerning stenosis:
  a. Minor disease (<50% stenosis by diameter) is associated with an increase in flow rate
  b. Minor disease (<50% stenosis by diameter) is associated with a decrease in flow rate
  c. Major disease (>75% stenosis by diameter) is associated with an increase in flow rate
  d. Major disease (>75% stenosis by diameter) is associated with a decrease in flow rate
  e. Total occlusion (100% stenosis) is associated with an increase in flow rate

Q5. Poiseuille's equation shows resistance to flow is inversely proportional to:
  a. Length
  b. Radius
  c. Radius squared
  d. Radius to the power of 4
  e. Velocity

Short-Answer Questions

Q1. Briefly describe the characteristics of laminar flow, turbulent flow and disturbed flow.

Q2. Describe three situations at which flow reversal may occur.

Q3. Briefly describe the change in flow rate and maximum velocity with increasing degree of stenosis.

Q4. What factors affect the return of venous flow to the heart?

FURTHER READING

Oates C. 2001. *Cardiovascular Haemodynamics and Doppler Waveforms Explained.* Cambridge: Greenwich Medical Media.

Thrush A, Hartshorne T. 2010. *Vascular Ultrasound, How Why and When* (3rd edn). Edinburgh: Churchill Livingstone Elsevier.

REFERENCES

Caro CG, Pedley TJ, Schroter RC, Seed WA. 1978. *The Mechanics of the Circulation.* Oxford: Oxford University Press.

Evans DH, McDicken WN. 2000. *Doppler Ultrasound: Physics, Instrumentation, and Signal Processing.* Chichester: Wiley.

Nichols WN, O'Rourke MF. 1999. *McDonald's Blood Flow in Arteries* (2nd edn). London: Edward Arnold.

Reneman RS, van Merode T, Hick P, Hoeks APG. 1985. Flow velocity patterns in and distensibility of the carotid artery bulb in subjects of various ages. *Circulation,* 71, 500–509.

Spencer MP, Reid JM. 1979. Quantification of carotid stenosis with continuous-wave (C-W) Doppler ultrasound. *Stroke,* 10, 326–330.

Spectral Doppler ultrasound

ABIGAIL THRUSH

SPECTRAL DISPLAY

A real-time spectral Doppler display is shown in Figure 9.1 displaying both arterial and venous flow. This displays time along the horizontal axis and the Doppler frequency shift or calculated velocity along the vertical axis. The brightness (or colour) of the display relates to the amplitude of each of the Doppler frequency components present, i.e. the relative proportion of the blood travelling with a particular velocity. The baseline indicated in the centre of the spectral display in Figure 9.1 corresponds to zero Doppler shift or zero velocity. The spectrum contains information about the speed and direction of the blood flow as well as the degree of pulsatility of the flow. Conventionally, it is arranged so that positive Doppler frequency shifts (blood flowing towards the transducer) are plotted above the baseline and negative Doppler shifts (blood flowing away from the probe) are plotted below the baseline, but the operator can invert this display as required. Both arterial and venous flow may be present within the path of the beam and when, as is usually the case, the flow in these vessels is in opposite directions, the Doppler waveforms from the arterial and venous flow appear on opposite sides of the baseline as seen in Figure 9.1. The Doppler ultrasound signal from a blood vessel will contain a range of Doppler frequency shifts. These arise in part from the range of blood velocities present within the region of the vessel being investigated but also arise from a phenomenon called 'intrinsic spectral broadening', which is described later in this chapter.

DOPPLER ULTRASOUND SYSTEMS

A spectral Doppler display is produced by performing spectral analysis of the Doppler signal. The Doppler signal can be obtained either from a stand-alone continuous-wave (CW) or pulsed-wave (PW) system or from systems combining Doppler ultrasound with imaging, known as duplex systems. Some of the advantages and disadvantages of these systems are discussed later.

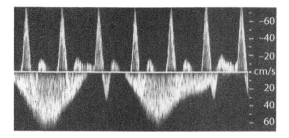

Figure 9.1 Spectral Doppler display showing arterial flow displayed above the baseline (negative value so flow is away from the transducer) and venous flow displayed below the baseline (positive value so flow is towards the transducer).

Continuous-wave Doppler

The CW transducer consists of at least two elements: one to continually transmit and the other to continually receive ultrasound. The frequencies of Doppler signals generated by moving blood are conveniently in the audible range and so can be output to a loudspeaker after amplification and filtering to remove the signal generated by the slow-moving vessel wall. Early Doppler systems had only audio outputs, and this approach is still taken in the simple pocket Doppler devices used in peripheral vascular applications. However, useful information can be provided if spectral analysis of the Doppler signal is performed. The fact that the CW system is continually detecting the blood flow means that CW systems do not suffer from the artefact of aliasing (Figures 7.14 and 9.2). However, the large sensitive region (Figure 7.7) means it is difficult to identify the source of the Doppler signal.

Pulsed-wave Doppler

Pulsed Doppler instrumentation has similarities to CW instrumentation. However, the transducer is excited with regular short pulses rather than continuously. In order to detect the signal from a specific depth in the tissue, a 'range gate' is used. This enables the system to only receive the returning signal at a given time after the pulse has been transmitted, and then for a limited time. The Doppler signal is, therefore, detected from a specific volume

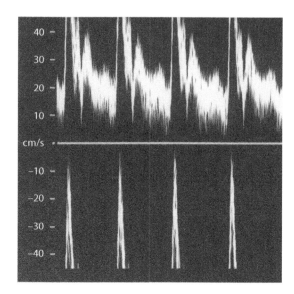

Figure 9.2 Spectral Doppler display showing aliasing of the signal (seen as wrapping around of the high velocities resulting in them being incorrectly displayed) due to undersampling of the Doppler signal.

within the body, known as the sample volume, at an identified range (Figure 9.3). The length of time over which the range gate is open is known as the gate length or sample volume length. The operator can control the depth and length of the sample volume, by varying the gate range and length. Thus, PW Doppler has the enormous advantage of allowing the operator to select the origin of the Doppler signal. However, PW Doppler does, as a consequence, suffer from the artefact of aliasing if the blood flow is undersampled. This can be seen in Figure 9.2 as the high velocities wrapping around the top of the display and shown as flow in the opposite direction (see Chapter 7).

Duplex systems

Initially, both the CW and PW Doppler were used without the aid of imaging and the vessels were located blindly, using only the detected Doppler signal to indicate the presence of the vessel. As imaging ultrasound developed, it become apparent that using Doppler ultrasound in conjunction with imaging, a combination known as duplex

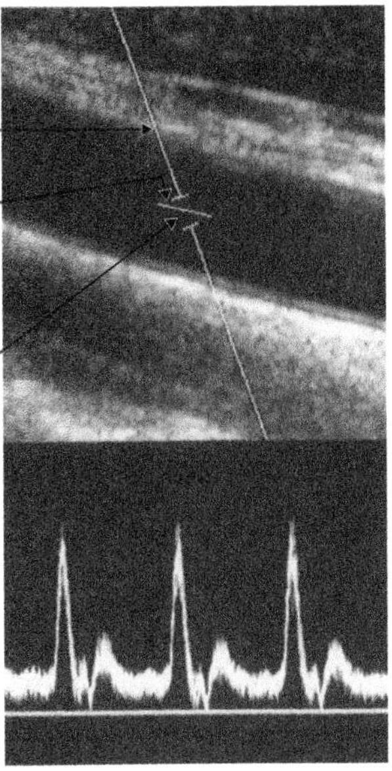

Path of Doppler ultrasound beam

Angle correction cursor

Sample volume

Figure 9.3 Ultrasound image of a vessel with the path of the spectral Doppler ultrasound beam, sample volume size and angle correction cursor displayed.

ultrasound, would enable better location of the relevant vessel. Duplex Doppler ultrasound enables precise location of the Doppler sample volume, e.g. at the centre of a vessel, as shown in Figure 9.3. It is also possible to estimate the angle of insonation, that is the angle between the path of the beam and the path of vessel, using the angle correction cursor, enabling the velocity of the blood to be calculated, using the Doppler equation.

Duplex scanners use arrays of piezoelectric elements to produce, independently, the B-mode imaging, colour flow and spectral Doppler beam, as described in Chapter 3. These multi-element transducers, capable of colour flow imaging and spectral Doppler measurements, include curvilinear-, linear- and phased-array transducers. When combining B-mode imaging with Doppler ultrasound, there is a conflict in the optimal angle of insonation. Ideally, the vessel walls should be imaged with a beam that is at right angles to the vessel whereas the optimal Doppler measurements are made when the beam is parallel to the vessel. Therefore, a compromise has to be reached.

Linear-array transducers are able to steer the beam, using beam-forming techniques (described later), by approximately 20°–30° to the left or right of the path perpendicular to the transducer face. This allows the Doppler beam to be set at an angle to the imaging beam enabling both the imaging and the Doppler recordings to be made in a vessel that runs parallel to the skin, such as vessels in the limbs and neck. Curvilinear- and phased-array transducers do not usually have the facility to steer the Doppler beam relative to the imaging beam so a suitable compromise for B-mode imaging and Doppler recordings has to be obtained by angling the transducer. Tilting the transducer enables a suitable angle of insonation to be obtained. In cardiac ultrasound, there will be imaging planes where the angle of insonation between the blood flow and the path of the beam is near zero degrees, i.e. they are parallel or almost parallel, leading to large Doppler shift frequencies being detected.

SPECTRAL ANALYSIS

The Doppler signal obtained from the blood flow can be analysed to find the frequency components in the Doppler signal using a mathematical process known as the Fourier transform. The range of frequency components relates to the range of velocities present within the blood flow. Figure 9.4 shows how the sonogram is produced by displaying consecutive spectra, with frequency displayed along the vertical axis and time along the horizontal axis. A complete spectrum is produced every 5–40 ms and each of these spectra is used to produce the next line in the sonogram. The brightness of the display at each point relates to the relative amplitude of each of the component frequencies, which in turn is an indication of the backscattered power for each value of Doppler frequency shift. If the angle of insonation is measured, using the angle-correction cursor, the vertical frequency scale can be converted to a velocity scale. In situations where the beam is parallel or almost parallel to the flow, such as cardiac ultrasound and transcranial Doppler measurements, ultrasound systems may assume the angle of insonation is zero rather than using the angle correction cursor to measure the angle.

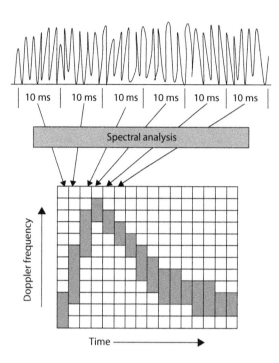

Figure 9.4 Schematic diagram showing how consecutive spectra produced by the spectrum analysis of the Doppler signal are used to construct the spectral Doppler display.

SPECTRAL DOPPLER CONTROLS AND HOW THEY SHOULD BE OPTIMIZED

Gain

The ultrasound backscattered from blood is of much lower amplitude than the signals returning from the surrounding tissue. As the received back-scattered signal from blood is small it will need to be amplified before it can be analysed. Increasing the spectral Doppler gain increases the brightness of the spectrum on the screen. However, the gain control increases the amplitude of not only the Doppler signal but also the background electronic noise. If the Doppler signal is of similar amplitude to the background noise, then no matter how much the gain is increased, only a poor Doppler sonogram will be obtained, as shown in Figure 9.5a. In these situations, it may be necessary to use a lower transmit frequency (resulting in less attenuation of the ultrasound pulse) or a higher transmit

power. If the gain is set too high, the Doppler system may become overloaded and may no longer be able to separate the forward and reverse flows accurately. This gives rise to an artefact whereby a mirror image of the true Doppler spectra appears in the reverse channel of the sonogram, as shown in Figure 9.5b. This is not true reverse flow but an artefact that can be eliminated by turning down the Doppler gain. Altering the Doppler gain may also affect the peak velocity detected. Increasing the gain may lead to a wider effective Doppler beam, resulting in the presence of a smaller angle of insonation at the edge of the beam. This will lead to higher detected frequencies.

Figure 9.5c shows the Doppler sonogram obtained from a flow rig where the gain has been increased over the time of the recording. On the left side where low gain has been used, the measured peak velocity is 128 cm s^{-1}, in the centre it measures 140 cm s^{-1} and on the right, where the gain has been set too high, the peak velocity measures 147 cm s^{-1}. On the far right of the image, where the gain is set too high the Doppler signal appears as saturated white and noise can be seen. Figure 9.5d shows the same effect on a Doppler recoding from a carotid artery with the gain increasing from left to right and the subsequent increase in measured velocity can be seen. The appropriate gain setting has been achieved in the centre of the displayed signal in Figure 9.5d.

Transmit power

Increasing the output power will increase the amplitude of the transmitted ultrasound, resulting in a higher-amplitude signal returning to the transducer. The output power of the transducer is altered by the scanner's transmit power control. On some ultrasound systems, the same control may be used to change the output power of all the three modalities: B-mode imaging, spectral Doppler and colour flow imaging. However, increasing the output power will also increase the patient's exposure to ultrasound. Therefore, the transmit power should only be increased after other controls, such as the gain, have been optimized to obtain the best Doppler spectrum. Changes in the output power should be reflected by changes in the displayed mechanical index (MI) and thermal index (TI) (see Chapter 16) displayed on the ultrasound image.

(a)

(b)

(c)

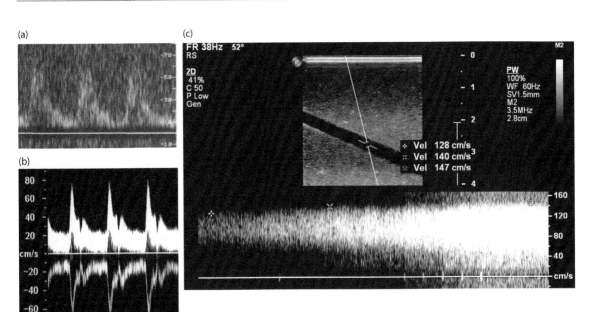

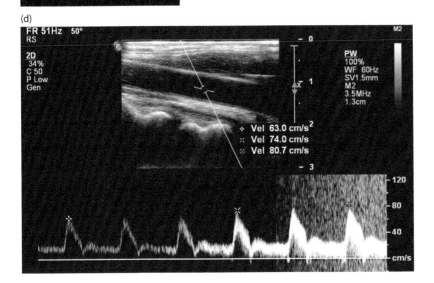

(d)

Figure 9.5 **(a)** Doppler spectral display showing that when the amplitude of the Doppler signal and the background noise are similar, increasing the gain does not help to improve the display. **(b)** Spectral Doppler display where the gain has been set too high, causing signal to break through on the reverse channel. **(c)** Spectral Doppler display obtained from a flow rig where the gain has been increased over the time of the recording and the effect this has on the measured peak velocity; **(d)** shows the same effect on a Doppler recording from a carotid artery with the gain increasing from left to right and the subsequent increase in measured velocity can be seen. The correct gain setting has been achieved in the centre of the displayed signal.

Transmit frequency

As high-frequency ultrasound is attenuated more than low-frequency ultrasound (see Chapter 2), the appropriate Doppler transmit frequency needs to be selected to ensure adequate penetration of the ultrasound. The frequency required will, therefore, depend on the depth of the vessel to be investigated. The Doppler transmit frequency is governed by the transducer that has been selected. Typically, ultrasound systems use broadband transducer technology, which means it is possible to operate the transducer over a range of different frequencies without too much loss in efficiency. Therefore, for example, a transducer may use a centre frequency of 5 MHz for B-mode imaging but use a lower frequency of 4 MHz for the Doppler measurements to ensure adequate returning backscattered signal from the blood. If the backscattered ultrasound signal from the blood is weak, it may be possible to improve the penetration of the beam by lowering the transmit frequency by 1 or 2 MHz. If this still does not provide sufficient penetration, it may be necessary to select a transducer that operates at a lower frequency range to enable adequate penetration.

Pulse repetition frequency (scale)

With pulsed Doppler ultrasound, the rate at which the pulses of ultrasound are transmitted is known as the pulse repetition frequency (PRF) and this can be controlled by the operator. On some scanners the PRF control is called the scale. Some, but not all, scanners display the actual values of the PRF used to produce the Doppler spectrum at the side of the display. The PRF can typically be set between 1.5 and 18 kHz, depending on the velocity of the blood that is to be detected. Most ultrasound scanners have application-specific pre-sets that, if selected, will set the controls, such as the PRF, to a suitable starting value. However, these values may need to be altered during the scan to obtain an optimal spectral display. If the PRF is set too low, aliasing will occur. This gives the appearance shown in Figure 9.2, where the high frequencies present are incorrectly displayed in the reverse channel. The PRF needs to be at least twice the maximum Doppler frequency required to be detected to prevent aliasing, i.e. $PRF = 2f_{d(max)}$ (see Chapter 7). This can be overcome by increasing the PRF. However, there will be an upper limit to the PRF that can be used, as the scanner will usually only allow one pulse to be 'in-flight' at a time in order to prevent confusion as to where a returning signal has originated from. If PRF_{max} is the upper limit of the PRF, the maximum velocity, v_{max}, which can be measured will be given by the following equation (also see Doppler Equation 7.1):

$$\frac{PRF_{max}}{2} = \frac{2v_{max}f_t \cos\theta}{c}$$

This can be re-written as:

$$v_{max} = \frac{PRF_{max}c}{4f_t \cos\theta}$$

For a depth of interest, d, and speed of sound, c:

$$PRF_{max} = \frac{c}{2d}$$

The 2, in the previous equation, arises from the fact that the pulse has to go to and return from the target. This gives the maximum detectable velocity, v_{max}:

$$v_{max} = \frac{c^2}{8df_t \cos\theta}$$

This shows that the maximum detectable velocity, without aliasing occurring, will be lower when making measurements at depth, or when the angle of insonation is small, giving a cos θ term approaching 1. It may be possible to overcome aliasing, once this maximum PRF is reached, by lowering the transmit frequency used or using a larger angle of insonation, giving a smaller cos θ term.

When measuring very high blood flow velocities, especially at depth, e.g. in an iliac stenosis in a large abdomen, some scanners will allow a 'high PRF' mode to be selected. This allows more than one pulse to be in-flight at a given time. The higher PRF allows higher velocities to be measured, but also introduces range ambiguity, i.e. a loss of certainty as to the origin of the Doppler signal.

If too high a PRF is used to detect slower-moving blood, the scale of the spectral display will not be fully utilized and the ability to identify changes in the Doppler frequency will be reduced. It is, therefore, important to set the PRF such that the Doppler waveform almost fills the display, without any wrap-around due to aliasing occurring.

Baseline

The baseline represents zero Doppler shift, i.e. zero velocity, and, therefore, demarcates the part of the display used for displaying forward flow (towards the transducer) from that used for reverse flow (away from the transducer). The position of the baseline can be changed by the operator to allow optimum use of all the spectral display, depending on the relative size of the forward and reverse flow velocities present. For example, the position of the baseline may be lowered to prevent aliasing of a positive Doppler shift signal, in the absence of flow in the opposite direction.

Invert

The invert control enables the operator to turn the Doppler display upside down, so that flow away from the transducer is displayed above the baseline and flow towards the transducer is displayed below. This can be indicated by showing negative velocity values on the vertical scale of the sonogram above the baseline. Typically, operators prefer to display arterial flow above the baseline.

Filter

The Doppler signal will contain not only the low-amplitude higher Doppler frequencies backscattered from the blood but also high-amplitude lower Doppler frequencies from the slow-moving tissue such as the vessel walls. These unwanted frequencies can be removed by a high-pass filter. As the name suggests, a high-pass filter removes the low-frequency signals while maintaining the high frequencies. There is, however, a compromise in the selection of the cut-off frequency to be used, as it is important not to remove the frequencies detected from the lower-velocity arterial or venous blood flow. Figure 9.6 shows the high-pass filter

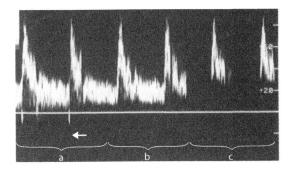

Figure 9.6 Doppler spectral display showing how the appearance changes with different high-pass filters. (a) The filter is set too low, allowing wall thump (shown by arrow) to be displayed. (b) The filter is set correctly and the wall thump has been removed. (c) The filter is set too high, removing the low velocities including diastolic flow component.

set at three different levels. The first (a) shows that the filter is set too low and the wall thump, generated by the slow-moving vessel wall, has not been removed. In the second situation (b), the filter has been correctly set to remove the wall thump and in the third situation (c) the filter has been set too high and the diastolic flow has been removed, giving a false impression of the waveform shape, which could lead to a misdiagnosis. The PRF and wall filter are often linked, so that increasing the PRF may automatically increase the wall filter.

Gate size and position

The size and position of the 'range gate' or 'sample volume' are selected by the operator and can typically be varied between 0.5 and 20 mm in length. The 'gate size' or 'sample volume length' may affect the appearance of the Doppler spectrum, as described later. It is important that care is taken to select the appropriate sample volume length, depending on whether the operator wishes to detect the velocities in the centre of the vessel only or velocities present across the whole vessel.

Beam-steering angle

The angle of insonation can be altered by changing the orientation of the transducer in relation to the vessel, e.g. by tilting the transducer. Linear-array transducers also have the facility to steer the Doppler beam 20° to the left or right of the centre or a range of angles between. The ultrasound beam is electronically steered by introducing delays between the pulses used to excite consecutive active elements. This is similar to the method used to focus the beam, described in Chapter 3, but uses a different sequence of delays. Figure 9.7 shows how the delay between excitation pulses results in the wavelets produced by each element interfering in such a way that a wave front is no longer parallel to the front of the transducer. The path of the beam can be steered left or right of the centre depending on the delays introduced. Steering the beam in this way is necessary when using a linear-array transducer to detect flow in a vessel that is parallel to the front face of the transducer. This enables a Doppler angle of insonation of 60° or less to be used while the imaging beam remains perpendicular to the vessel wall, optimal for B-mode imaging. However, a steered Doppler beam has a lower sensitivity than a beam that is perpendicular to the transducer face, which may result in a noisy or lower-amplitude Doppler signal and therefore require increased gain. The angle of insonation of the Doppler beam should be optimized to 60° or less to enable a good Doppler signal to be acquired and to minimise errors in velocity measurements (discussed later in the chapter).

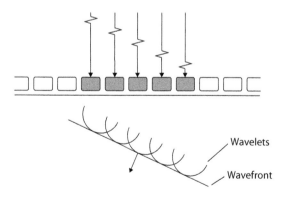

Figure 9.7 The Doppler beam produced by linear array transducers may be steered by introducing delays between the pulses used to excite consecutive active elements in the array.

Doppler angle cursor

In order to calculate the velocity of blood from the measured Doppler shift frequencies, the angle of insonation between the Doppler beam and the direction of flow must be known. Duplex scanners have the facility to allow the operator to line up an angle cursor with either the vessel wall, as seen on the B-mode image, or the direction of flow, as seen on the colour flow image. Knowing the angle of insonation, the scanner can then convert the Doppler shift frequencies to velocities and a velocity scale will be seen alongside the spectral display. Errors in the alignment of the angle-correction cursor can lead to significant errors in velocity measurement (discussed later in the chapter).

Focal depth

In many systems, the Doppler beam is focused and, depending on the ultrasound system, this focus may be fixed or adjustable. In some systems the Doppler beam focal depth automatically follows the sample volume, when the operator moves it.

Greyscale curve

The greyscale curve governs the relative display brightness assigned to the different amplitude signals detected. The relative amplitude signal detected at each frequency relates to the relative proportion of red blood cells moving at each given velocity. The Doppler spectra may be displayed using different greyscale curves or even colour scales.

FACTORS THAT AFFECT THE SPECTRAL DOPPLER DISPLAY

Both the shape of the Doppler waveform and the velocity measurements made from it are used to diagnose disease. However, the spectra may also be affected by factors other than disease (Thrush and Hartshorne 2010). It is important for the operator to understand these effects in order to be able to correctly interpret the spectral Doppler display, and these factors are discussed in the following sections.

Blood flow profile

The velocity profile across a vessel, at a given point of time, may be blunt, parabolic or partway in between (see Chapter 8). If the width of the Doppler beam and sample volume length are such that they cover the entire vessel, i.e. completely insonate the vessel cross section, then signals from flow across the full width of the vessel will be detected. With a blunt flow profile, all the blood is travelling at a similar velocity, therefore the velocity spectrum would display a narrow spread of velocities as shown in Figure 9.8a. However, the spectral Doppler display for complete insonation of parabolic flow would demonstrate the wider range of velocities present in the vessel as seen in Figure 9.8b.

Non-uniform insonation

Multi-element array transducers typically produce very narrow beams. When an array probe is aligned along the length of a section of vessel, it is the beam width in the elevation plane (perpendicular to the scan plane) that is important. If a beam is narrow in this dimension, then the beam will not insonate the entire vessel and so the slower-moving blood on the lateral walls of the vessel will not be detected (as shown in Figure 9.9a). Therefore, the spectrum will no longer represent the true relative proportions of blood moving at the slower velocity in the presence of parabolic or near-parabolic flow.

Sample volume size

The sample volume size and position will also affect the proportion of the blood velocities within the vessel that will be detected. A large sample volume will enable the flow near the anterior and posterior walls to be detected, but, as discussed, the narrow beam width in the elevation plane may mean that the flow near the lateral walls will remain undetected. If only the fast flow in the centre of the vessel is to be measured, then a small sample volume should be selected (Figure 9.9b). It is also important to use a small sample volume when assessing the degree of spectral broadening, i.e. the width of the Doppler spectrum, as an indication of the presence of disturbed flow within a diseased vessel.

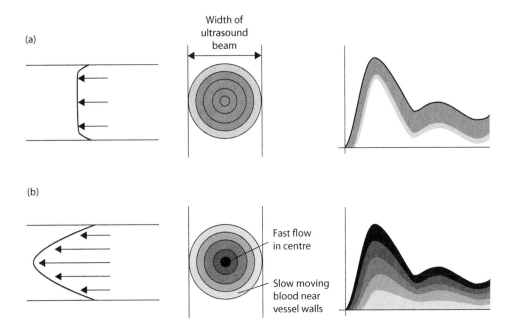

Figure 9.8 A schematic diagram showing how the velocities displayed in the Doppler spectrum will depend on the velocity profile within the vessel shown for idealized **(a)** blunt flow and **(b)** parabolic flow.

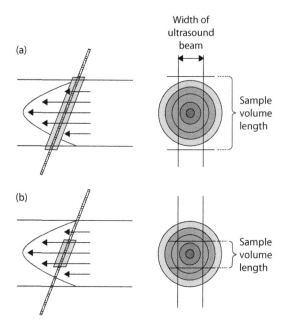

Figure 9.9 Effect of **(a)** incomplete insonation and **(b)** sample volume size on the detected Doppler signal.

Intrinsic spectral broadening

The Doppler beam is produced by a sub-group of elements within the array (as shown in Figure 9.10), producing a Doppler aperture. This means that, in reality, the Doppler beam insonates the blood at a range of different angles due to the shape of the beam. For example, in the diagram shown in Figure 9.10, the element on the far left produces an angle (θ_3) and the element on the far right produces a smaller angle (θ_1). The detected Doppler frequency is proportional to cosine (θ), so that the highest Doppler frequency is detected for the element on the far right and the lowest Doppler frequency is detected for the element on the far left. The overall effect is that there is spreading of the range of Doppler shift frequencies detected which is due to the beam shape rather than the blood flow. This effect is known as intrinsic spectral broadening (Thrush and Evans 1995; Hoskins 1996; Hoskins et al. 1999). It can be demonstrated by making ultrasound velocity measurements from a single moving target, such as a string driven by a motor at constant speed. The Doppler spectra produced using a 5 MHz linear-array transducer insonating

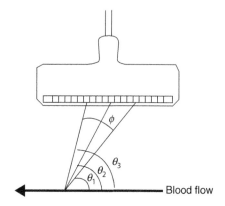

Figure 9.10 A number of elements of a linear-array transducer are used to produce the Doppler beam, leading to a range of angles of insonation. Typically the angle produced by the centre of the beam, θ_2, is used to calculate the velocity.

moving string is shown in Figure 9.11. This demonstrates that, instead of the Doppler spectra displaying a single velocity for the moving string, a spread of velocities is displayed. As the string (unlike blood) is moving at a single velocity, the spreading of the detected velocities seen in the spectra is due to intrinsic spectral broadening, i.e. due to the ultrasound system rather than the motion being detected. The degree of spectral broadening

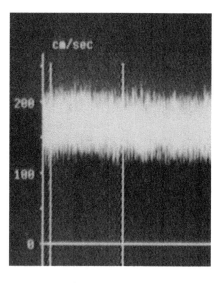

Figure 9.11 Doppler spectrum obtained from a moving string test object showing a range of velocities despite the fact that the string can only move with a single velocity.

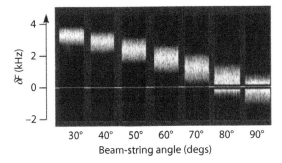

Figure 9.12 The Doppler shift frequency δF (kHz) detected from a moving string test object with constant velocity using an Acuson system over a range of angles of insonation. (Reprinted from Hoskins PR et al. 1999. *Ultrasound in Medicine and Biology*, 25, 391–404, with permission.)

will depend on the angle of insonation used, with larger angles of insonation generating a larger degree of intrinsic spectral broadening. Figure 9.12 shows the spread of Doppler frequencies obtained, using an Acuson scanner with a 7 MHz transducer and the sample volume set at 2–3 cm depth, from a moving string test object over a range of beam-to-string angles of insonation. It can be seen that the degree of spectral broadening increases as the angle of insonation increases towards 90°. This spectral broadening effect can lead to errors when making velocity measurements, or estimating the range of velocities present due to disturbed or turbulent flow in the presence of arterial disease.

Equipment setup

If an inappropriately low PRF is used, this will result in aliasing of the signal. This will alter the appearance of the waveform shape (see Figure 9.2) and lead to an underestimate of the peak velocity.

The high-pass filter is used to remove unwanted low-frequency signals arising from the slow-moving vessel walls. However, if the filter is set too high the waveform shape may be significantly altered (see Figure 9.6), e.g. the filter may remove the low frequencies detected during diastole. If the gain is set too high, a mirror image of the spectrum will be seen (see Figure 9.5b). If the gain is set too low, all the detected velocities may not be adequately displayed.

EFFECT OF PATHOLOGY ON THE SPECTRAL DOPPLER DISPLAY

Changes in detected Doppler frequency shift

The velocity of blood increases as blood passes through a narrowing (as described in Chapter 8) and velocity measurements or velocity ratios are often used to quantify the degree of narrowing of a vessel. Velocity ratios compare the velocity in the normal vessel proximal to a stenosis with the highest velocity at or just beyond the stenosis. A significant stenosis or occlusion may lead to a reduction in flow and this will be indicated by the presence of untypically low velocities proximal or distal to the site of disease.

Changes in spectral broadening

The presence of disturbed or turbulent flow can lead to an increase in spectral broadening which may be used as an indicator of disease. However, spectral broadening should be interpreted cautiously as intrinsic spectral broadening can be introduced due to properties of the ultrasound scanner.

Changes in waveform shape

The shape of the Doppler spectrum will depend on which vessel is investigated. For example, the waveform shape detected in the internal carotid artery (see Figure 8.15), which supplies blood flow to the brain, is very different from the waveform shape detected from the femoral arteries (see Figure 8.6), which supply the leg. Significant disease either proximal or distal to the measurement site will also affect the waveform shape and can, therefore, provide a useful indication of the presence and possible site of disease (see Chapter 8).

ARTEFACTS

Some of the artefacts that spectral Doppler suffers from are of the same origin as imaging ultrasound artefacts in that the ultrasound beam does not follow the expected path, or that the ultrasound has been attenuated. This leads to the following artefacts:

- *Shadowing*: This is due to highly reflecting or attenuating structures, such as bowel gas or calcified vessel wall, overlying the blood flow, leading to a loss of Doppler signal.
- *Multiple reflections*: In the presence of a strongly reflecting surface, e.g. a bone-tissue or air-tissue interface, multiple reflections may occur, leading to the appearance that Doppler signals have been detected outside the vessel. This will affect both the spectral Doppler and the colour flow. An example of where this is often seen is when imaging the subclavian artery as it passes over the lung, leading to the appearance of a second vessel lying within the lung. This is known as a mirror image.
- *Refraction*: This occurs if an ultrasound beam passes a boundary between two media with different propagation speeds, at an angle of less than 90°. This can lead to misregistration of both the image and the Doppler sample volume but this can be difficult for the operator to identify.

Other artefacts seen in spectral Doppler relate to production of the Doppler spectrum. These artefacts include the following:

- *Aliasing*: This occurs due to undersampling of the blood flow (see Figure 9.2) and can be corrected by increasing the Doppler PRF.
- *Angle dependence*: The Doppler shift frequencies detected are dependent on the angle of insonation by the $\cos \theta$ term. The larger the angle of insonation the smaller the Doppler shift frequencies detected, possibly leading to a poorer-quality Doppler spectrum. As the angle of insonation approaches 90°, the Doppler frequency may be very small and, therefore, may be removed by the high-pass filter.
- *Intrinsic spectral broadening*: This occurs due to the geometry of the ultrasound beam and the vessel (Figures 9.10, 9.11 and 9.12).
- *Range ambiguity*: This occurs when more than one pulse is in-flight at a time, such as when a high PRF is used to investigate high-velocity

flow at depth, as the origin of the Doppler signal is no longer certain.
- *Inverted mirror image of the Doppler spectrum*: This can occur if the gain is set too high (see Figure 9.5b).

MEASUREMENTS AND THEIR POTENTIAL SOURCES OF ERRORS

Velocity

Measurements of blood velocities are often used to quantify disease. Duplex imaging allows an estimate of the angle of insonation (θ) between the Doppler ultrasound beam and the blood flow. The velocity of the blood (v) can then be estimated from the measured Doppler shift frequency (f_d), using the Doppler equation, as the transmitted frequency of the Doppler beam (f_t) is known and the speed of sound in tissue (c) is assumed to be constant (1540 m s^{-1}):

$$ v = \frac{f_d c}{2 f_t \cos \theta} $$

There is often a spread of velocities present within the Doppler sample volume due to the groups of blood cells moving at different velocities. The effect of intrinsic spectral broadening will lead to a further spreading of the measured velocities. Therefore, when measuring velocity, a choice has to be made whether to use the maximum or the mean Doppler frequency. Peak systolic velocity, the maximum velocity at peak systole, and velocity ratios are used to quantify vascular disease (Thrush and Hartshorne 2010). The velocity ratio is given by the maximum peak systolic velocity within a stenosis, v_{sten}, and in the normal vessel proximal to the stenosis, v_{prox}, as follows:

$$ \text{Velocity ratio} = \frac{v_{sten}}{v_{prox}} $$

The ultrasound scanner can also calculate the mean velocity by finding the average of all the velocities detected at a given instant in time. As well as spot measurements of velocity at given points in time, many ultrasound systems are also capable of providing velocity measurements averaged over time. One such measurement is the mean velocity averaged over a number of complete cardiac cycles, usually known as time-average velocity (TAV). This can be used to estimate blood flow as described later in the chapter.

Errors in velocity measurements due to the angle of insonation

An estimate of the angle of insonation is required to convert the detected Doppler shift frequency into a velocity measurement. Any inaccuracy in placing the angle-correction cursor parallel to the direction of flow will lead to an error in the estimated angle of insonation. This, in turn, will lead to an error in the velocity measurement. As the velocity calculation depends on the $\cos \theta$ term, the error created due to cursor misplacement will be greater for larger angles of insonation. Figure 9.13 shows the relationship between the percentage error in the velocity measurement as the angle of insonation increases, where there is a 5° error in placement of the angle-correction cursor. Ideally, the angle of insonation should be kept at or below 60° in order to minimise the effect of errors due to imperfectly positioning the angle-correction cursor on an image. However, estimating the angle of insonation is not always straightforward, especially

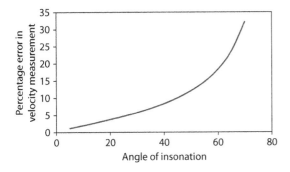

Figure 9.13 Graph showing the relationship between the percentage error in the velocity measurement as the angle of insonation increases, for a 5° error in the placement of the angle-correction cursor.

in the presence of disease. Some of the limitations are discussed in the following text.

DIRECTION OF FLOW RELATIVE TO THE VESSEL WALLS

The direction of the blood flow may not be parallel to the vessel wall, especially in the presence of a stenosis. Therefore, lining the angle-correction cursor to be parallel to the walls may lead to large errors in estimating the true angle of insonation, which, in turn, will lead to errors in the velocity estimation. If there is a clear image of the flow channel through a narrowing, it may be possible to line the angle cursor up with the flow channel. However, the maximum velocity caused by a stenosis may be situated just beyond the stenosis, and the direction of flow may be less obvious there.

LOCATION OF SITE OF MAXIMUM VELOCITY AND DIRECTION OF FLOW FROM COLOUR IMAGE

The advent of colour flow imaging has enabled better assessment of the direction of the blood flow, especially if the vessel lumen is unclear on the B-mode image. The colour image may also be used to identify the site of maximum velocity, although this can be misleading since the colour image displays the mean velocity of the blood relative to the direction of the Doppler beam, rather than the actual blood velocity. As the velocity estimated is angle dependent, the point at which the highest velocity is displayed on the colour flow image may not be the site of the true highest velocity. Instead the highest velocity displayed may be the site at which the angle of insonation is smallest. If the direction of blood flow has changed relative to the Doppler beam, within the imaged area, there will be different angles of insonation at different sites in the colour flow image. Ideally, both the colour and the spectral Doppler assessment of velocity changes across a stenosis should be used to locate the site of the maximum velocity.

OUT-OF-PLANE ANGLE OF INSONATION

It is important to remember that the interception of the ultrasound beam with the blood flow occurs in a three-dimensional space and not just in the two-dimensional plane shown on the image. Therefore, the transducer should be aligned with a reasonable length of the vessel, seen in longitudinal section on the image (see Figure 9.3). This ensures that the angle between the beam and the flow is approaching zero in the non-imaging plane, and this results in the minimum error.

DOPPLER ULTRASOUND BEAM APERTURE CREATES A RANGE OF ANGLES

The wide aperture used by multi-element transducers used to produce the Doppler beam means that the beam produces a range of angles of insonation (see Figure 9.10). Ideally, if the maximum velocity is to be measured, then the angle produced by the edge of the beam that gives the smallest angle (θ_1) should be used to estimate the velocity. However, the majority of modern scanners use the angle produced by the centre of the beam (θ_2) to calculate the velocity. This can lead to an overestimate in the velocity. Because the velocity measurement is dependent on the $\cos\theta$ term, the size of this error is also angle-dependent, becoming greater the larger the angle of insonation. Figure 9.14 (Hoskins et al. 1999) shows an example plot of the maximum velocity measured, using an ultrasound scanner, from a moving string test object over a range of angles of insonation. The solid line on the graph shows the true velocity of the moving string. It can be seen that the velocity of the string has been overestimated by the scanner, with the size of this systematic overestimation becoming greater as the angle of insonation increases.

The errors in velocity measurements obtained may also depend on the position of the beam on the screen. When the beam is positioned near the edge of the transducer, the number of elements used to form the beam may be smaller than when the beam is positioned in the centre of the image. This may result in different-size errors caused by intrinsic spectral broadening. Figure 9.15a and b show two measurements made at the same point in a flow rig with constant flow. It can be seen that the measurement made with the beam positioned close to the edge of the transducer gives a lower velocity measurement (111 cm s^{-1}) than the measurement made with the beam placed in the centre of the field of view (140 cm s^{-1}). It can be seen that the sensitivity of the Doppler beam produced at the edge of the transducer is also lower, compared to that produced in the centre, and more gain is required, resulting in noise seen on the spectral Doppler display. The same effect can be seen in Figure 9.15c and d, where the peak systolic velocity has been measured at the same point in a carotid artery with the Doppler beam at the edge (c) and the Doppler beam in the centre (d) of the transducer, resulting in velocity measurements of 68 and 90 cm s^{-1}, respectively. In order to minimise these differences in measurement, ideally measurements should be made with the Doppler beam in the centre of the field of view. The error produced due to spectral broadening can also vary with changes in the sample volume depth as this may result in changes in the active aperture.

The size of errors due to intrinsic spectral broadening is potentially large and may vary between models of scanner and between manufacturers (Fillinger and Baker 1996; Alexandrov et al. 1997; Kuntz et al. 1997). Unfortunately, manufacturers of ultrasound scanners provide little data on possible errors in velocity measurements.

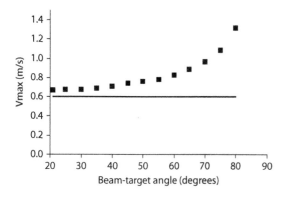

Figure 9.14 Errors in velocity measurements due to intrinsic spectral broadening: an example plot showing the maximum velocity, measured using an ultrasound scanner, of a moving string test object over a range of angles of insonation. The solid line shows the true velocity of the string. (Graph courtesy of PR Hoskins.)

Optimizing the angle of insonation

The B-mode and the colour flow images are used to estimate the direction of flow in the area

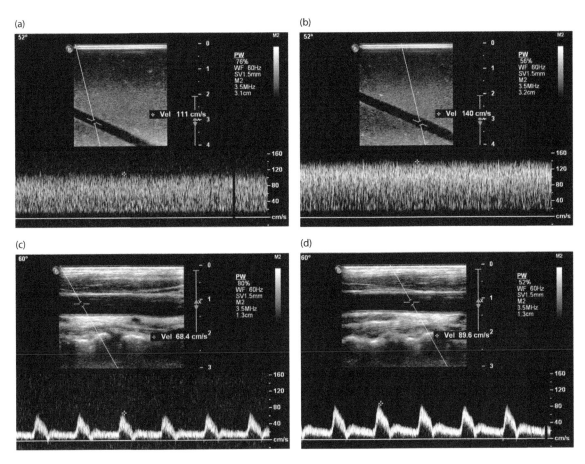

Figure 9.15 Spectral Doppler displays showing two measurements made at the same point in a flow rig with constant flow. The measurement made with the beam positioned close to the edge of the transducer (a) gives a lower velocity (111 cm s⁻¹) than the measurement made with the beam placed in the centre (b) of the field of view (140 cm s⁻¹). Panels (c) and (d) show the same effect where the peak systolic velocity has been measured at the same point in a carotid artery with the Doppler beam at the edge (c) and the Doppler beam in the centre (d) of the transducer, resulting in velocity measurements of 68 and 90 cm s⁻¹, respectively.

to be investigated so that the spectral Doppler beam can be steered appropriately. The angle of insonation is measured by lining up the angle-correction cursor with the estimated direction of flow. In some clinical settings, such as cardiac or transcranial Doppler, angles of insonation used are often at or approaching zero degrees resulting in smaller errors due to misalignment of the angle correction cursor (see Figure 9.13). In these situations, the angle of insonation is often assumed to be zero rather than measured by the operator. There are many possible pitfalls when making

velocity measurements and no single method of estimating the angle of insonation is completely reliable. The various possibilities and their advantages and disadvantages are discussed in the following sections.

VELOCITY RATIO MEASUREMENTS

Ideally, the angle of insonation used to make the velocity measurement proximal to and at the stenosis should be similar. This will result in the two velocities having similar systematic errors that will cancel out when finding the ratio.

ABSOLUTE VELOCITY MEASUREMENTS

There are two schools of thought about selecting the angle of insonation at which to make absolute velocity measurements:

1. *Always set the angle of insonation to 60°*: This ensures that any error in alignment of the angle-correction cursor only leads to a moderate error in the velocity estimate (see Figure 9.13) and that the errors caused by intrinsic spectral broadening are kept reasonably constant between measurements.
2. *Always select as small an angle of insonation as possible*: This ensures that any error in the alignment of the angle-correction cursor produces as small an error in velocity estimation as possible. The error due to intrinsic spectral broadening will also be minimised. However, this error will be different for measurements made at different angles of insonation. This makes comparison between measurements made at different angles less reliable.

In the Joint Recommendations for Reporting Carotid Ultrasound Investigations in the United Kingdom, Oates et al. (2009) recommend that the Doppler angle be in the range 45°–60° to minimise the effects on velocity measurement relating to the angle of insonation. The blood velocity may need to be measured at a few points through and beyond the stenosis to ensure the highest velocity has been obtained. Doppler criteria developed over the years may not have been produced with a full understanding of all these possible sources of error. Different models of ultrasound systems may produce different results for the same blood flow. However, despite these sources of error, velocity measurements have been successfully used to quantify vascular disease for the past two decades. A greater understanding of the sources of error in velocity measurement may lead to improvements in accuracy.

Measurement of volume flow

Volume flow is a potentially useful physiological parameter that can be measured using duplex ultrasound. Flow can be calculated by multiplying the time-average velocity (TAV) obtained from the

Doppler spectrum, by the cross-sectional area of the vessel, obtained from the B-mode image:

$$Flow = TAV \times cross\text{-}sectional\ area\ of\ the\ vessel$$

One method of estimating the cross-sectional area is to measure the diameter of the vessel, d, and calculate the area, A, assuming the vessel is cylindrical as follows:

$$A = \frac{\pi d^2}{4}$$

Some scanners allow the operator to trace around the vessel perimeter to calculate the area but this method relies on a good image of the lateral walls and a steady hand and is therefore prone to error (see Chapter 6). Figure 9.16 shows how the vessel diameter has been obtained from the image and the velocity of the blood has been measured by placing the sample volume across the width of the vessel and estimating the angle of insonation from the image. Although the measurement of flow is relatively simple to perform there are many errors

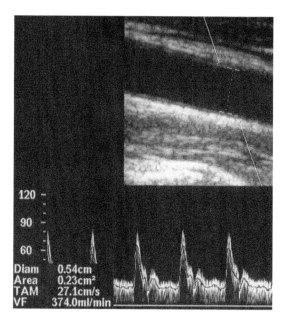

Figure 9.16 Image and spectral Doppler display showing the technique used to measure time-averaged mean velocity (TAM) and volume flow (VF).

relating to both the measurement of the TAV and to the cross-sectional area (Evans and McDicken 2000). These errors limit the value of one-off absolute flow measurements but serial measurements of flow may provide useful information on flow changes. Volume flow measurements may be useful where large changes in volume flow are diagnostically significant such as in the assessment of arteriovenous fistulas created for haemodialysis access.

ERRORS IN DIAMETER MEASUREMENT

As the measurement of flow is dependent on the diameter squared, any error in the diameter will produce a fractional error in the flow measurement that is double the fractional error in the diameter measurement. The accuracy of the diameter measurement depends on the resolution of the image and the accuracy of the callipers. Calculation of the cross-sectional area from a diameter measurement assumes that the vessel lumen is circular, which may not be the case in the presence of disease. Also, the arterial diameter varies by approximately 10% during the cardiac cycle so, ideally, several diameter measurements should be made and the average found.

ERRORS IN TAV MEASUREMENT

Incomplete insonation of the vessel will lead to errors in the mean velocity measurements due to an underestimation of the proportion of slower-moving blood near the vessel wall (see Figure 9.9). This is the case even if the sample volume is set to cover the near and far wall of the vessel, as the out-of-imaging-plane flow will not be sampled. Incomplete insonation of the vessel can lead to an overestimate in the measured value of TAV or flow of up to approximately 30%. The use of high-pass filters, if set too high (see Figure 9.6), can also lead to an overestimate in the mean velocity since the low-velocity blood flow would be excluded. Aliasing would lead to underestimation of the mean velocity due to the incorrect estimation of the high velocities present within the signal.

Waveform indices

The presence of significant disease can lead to alterations in the spectral Doppler waveform shape and may indicate whether the disease is proximal or distal to the site from which the Doppler waveform is obtained (see Chapter 8). Over the years different methods of quantifying the waveform shape have been developed (Evans and McDicken 2000) and modern scanners incorporate the facilities to calculate various indices to help identify changes in the waveform shape.

PULSATILITY INDEX

The pulsatility index (PI) can be used to quantify the degree of pulse-wave damping at different measurement sites. It is defined as the maximum, either frequency or velocity, of the waveform, S, minus the minimum, D (which may be negative), divided by the mean, M, as shown in Figure 9.17:

$$PI = \frac{(S-D)}{M}$$

Damped flow, beyond significant disease, will have a lower PI value than a normal pulsatile waveform.

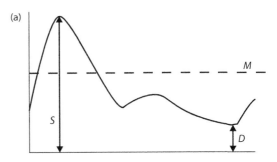

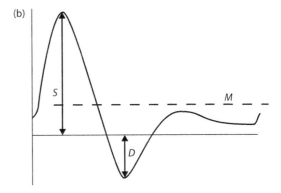

Figure 9.17 Schematic diagram showing the quantities used in estimation of resistance index and pulsatility index, for **(a)** a waveform with forward flow only, **(b)** a waveform with a period of reverse flow.

POURCELOT'S RESISTANCE INDEX

The resistance index (RI) is defined as (see Figure 9.17):

$$RI = \frac{(S - D)}{S}$$

SPECTRAL BROADENING

Over the years, there have been several definitions of spectral broadening to quantify the range of frequencies present within a spectrum (Evans and McDicken 2000) and one such definition is:

$$SB = \frac{(f_{max} - f_{min})}{f_{mean}}$$

When using these indices, it is important for the operator to understand how the ultrasound system calculates these values. Increased spectral broadening indicates the presence of arterial disease but some broadening can also be introduced by the scanner, as previously discussed.

Manual versus automated measurement

Duplex systems may enable the user to choose either to manually trace the maximum velocity over a cardiac cycle or to allow the system to automatically trace the maximum velocity. Manual tracing of the velocity can be both awkward and time consuming. Automatic measurements are usually quicker to implement but may be inaccurate in the presence of noise. Noise can include random background noise, electrical spikes or signals from other vessels, all of which may result in an incorrect estimation of the velocity. Therefore, the automated maximum trace should be displayed alongside the Doppler spectrum so that any large discrepancies can be seen and a judgement made on the accuracy of the measurement. Automatic tracing of the maximum velocity also allows automated measurement of peak systolic velocity and calculation of various indices which are displayed on the screen, sometimes in real time.

QUESTIONS

Multiple Choice Questions

Q1. Errors in peak systolic velocity measurements may occur due to:
  a. Small angle of insonation
  b. Large sample volume
  c. High Doppler gain setting
  d. Incorrect angle correction cursor placement
  e. A low filter setting

Q2. Aliasing:
  a. Is due to the incorrect sampling rate used to measure the blood velocity
  b. Can affect pulsed-wave Doppler measurements
  c. Can affect continuous-wave Doppler measurements
  d. Is more likely to occur when making measurements at depth
  e. Will not affect peak systolic velocity measurements

Q3. Aliasing may be overcome by:
  a. Decreasing the pulse repetition frequency (scale) of the Doppler pulse
  b. Increasing the angle of insonation
  c. Using a lower transmit frequency
  d. Changing the baseline
  e. Using 'high PRF mode', i.e. having more than one pulse in flight at a time

Q4. Volume flow measurements are typically calculated using:
  a. Peak systolic velocity measurement
  b. End diastolic velocity measurement
  c. Peak systolic velocity and end diastolic velocity measurements
  d. Time-averaged mean velocity measurements
  e. Time-averaged maximum velocity measurements

Q5. Incomplete insonation of a vessel can be due to:
  a. High gain setting

b. Small sample volume setting
c. Vessel wider than out of imaging plane Doppler beam width
d. Low pulse repetition frequency
e. Low transmit frequency

Q6. Intrinsic spectral boarding:
a. Is caused by the size of the aperture (group of elements) used to form the Doppler beam
b. Will be reduced by a reduction in the high-pass filter
c. Can be reduced by using a smaller angle of insonation
d. May lead to errors in peak systolic velocity measurement
e. May lead to errors in time-averaged mean velocity measurement

Short-Answer Questions

Q1. What are the advantages and disadvantages of PW over CW Doppler systems?

Q2. What is aliasing and how may it be overcome? Are there situations where aliasing cannot be prevented?

Q3. What is the sample volume length? How might the size and position of the sample volume affect the appearance of the Doppler spectral display?

Q4. What is intrinsic spectral broadening and what measurement may it affect?

Q5. When making velocity measurements, how is the angle of insonation measured? Discuss what size of angle of insonation should be used when making velocity measurements. Why is this important?

REFERENCES

Alexandrov AV, Vital D, Brodie DS, Hamilton P, Grotta JC. 1997. Grading carotid stenosis with ultrasound. An interlaboratory comparison. *Stroke*, 28, 1208–1210.

Evans DH, McDicken WN. 2000. *Doppler Ultrasound: Physics, Instrumentation, and Signal Processing*. Chichester: Wiley.

Fillinger MF, Baker RJ. 1996. Carotid duplex criteria for a 60% or greater angiographic stenosis: Variation according to equipment. *Journal of Vascular Surgery*, 24, 856–884.

Hoskins PR. 1996. Measurement of maximum velocity using duplex ultrasound systems. *British Journal of Radiology*, 69, 172–177.

Hoskins PR, Fish PJ, Pye SD, Anderson T. 1999. Finite beam-width ray model for geometric spectral broadening. *Ultrasound in Medicine and Biology*, 25, 391–404.

Kuntz KM, Polak JF, Whittemore AD, Skillman JJ, Kent KC. 1997. Duplex ultrasound criteria for identification of carotid stenosis should be laboratory specific. *Stroke*, 28, 597–602.

Oates CP, Naylor AR, Hartshorne T et al. 2009. Joint recommendations for reporting carotid ultrasound investigations in the United Kingdom. *European Journal of Vascular Surgery*, 37, 251–261.

Thrush AJ, Evans DH. 1995. Intrinsic spectral broadening: A potential cause of misdiagnosis of carotid artery disease. *Journal of Vascular Investigation*, 1, 187–192.

Thrush AJ, Hartshorne TC. 2010. *Vascular Ultrasound: How, Why and When*. London: Churchill Livingstone.

Colour flow

PETER R HOSKINS AND ALINE CRITON

INTRODUCTION

Doppler ultrasound remained a minority imaging methodology until the introduction of colour Doppler in 1982. In this technique, the motion of the blood is colour coded and superimposed on the B-mode image. This allows rapid visualisation of the flow patterns in vessels, allowing high-velocity jets in arteries and in cardiac chambers to be seen. It quickly became apparent that the ability to visualise flow patterns, such as the presence of intracardiac jets, was of great value. In addition, it considerably speeded up the placement of the Doppler sample volume in spectral Doppler investigation, hence reducing scanning time.

Prior to the introduction of commercial colour flow systems, several approaches were described to provide an image showing the pattern of blood flow. These often relied on manual scanning of the probe over the skin to build up a two-dimensional (2D) image. These systems, which used electronic or mechanical sweeping of the beam, were not real-time, having only a maximum of a few frames per second. These are reviewed in Evans and McDicken (2000), Wells (1994) and Cobbold (2007). General reviews of colour flow ultrasound technology and applications are provided by Hoskins and McDicken (1997), Evans (2010) and Evans et al. (2011).

As noted in Chapter 3 beam forming classically involves single-line methods which result in limited frame rates. Introduction of zone and full-field beam forming has resulted in increased frame rate and opened up the possibility of more sensitive Doppler techniques. At the time of writing the vast majority of clinical practice involving colour flow ultrasound is based on imaging of blood flow using ultrasound systems with classic beam forming. Other techniques which are available commercially are described in Chapter 11. These include Doppler tissue imaging, B-flow, high frame rate colour flow and vector-flow techniques. This chapter will concentrate on colour flow ultrasound imaging with an emphasis on classic beam forming.

Terminology

The term 'colour flow' originally referred only to an ultrasound system in which a 2D image of mean Doppler frequency from blood was displayed using colour coding. For many years now other quantities have been displayed in colour, such as the power of the Doppler signal. In order to provide a consistent terminology, the descriptions that follow are used in this chapter:

- *Colour flow*: Imaging of blood flow. This is the generic term, which is used in this chapter, and it encompasses the three modalities that follow.
- *Colour Doppler*: Image of the mean Doppler frequency from blood, displayed in colour superimposed on a B-mode image.
- *Power Doppler*: Image in which the power of the Doppler signal backscattered from blood is displayed in colour.
- *Directional power Doppler*: Image in which the power of the Doppler signal is displayed, including separate colour coding of blood velocities towards and away from the probe.

2D image production

Production of a 2D colour flow image includes elements of B-mode image formation and pulsed Doppler techniques. As in B-mode image formation, the image is built one line at a time, by transmitting ultrasonic pulses and processing the sequence of returned echoes. However, unlike B-mode image formation in which echo amplitude information is processed to form the image, the echoes are demodulated to produce a Doppler shift signal. In the pulsed-wave spectral Doppler systems described in Chapter 9, Doppler information was obtained from only a single sample volume. In a colour flow system, each line of the image is made up of multiple adjacent sample volumes.

As colour flow is a pulsed-wave Doppler technique, the Doppler shift information for each line is obtained from several transmission pulses. Unlike spectral Doppler, which relies on the fast Fourier transform (FFT) to extract the whole spectrum of frequencies that are present, colour flow imaging uses a technique known as 'autocorrelation', which was introduced in Chapter 7. This calculates the mean frequency detected within each sample volume, which is then colour coded on the display. For the mean Doppler frequency to be detected, at least two pulses are required to be transmitted along each line. However, as a more accurate estimate of the detected mean frequency is obtained

when more pulses are used, a typical colour scanner may use about 10 pulses. The requirement of several pulses per line to produce the colour image, compared with a minimum of one in the B-mode image, means that the frame rate for a comparable number of scan lines is much less in the colour image than it is in the B-mode image. For example, if 10 pulses were to be used for each colour line, and one pulse for each B-mode line, then the maximum frame rate for the colour image would be one-tenth that of the B-mode image. There is, therefore, a compromise between the size of the colour image (number of scan lines), accuracy of the frequency estimate (number of pulses per line) and the rate at which the image is updated (frame rate). In order for a sonographer to appreciate the pulsatile nature of blood flow, it is preferable for the machine to maintain a frame rate above 10 frames per second. If the whole field of view were to be used, then the maximum achievable colour frame rate would be only a few frames per second. In order to improve the colour frame rate, the colour-coded flow is only displayed in a limited region of interest called the 'colour box' within the displayed B-mode image. The reduced depth and number of scan lines that this provides enable fewer ultrasound pulses to be sent per colour frame, and hence allows the use of higher colour frame rates. The width and depth of the colour box are under operator control, and the frame rate may be increased by narrowing the box, or by decreasing the box depth. In order to further increase frame rate, the line density may also be reduced on some systems. Typically, frame rates of 10–15 frames per second can be achieved in peripheral arterial applications, though this may fall to five frames per second or less in venous applications, where a large number of pulses are needed for each colour line in order to measure relatively low Doppler shifts. Low frame rates may also result for abdominal and obstetric applications, where the vessel depth is large. Typical colour box shapes and relative sizes are shown in Figure 10.1.

Phase- and time-domain techniques

The common theme of colour flow techniques is that the colour image is derived by consideration of the motion of the blood. There are two basic classes of instrument, dependent on whether they

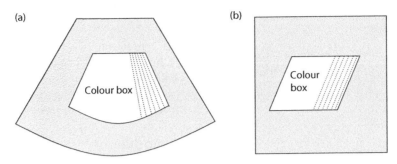

Figure 10.1 Colour box shapes for **(a)** sector and **(b)** linear-array transducers. Within the colour box, the image is built up as a series of colour lines. Each line consists of a series of adjacent sample volumes.

determine the presence of motion by analysis of the phase shift or the time shift, as described in Chapter 7. Few commercial machines use the time-domain approach, probably as this is computationally more demanding, and hence more expensive to implement. Virtually all modern commercial colour flow systems employ the phase-shift approach using autocorrelation detection, and the sections immediately following refer to this approach.

COLOUR FLOW SYSTEM COMPONENTS

A colour flow system will independently process the received B-mode and colour flow echoes (Figure 10.2). In addition, a spectral Doppler display can be obtained from a single sample volume as selected by the operator. A small number of pulses, typically 2–20, are transmitted and received for each colour line that is produced. Each line is divided into a large number of sections each of which represents a different sample volume. The Doppler signal from all of the gates is processed simultaneously. This situation is different to pulsed-wave spectral Doppler, where only one gate is considered. Figure 10.3 shows the essential components of the colour flow processor. The functions of the components are described in the following sections.

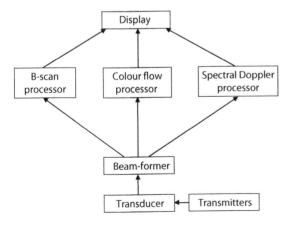

Figure 10.2 Components of a colour system. For a colour flow system, three types of information are processed: B-mode, colour flow and spectral Doppler.

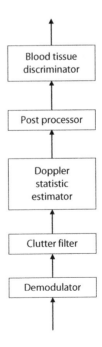

Figure 10.3 Components of the colour flow processor.

Table 10.1 Typical B-mode and colour flow transmit frequencies

Application	B-mode frequency (MHz)	Colour frequency (MHz)
Peripheral vascular	7–12	4–6
Abdominal/obstetrics	2–5	2–4
Transcranial	1.5–2.5	1–2

Doppler transmitter

Colour flow imaging uses the pulse-echo technique. Modern colour flow systems do not use the same pulses that are used for the B-mode image, but instead use separate lower-frequency pulses, as shown in Table 10.1.

Transducer

Any transducer used for B-mode imaging can, in principle, be used for colour flow. Commercial colour flow scanners typically use linear-, curvilinear- or phased-array transducers. The use of mechanically swept systems is possible, but is more problematic as the vibration that is produced can be picked up by the colour flow system, so that careful attention to design is needed in order to reduce false colour display.

Beam-former

This component of the system is the same as the B-mode beam-former discussed in detail in Chapter 3. The beam-former controls all aspects concerned with focusing and sweeping the beam through the tissue to produce a 2D colour image.

Demodulator

The demodulator extracts the Doppler shift frequencies as discussed in Chapter 7. This process is invisible to the user, with no relevant user controls.

Clutter filter

Clutter refers to signals from stationary and slowly moving tissue. It was noted in Chapter 7 that the signal from tissue is some 40 dB higher than the signal from blood. If it is moving blood that is of interest, then it is necessary to remove, as far as possible, the clutter signal, and so a clutter filter is present in colour flow systems. The clutter filter is analogous to the wall thump filter of spectral Doppler. Early colour flow systems used relatively unsophisticated clutter filters and were unable to detect low-velocity flow (Figure 10.4). The detection of low velocities by modern colour flow systems is linked to the use of a sophisticated clutter filter (Figure 10.5).

- Moving target indicator
- Delay line cancellor

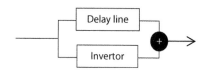

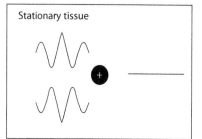
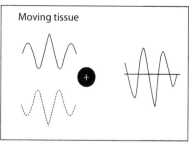

Figure 10.4 Simple clutter filter. The previous received echo is delayed and added to the current received echo. If the tissue is stationary, consecutive echoes will cancel out, whereas if there is blood or tissue motion, consecutive echoes will not cancel out. This simple method is also called a 'delay line cancellor' or a 'moving target indicator'.

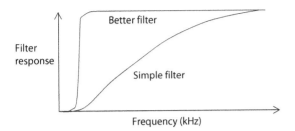

Figure 10.5 Clutter filter response as a function of Doppler frequency. The ideal filter completely suppresses the clutter, and allows through the Doppler frequencies. The simple delay line canceller of Figure 10.4 causes suppression of low Doppler frequencies, which will prevent the display of low velocities; however, there is suppression of higher blood velocity signals also. The more complex clutter filter has better low-frequency behaviour.

Mean-frequency estimator

The requirement for real-time scanning means that Doppler frequencies must be estimated in a much shorter time than is available for spectral Doppler; usually 0.2–2 ms for colour flow, as opposed to 5–40 ms for spectral Doppler. When colour flow systems were introduced in 1982, the processing capabilities available at that time meant that it was not possible to use an FFT approach. This would have meant calculating the full Doppler spectrum followed by extracting the mean frequency. The realisation by Namekawa et al. (1982) and Kasai et al. (1985) that a computationally simple algorithm termed 'autocorrelation' could be used to calculate mean frequency directly was the breakthrough that led to the introduction of real-time colour flow systems into commercial use. In addition to the short time that is available for calculation in colour flow, the number of pulses used for generation of each line is much less than in spectral Doppler; 2–20 in colour flow as opposed to 80–100 in spectral Doppler. The number of pulses used for generation of each colour line is called the 'ensemble length'.

The autocorrelator provides simultaneous estimates of three quantities:

- *Power:* Proportional to the square of the amplitude of the Doppler signal.

- *Mean Doppler frequency:* The mean or average Doppler frequency.
- *Variance:* A quantity related to the variability of the Doppler signal. It is defined by the square of the standard deviation of the Doppler signal amplitude, estimated over the ensemble length.

Since the introduction of the autocorrelation technique, there have been many different techniques used to estimate the single quantity that can be displayed on a colour flow image. Some of these techniques have been alternative ways of estimating mean frequency. Other techniques have been for estimation of different quantities, such as maximum Doppler frequency. The descriptions of these techniques are beyond the scope of this book, and are reviewed in Evans and McDicken (2000) and in Evans (2010). Most modern colour flow scanners use a technique called '2D autocorrelation' or variants of this. The 2D autocorrelation technique is described by Loupas et al. (1995a, b), and is an extension of the basic autocorrelation technique.

Post-processor

Even under conditions where the blood or tissue velocity is unchanging, the estimated mean Doppler frequency will change to some extent in a random manner. On the colour flow image, this variation manifests itself as a speckle pattern called 'colour speckle'. The cause of this speckle is associated with the variation in echo amplitude received at the transducer arising from the variation in the detailed position of each of the red cells within the sample volume from one pulse to the next. This is the same reason that there is a speckle pattern on B-mode images and on spectral Doppler waveforms. This speckle pattern can mask changes in the displayed colour. However, it is possible to reduce the degree of noise by averaging over several frames. This is the same frame-averaging technique as used in B-mode imaging, and it gives rise to a persistence effect.

Blood-tissue discriminator

For each pixel of the image, it is possible to estimate the echo brightness level for the B-mode image and also the mean Doppler frequency for the colour flow image. However, it is only possible to display one of these in the final composite image. The function of the blood-tissue discriminator is to ensure that colour is displayed only in regions of true blood flow and not in the presence of moving tissue. There are several methods by which discrimination between the signals from blood and moving tissue can be achieved. These include:

- *B-mode amplitude threshold*: If the amplitude of the B-mode image is large, it is likely that the signal arises from a region of tissue. A threshold based on the amplitude of the B-mode data is used to suppress the display of colour in regions of the image where the B-mode amplitude exceeds the threshold (Figures 10.6 and 10.7). See the description of the 'colour write priority' control later in the text.
- *Doppler signal amplitude threshold*: In regions of tissue that are moving slowly with respect to the transducer, the Doppler frequency shifts from

the tissue will be of high amplitude, but of low Doppler frequency. The clutter filter acts to eliminate these signals. After the clutter filter, the signal from blood is much stronger than the signal from tissue. The use of a simple threshold on the Doppler amplitude is used to determine whether blood or tissue signals are displayed for a given pixel (Figures 10.6 and 10.7). See the description of the 'colour gain' control later in the text.

- *Flash filter*: Rapid motion of either the tissue or of the transducer produces a Doppler shift, which may be displayed as a region of colour; these are called 'flash artefacts'. The threshold method described previously is often insufficient to remove such flashes. Manufacturers have developed more sophisticated removal methods, based on the detection of very rapid changes in the Doppler signal level, such as are produced by motion of the transducer with respect to the patient, breathing motions, cardiac motion and bowel motion.

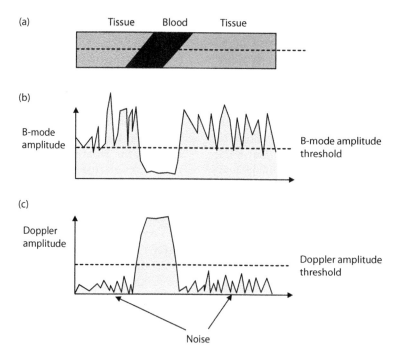

Figure 10.6 Operation of the blood-tissue discriminator. (a) A single image scan line through a region of tissue and flowing blood is shown. (b) Amplitude of B-mode echoes with depth. For values of the amplitude above the threshold, colour is not displayed, and for values below the threshold colour data are displayed. (c) Amplitude of the Doppler signal with depth; for values of the amplitude above the threshold, colour is displayed, and for values below the threshold colour data are not displayed.

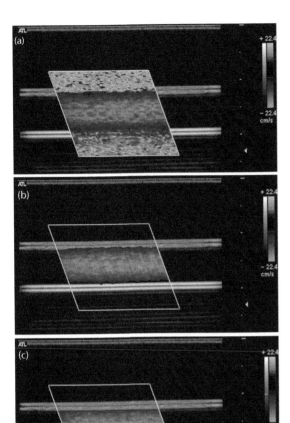

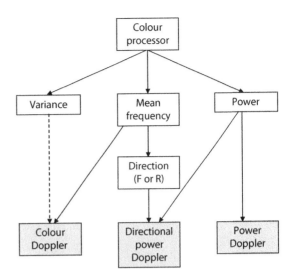

Figure 10.8 Relationship between calculated quantities and colour flow display modes. The three quantities estimated by the colour flow system are the mean Doppler frequency, the Doppler power and the variance. The colour Doppler image is a display of the mean frequency in which the variance may be added, if required. The power Doppler image is a display of the Doppler power. For directional power Doppler, flow direction is obtained from mean frequency, and used to colour code the power Doppler image.

Figure 10.7 Operation of the blood–tissue discriminator. Colour Doppler images are acquired from a flow phantom with a blood mimic flowing through a tube. **(a)** No blood-tissue discriminator present – there is colour throughout the colour box consisting of a uniform colour in the region of flow, surrounded by random noise in the surrounding regions. **(b)** Application of the Doppler amplitude threshold using the colour gain removes most of the colour from the region of no flow; however, there is some 'bleeding' of the colour into the tube. **(c)** Application of the B-mode amplitude threshold using colour write priority removes the bleeding.

COLOUR FLOW MODES

The three outputs from the autocorrelator (mean frequency, variance and power) can be colour coded and displayed, either alone or in combination with each other. This produces a number of possible colour modes, which can be selected. In practice, only a few are sensible (Figure 10.8). These are considered in the following sections.

Colour Doppler

The earliest commercial colour flow systems concentrated on this mode, in which the mean frequency in each pixel is colour coded. Although in principle any colour scale could be used for the display, most manufacturers adopt a red-based scale for blood flowing in one direction, and a blue-based scale for blood flowing in the opposite direction (Figure 10.9).

Where variance is displayed alone, it is shown as a green colour, together with the red or blue representing the mean frequency, to produce a composite display. This option was widely available on

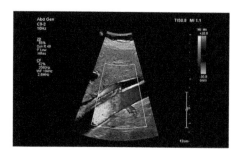

Figure 10.9 Colour Doppler image of flow in the aorta with a blue-red scale.

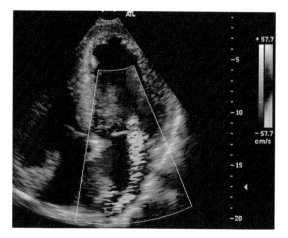

Figure 10.10 Colour Doppler image of tricuspid regurgitation with variance admixed. Regions of green colouration associated with high variance are seen.

early colour flow machines, where it was thought that variance was related to turbulence produced by narrowed arteries or cardiac valves. Figure 10.10 shows a cardiac jet, where the green colouration in the jet is seen. This mode is not used much outside cardiac applications.

Power Doppler

Display of the power of the Doppler signal had been a feature of early colour flow systems; however, the same instrument settings were used as for colour Doppler and there was not much improvement over the colour Doppler image. It was only when the processing was optimized for power Doppler (Rubin et al. 1994) that it became popular, largely due to improved sensitivity over colour Doppler.

The first step in this optimization is concerned with the display of noise. In colour Doppler, the noise is present on the image if the colour gain is set too high, and its appearance is a multicoloured mosaic in which the true image from the vessel of interest may be difficult to observe. However, for power Doppler, the noise appears as a low-level uniform hue in which it is easily possible to observe the vessel of interest. Consequently, the first step in optimization is to reduce the level of the threshold used to differentiate between the signal from blood and the signal from noise. The second step recognises that it is not possible to follow the changes in blood flow with time using power Doppler, so it is not necessary to have good time resolution. Consequently, for power Doppler, the degree of averaging over successive acquired frames (persistence) is much higher than for colour Doppler. This averaging process acts to reduce the colour noise level, and hence make it easier to distinguish small vessels with low-signal levels. In order to make the best use of the increase in sensitivity, the first manufacturers to use power Doppler displayed colour throughout the colour box. This is shown in Figure 10.11, where small vessels can be seen in yellow above the noise, which is shown in red. The disadvantage of this mode is that the underlying tissue anatomy cannot be seen. Most manufacturers now prefer to display the power Doppler image superimposed on the B-mode image (Figure 10.12). A commonly used colour scale is a 'heated body scale', in which the displayed colour changes through black, red, orange and yellow as the Doppler power increases.

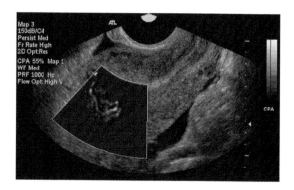

Figure 10.11 Power Doppler image of arteriovenous fistulae, in which there is colour throughout the colour box.

Figure 10.12 Power Doppler image of the kidney taken intraoperatively consisting of colour superimposed on the B-mode image.

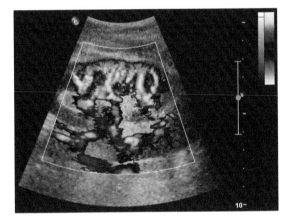

Figure 10.13 Directional power Doppler from a transplanted kidney.

Directional power Doppler

In this mode, the directional information on blood flow is obtained from the mean-frequency data, and used to colour code the power Doppler data (Figure 10.13). This mode is supposed to combine the enhanced sensitivity of power Doppler with the directional capability of colour Doppler.

COLOUR CONTROLS

There are a large number of instrument settings, which affect the displayed colour image. Thankfully in most modern colour flow systems, the default values of the settings are pre-programmed by the manufacturer for particular clinical applications, which the operator can recall

from the applications list. The operator must then adjust a small number of controls for individual patients. In the list that follows, the controls are classified according to three categories: controls which affect colour image acquisition, controls which affect Doppler signal extraction and frequency estimation, and controls which affect the display of the colour flow signals. The section finishes with a description of the use of controls in clinical practice.

Controls affecting the acquisition of colour flow images

POWER OR ACOUSTIC OUTPUT

The amplitude of the ultrasound pulses used for generation of the colour flow images may usually be adjusted over a wide range of values. The sensitivity of the instrument will improve as the power increases; however, in order to maintain patient exposure within safe limits, it is best if other controls, such as colour gain, are also used to obtain the desired image quality.

PULSE REPETITION FREQUENCY

The pulse repetition frequency (PRF) is the total number of pulses which the transducer transmits per second. It is limited largely by the maximum depth of the field of view; the transmit-receive time is less for smaller depths and a higher PRF is then possible. The value of the PRF selected in the various system presets, such as arterial or venous, will depend on the expected velocities present in the region of interest, but PRF may need to be altered by the operator, e.g. to prevent aliasing or to enable the detection of low flow. On modern systems, there is not a single control labelled 'PRF'. Instead, PRF is usually determined automatically from various controls, including the colour box size and the velocity scale.

The total number of pulses transmitted is divided between the B-mode image, the colour flow image and spectral Doppler. Maximum PRF for the colour image is achieved when the spectral display is switched off, and the colour box depth and width are reduced as much as possible. This will result in a colour image with a high frame rate.

STEERING ANGLE

This control is applicable to linear-array systems, where it is possible to steer the colour beam in a variety of directions with respect to the B-mode scan lines. Most systems provide three angles (e.g. −20°, 0°, +20°), though some provide a choice of five or more directions between this range. Steering the colour beams is desirable in colour flow imaging as many peripheral vessels run parallel to the skin surface, and an ultrasound beam perpendicular to the skin would provide zero Doppler signal. However, optimum B-mode imaging of the vessel is provided in this situation. By a combination of beam steering and probe angulation, it is usually possible to obtain a beam-vessel angle in the range 40°–70°, which is sufficient for adequate colour flow image production, as well as good B-mode visualisation of vessel walls. The power Doppler image is much less dependent on the beam-vessel angle, as explained in the following text, and it is generally not necessary to steer the beam away from the 0° direction.

FOCAL DEPTH

Due to frame-rate considerations, it is usual to only have one transmit focal depth for colour flow images. In some systems, the default is set automatically at the centre of the displayed field of view; in others, it is necessary to set this manually at the depth of interest.

BOX SIZE

The depth and width of the colour box are set by the user. The box depth directly influences the PRF, with higher PRF, and hence higher frame rates, being possible for box depths that are nearer to the surface. Higher frame rates may also be achieved by restriction of the width of the colour box as this reduces the number of Doppler lines required.

LINE DENSITY

It is not necessary for the line density (number of Doppler lines per centimetre across the image) of the colour image to be the same as the B-mode image. Reduction in line density increases frame rate; however, this is done at the expense of reduced lateral spatial resolution of the colour image.

GATE LENGTH

The gate length will determine the number of cycles in the transmitted pulse, and so alters the sample volume size. This improves sensitivity, but decreases axial resolution.

DEPTH OF FIELD

Reducing the image depth enables higher PRF to be used, and therefore higher frame rates to be achieved.

Controls affecting the extraction and estimation of Doppler frequencies

FILTER CUT-OFF

It is common practice to set the filter cut-off frequency as a certain fraction of the total displayed frequency scale, rather than as an absolute value of, say, 200 Hz. This means that as the frequency scale increases, so does the level of the clutter filter. Observation of low blood velocities requires that the frequency scale is set to low values.

There are usually three or four clutter filter options to choose from. Selection of too low a filter level results in breakthrough of the clutter signal from slowly moving tissue.

ENSEMBLE LENGTH

The term 'ensemble length' is used to refer to the number of pulses used to generate each colour line. Provided that flow is steady during the time spent measuring the Doppler shifts along one scan line, the variability of estimated mean frequency decreases as the ensemble length (number of pulses per estimate) increases. Low variability is required for accurate estimation of low velocities. In cardiology, it is higher blood velocities that are mainly of interest, whereas in radiology low venous blood velocity may be of more interest. Consequently, the ensemble length is partly determined by the selected application, with longer ensemble lengths used in radiology applications than in cardiology applications. Visualisation of low velocities is best achieved by adjustment of the velocity scale, so in many systems the ensemble length is directly linked with the velocity scale. For example, a 10-pulse ensemble will take longer to complete when a low PRF is used than when a high

PRF is selected. As the velocity scale is reduced to enable better visualisation of low velocities, the increased ensemble length will result in a reduction in frame rate.

BASELINE

If aliasing is a problem, then one method of dealing with this is to shift the baseline to enable higher positive velocities to be presented. This is identical to the technique used in spectral Doppler.

PERSISTENCE OR FRAME AVERAGING

Persistence refers to the averaging of Doppler shift estimates from current and previous frames. If flow is stable over the averaging period, then strong frame averaging will result in reduced colour noise, enabling better visualisation of the true flow pattern. If the degree of frame averaging is kept fixed through the cardiac cycle, then rapidly changing flow patterns will not be properly visualised. Some commercial systems attempt to overcome this by automatic adjustment of the level of frame averaging. For example, if the measured velocity is high, the persistence will be low, enabling visualisation of the high-velocity pulsatile flow patterns in arteries; when the velocity is low the persistence will be high, allowing the (usually) less pulsatile flow in veins to be observed.

Controls affecting the display of the colour flow signals

COLOUR GAIN

Colour is displayed if the amplitude of the Doppler signal amplitude is above a threshold value (see Figure 10.6). The level of the threshold can be adjusted by use of the colour gain control. If the gain is too low then no colour is displayed, whereas if the gain is set too high, then noise may be displayed as a mosaic pattern throughout the image. This control is adjusted for each patient in a similar manner to spectral Doppler gain. Figure 10.14 shows flow in the common carotid artery where the gain is too high, correct and too low.

COLOUR WRITE PRIORITY

This control ensures that pixels with high B-mode echo values, likely to arise from tissue, are not

(a)

(b)

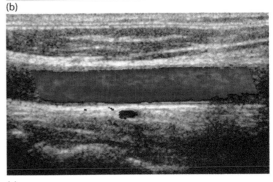

(c)

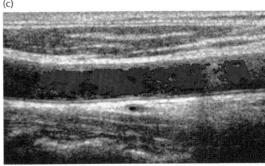

Figure 10.14 Flow in the common carotid artery at different gain settings. **(a)** Gain too high with noise in the tissue; **(b)** gain correct with colour contained within the artery; **(c)** gain too low with inadequate colour coverage in the artery.

displayed in colour. The colour write priority enables the operator to adjust the B-mode echo amplitude threshold, above which colour is not displayed and below which colour data are displayed.

POWER THRESHOLD

This is a threshold on the calculated power value, with no display of colour if the power is below the threshold.

FLASH FILTER

This is the process whereby the colour flashes from transducer or tissue motion are removed. Few details of these are available from manufacturers; however, one possibility is that flash filters are based on the detection of very rapid changes in the Doppler signal level. These could be produced by motion of the transducer with respect to the patient, breathing, cardiac motion and bowel movements. The operator usually has the choice of turning the flash filter on or off.

Use of controls

The operator chooses the probe and the application from the pre-set menu. This provides default values relevant to the typical patient for the application selected. It is usual to start the examination using B-mode in order for the operator to familiarise themselves with the anatomy, then progress to colour flow. The operator adjusts the size of the colour box to cover the desired region. With a linear-array transducer, it is also possible for the operator to steer the colour box in order to optimize the colour Doppler angle of insonation. The mode is selected (Doppler or power), and the scale and baseline are adjusted to enable display of the blood velocity range. The colour gain is adjusted so that as much of the vessel as possible is filled with colour, but at the same time avoiding excess noise in the tissue. A further refinement for colour Doppler is to adjust the probe angulation and steering angle to ensure that a Doppler angle away from 90° is obtained. This avoids colour drop-out due to the action of the clutter filter on the low Doppler shifts obtained near to 90°. This limited sequence of control adjustments is often all that is needed in a colour flow or Doppler tissue examination, although the user can alter other controls, if necessary.

FEATURES OF COLOUR FLOW

Penetration

The penetration depth is the maximum depth at which Doppler signals can be distinguished from noise. Improved penetration can simply be achieved by turning up the output power. High-output power is recognised as being hazardous to the patient, as described in Chapter 16. For targets at the deepest depths, the returning ultrasonic signal and hence detected Doppler signal is of small amplitude due to the effect of attenuation within the tissue. The task of the Doppler system in this situation is to distinguish the true Doppler signal from noise, and good machine design uses low-noise components. A standard signal-processing method to improve the detection of the signal is to combine a large number of measurements by some form of averaging. In this process, the signal size increases in comparison with the random noise that tends to cancel out, hence the signal can be more easily detected from the noise. For Doppler systems, the averaging can involve the use of a larger ensemble length or frame averaging; however, both of these are done at the expense of a lower frame rate.

Display of low velocities

The most important components determining the visualisation of low velocities are the clutter filter and the PRF, as previously noted. Optimization of machine settings in order to detect low velocities is achieved by increase in the ensemble length, by the use of persistence and by the decrease of the PRF and clutter filter. This is performed automatically by selection of the clinical protocol and adjustment of the 'velocity scale', but some systems may allow access to these controls directly by the operator in a hidden menu.

Display of flow in small vessels

The first requirement to display flow in small vessels is that the spatial resolution of the B-mode and colour flow images is adequate. It is the display of flow in small vessels where the superior characteristics of power Doppler over colour Doppler are demonstrated. Figure 10.15 compares colour Doppler and power Doppler images in a simple flow test device consisting of a 1 mm diameter vessel embedded in tissue-mimicking material. The image of flow in the vessel should ideally be a continuous line of colour.

In Figure 10.15a and b the persistence is set to zero. The colour Doppler image is not continuous, instead showing drop-out at several locations.

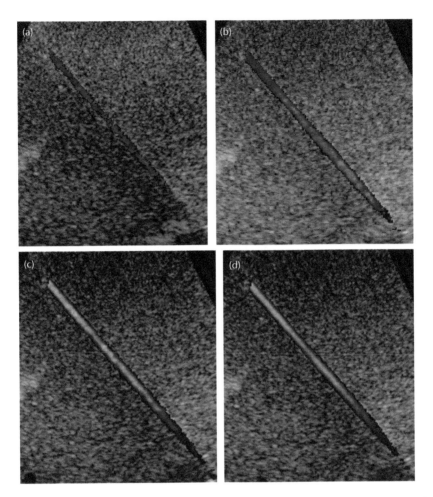

Figure 10.15 Effect of colour imaging mode and persistence. Images are shown in a 1 mm diameter vessel taken using a C5–2 curvilinear probe. With persistence turned off, **(a)** colour Doppler shows drop-out, **(b)** power Doppler shows a continuous line of colour with no drop-out. With persistence at maximum a continuous line of colour with no drop-out is shown for **(c)** colour Doppler and **(d)** power Doppler.

This is associated with variations in the calculated mean frequency produced by the autocorrelator. Values of mean frequency that are low will trigger the blood-tissue discriminator, and colour will not be displayed. The calculated Doppler power is less variable than the calculated mean frequency, so the power Doppler image of the vessel demonstrates less drop-out.

In Figure 10.15c and d the persistence is increased to maximum for both colour and power Doppler. With this increased persistence both mean frequency and power demonstrate less variability, resulting in fewer values falling below the blood-tissue discriminator threshold.

From this simple example it can be seen that when the same machine settings are used for power and colour Doppler, the penetration depth is similar for the two modalities. The improved detection of small vessels using power Doppler in clinical practice is due to the use of higher frame averaging, and also to the inherently less confusing nature of the power Doppler image, as there is no aliasing effect and only a limited angle dependence (Figure 10.16).

Display of complex flow patterns

The ideal colour Doppler display would provide images in which the displayed colour was related

(a)

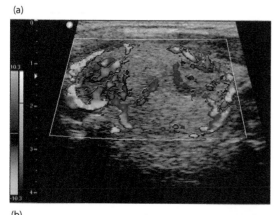

(b)

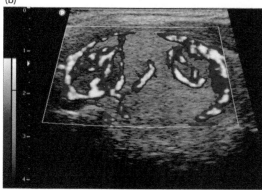

Figure 10.16 Colour **(a)** and power **(b)** Doppler images demonstrating thyroid nodules. The power Doppler image shows the anatomy more clearly than the colour Doppler image.

(a)

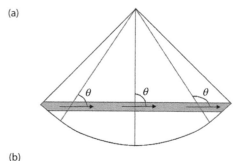

(b)

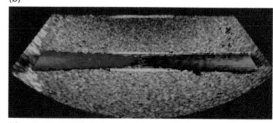

(c)

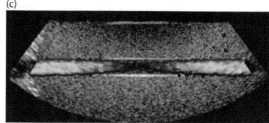

Figure 10.17 Angle dependence of colour and power Doppler images. **(a)** A straight tube is shown in which the flow is identical at all points along the tube. For the sector scanner, the angle θ between the beam and direction of motion varies from the left to the right side of the image. **(b)** The colour Doppler image shows variation in the displayed colour throughout the length of the tube, with no colour shown at 90°. **(c)** The power Doppler image shows little variation of displayed colour, except for 90°, when no colour is displayed.

to the velocity of the blood in the scan plane. Similarly, the ideal power Doppler image would provide a display in which the colour was related to the presence or absence of moving blood. There are two phenomena that limit the ability of the technology to provide these ideal displays: angle dependence and aliasing.

The Doppler shift arises primarily from blood motion in the direction of the ultrasound beam; this leads to the cosine dependence on the angle between the beam and the direction of motion, when Doppler frequency is calculated (see Chapter 7). Consequently, colour Doppler demonstrates an angle dependence which can be demonstrated using a flow phantom as shown in Figure 10.17a. The flow on the left side of the image is displayed as towards the transducer, in red, and the flow on the right is displayed as away from the

transducer, in blue (Figure 10.17b). At first glance, the image gives the appearance that the flow is changing direction midway across the image; however, careful consideration of the changing angle of insonation allows the observer to establish that the flow is all in one direction. The flow in the centre of the image is not detected due to the poor Doppler angle resulting in small Doppler shift frequencies that are removed by the clutter filter. The power Doppler image maintains a uniform colour over a wide range of angles (Figure 10.17c); however,

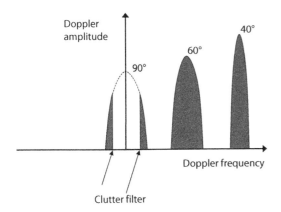

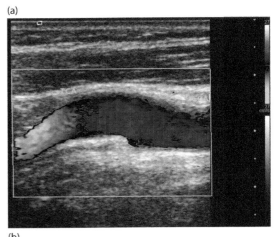

Figure 10.18 Effect of a change in angle on the Doppler spectrum, at constant blood velocity. As the angle increases from 40° to 60°, the Doppler frequency shift decreases, but the power of the Doppler signal, represented by the area under the curve, remains approximately constant. However, at angles close to 90°, the Doppler frequencies are low and may be removed by the clutter filter, which causes a reduction in the Doppler power. In the figure, there is only partial removal of the Doppler signal at 90°.

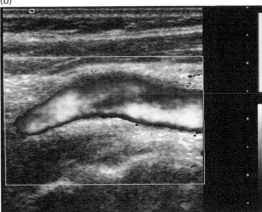

Figure 10.19 Display of a tortuous carotid artery using colour Doppler and power Doppler. There are large changes in blood flow direction along the course of the carotid artery. This results in large changes in the displayed colour for the colour Doppler display (a), but only minor changes in colour for the power Doppler display (b).

when the angle approaches 90°, the power signal may be lost. This may be understood with reference to Figure 10.18, which shows the received Doppler signal at different angles. As the angle increases the Doppler frequencies fall, but the total Doppler power, indicated by the area under the curve, remains constant. Near 90°, there is some loss of signal due to the clutter filter, and the power reduces. The angle dependence of directional power Doppler is similar to that for power Doppler, with the difference that flows towards and away from the transducer are coded with different colours. Display of tortuous vessels is often confusing using colour Doppler due to the angle dependence (Figure 10.19), whereas the corresponding power Doppler image has a uniform hue and is less confusing.

Increase in blood velocity results in increase in Doppler frequency shift up to a maximum value set by the Nyquist limit (PRF/2). For frequency shifts above the Nyquist limit, there are two consequences of aliasing; the Doppler frequencies are inaccurately calculated, and the direction of flow is inaccurately predicted.

As the Doppler frequency is not estimated in power Doppler, aliasing will not affect the displayed image. An alternative way of understanding this is illustrated in Figure 10.20, which shows that the Doppler shift increases up to a critical velocity. Above this critical velocity aliasing occurs. However, in all cases the area under the curve, which represents the Doppler power, is the same. However, directional power Doppler does suffer from aliasing as directional information is calculated.

Both angle dependence and aliasing can occur in the same image. A flow model of a diseased

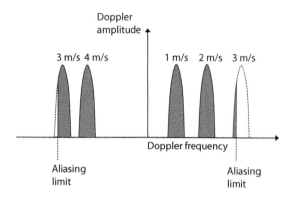

Figure 10.20 Colour aliasing explanation. As velocity increases from 1 to 2 m s^{-1} the Doppler shift increases. However, when the aliasing limit (or Nyquist limit) is reached, at approximately 3 m s^{-1}, in the figure, the Doppler frequency is not estimated correctly. The Doppler signal for the velocity of 3 m s^{-1} consists of two components, one with a positive Doppler frequency and one with a negative Doppler frequency. The combined Doppler power, represented by the area under the curve of the combined signals, remains unchanged. Consequently, the power Doppler image is insensitive to aliasing.

artery may be used to illustrate these effects (Figure 10.21a). In Figure 10.21b, the increase in mean Doppler frequency within the region of the stenosis is seen as an orange-coloured area, and there is jet formation and recirculation in the post-stenotic region. Increase in flow rate (Figure 10.21c) leads to aliasing, where the previous orange-coloured region is now coloured green. The corresponding power Doppler images are uniformly coloured. At the low flow rate (Figure 10.21b), there is a gap in the region of recirculation, where the velocity is low. The gap is filled when the higher flow rate is used (Figure 10.21c). The practical consequence of this last observation is that considerable care must be taken when using poor filling of the vessel in the power Doppler image as evidence of thrombus.

Display of rapidly changing flow patterns

The ability of the colour Doppler image to follow faithfully the changing flow pattern is determined by frame rate and persistence. As previously noted, the frame rate is maximised by the use of a small ensemble length, restriction of the colour box size and (in some systems) by the simultaneous acquisition of multiple beams. For display of lower velocities, a larger ensemble length is required, and the frame rate reduces accordingly. A degree of persistence is acceptable and also desirable as the effect of noise is reduced, and the visualisation of small vessels is improved. Although it is possible to observe changes in blood flow during the cardiac cycle using the colour flow image, this task is best performed using spectral Doppler.

In power Doppler, there is no information on the dynamic nature of blood flow, and persistence is set high to obtain maximum noise reduction.

ARTEFACTS

Many of the features of B-mode images are applicable to colour flow images, as the propagation of the ultrasound pulse through the tissue will obey the same physics, whether it is used to produce a B-mode image or a colour flow image. Some of these artefacts have been described earlier in this chapter. The purpose of this section is to list all the major artefacts in one location. Papers which provide further useful reading on colour flow artefacts are those by Hoskins and McDicken (1997), Nilsoon (2001), Kamaya et al. (2003), Arning and Eckert (2004), Campbell et al. (2004) and Rubens et al. (2006).

Shadowing

There is reduction in the amplitude of the Doppler signal whenever there is attenuation of the ultrasonic pulse. Hence, colour signal is lost when there is an intervening high-attenuation region or a region of high reflectivity, such as a calcified area or bowel gas. This is similar to the production of shadows on a B-mode image.

Ghost mirror images

Ghost mirror images may be produced by partial reflection of the beam from a highly reflecting surface.

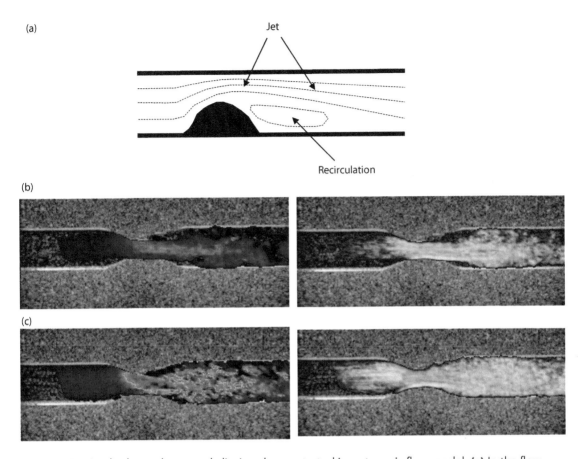

Figure 10.21 Angle dependence and aliasing demonstrated in a stenosis flow model. **(a)** In the flow model, there is a localised narrowed region that mimics a stenosis. A blood-mimicking fluid is pumped from left to right through the tube. As the fluid passes through the stenosis, there is increase in velocity with the formation of a jet. In the post-stenosis region there are regions of recirculation, vortex shedding and turbulence. **(b)** Low flow rate. For the colour Doppler image (left), there is alteration of colour throughout the image; however, the power Doppler image (right) is of a uniform colouration. The jet is clearly seen; however, the fluid in the region of recirculation has low velocity, and the Doppler shift frequencies are suppressed by the clutter filter, resulting in an absence of colour in this region. **(c)** The flow rate is doubled, resulting in a doubling of velocities. There is aliasing at the narrowest point within the stenosis and in the post-stenotic region for colour Doppler (left). The power Doppler image (right) is unaffected by aliasing and remains of a uniform colouration. The Doppler frequencies from the region of recirculation are now high enough to be not suppressed by the clutter filter, and both colour and power Doppler images no longer demonstrate a flow void.

Angle dependence

The displayed colour is dependent on the angle between the beam and the direction of motion as illustrated in Figures 10.17 and 10.19:

- *Colour Doppler*: Displayed colour depends on the cosine of the angle.

- *Power Doppler*: Little angle dependence on angle except near 90°, where the Doppler frequencies fall below the clutter filter if the velocity is too low.
- *Directional power Doppler*: Similar to power Doppler, except it is noted that flows towards and away from the transducer are coded in different colours.

Aliasing

As explained in Chapter 7, the maximum Doppler frequency shift that can be estimated is equal to PRF/2. Higher blood or tissue velocities will be displayed colour-coded with opposite direction:

- *Colour Doppler and directional power Doppler*: Both suffer from aliasing
- *Power Doppler*: Does not suffer from aliasing

Drop-out

This is loss of colour due to the variable nature of the calculated mean frequency or power. If this is high, it is possible that the estimated mean frequency or power will fall below the threshold value used in the blood-tissue discriminator. When this happens, the system does not display colour. For colour flow, the effect is most marked at low velocities and in small vessels.

Noise

There are several types of noise present on the colour image:

- *Electronic noise*: This is produced within the colour flow system electronics. If the colour gain is set too high, the noise will be displayed as colour in regions of tissue in which there is no flow.
- *Clutter breakthrough 1*: Moving tissues (cardiac motion, vessel wall motion, bowel movement) produce Doppler shifts which may be above the level of the clutter filter producing patterns of colour not associated with blood flow within regions of tissue.
- *Clutter breakthrough 2 (twinkling artefact)*: Random colour signals often with a long tail may be observed arising at heavily calcified regions such as kidney stones (Rahmouni et al. 1996; Kamaya et al. 2003). This artefact is thought to be caused by phase jitter within the Doppler system. When the echo amplitude is especially high, as it is from calcifications, this phase jitter breaks through and appears as a false colour signal. The artefact may be removed by suitable adjustment of the colour-write priority.

- *Audio sound*: Sound that is produced within the body is indistinguishable from the Doppler shifts produced from blood and tissue. The sound is detected and displayed as regions of colour within tissues. This most obviously occurs during scanning of the neck when the patient speaks; the audio waves pass through the adjacent tissues and are detected by the colour flow system. Sound in the form of 'bruits' arising from turbulent flow in diseased arteries also gives rise to colour noise within tissues.
- *Flash artefacts*: These are false areas of colour on the colour flow image, which are produced when there is movement of the transducer with respect to the tissue. Some systems are able to remove these artefacts by use of a 'flash filter'.
- *Speckle*: The variation in the autocorrelator estimate of mean frequency and power gives rise to a noise superimposed on the underlying colour and power Doppler images; this noise is called 'colour speckle'. This speckle pattern may be reduced by the use of persistence.

Colour display at vessel-tissue boundaries

Ideally the power Doppler image would show a uniform colour up to the edge of the vessel. For colour Doppler, it is known that the blood velocities are low at the edge of the vessel, so that the displayed colour should show this. In practice, there are a number of effects that will lead to incorrect display of colour:

- *Partial volume effect*: At the edge of vessels, the colour sample volume is located partially within the vessel and partially in the tissue. This effect will lead to reduction in Doppler signal amplitude which will cause a change in displayed colour seen on power Doppler images. For colour Doppler, it is the mean frequency in that part of the sample volume that is located within the vessel which is displayed, so the displayed colour is not affected (Figure 10.22).
- *Image smoothing*: If there is any smoothing in the colour image, by averaging of adjacent pixels or by interpolation, this leads to false colours at the edge of vessels for both colour Doppler and power Doppler.

(a)

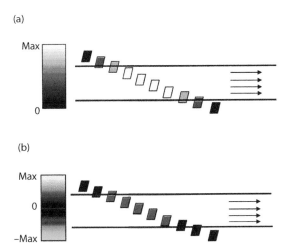

(b)

Figure 10.22 Partial volume effect. In the vessel, it is assumed that all of the blood is moving at the same velocity. **(a)** Power Doppler – change in colour at the edge of a vessel. **(b)** Colour Doppler – no effect on the displayed colour.

- *Clutter filter and blood-tissue discrimination*: Both of these will act to prevent the display of colour at the edge of vessels as the velocities (and hence Doppler frequency shifts) are low, and the tissue signal strength is high.

MEASUREMENTS

This short section covers measurements, which are occasionally made from the colour flow image by clinical users, though it is noted that the vast majority of quantitative measurements are made using spectral Doppler. In research studies, quantitative analysis of the colour flow image using offline computer analysis is widely performed; however, these techniques have yet to impact on clinical practice.

Single-site velocity measurement

Some colour flow systems have the capability of showing the mean-frequency value at a specific location chosen by the operator. This can be converted to a velocity using the same angle-correction techniques that are used in spectral Doppler. This information is occasionally useful in clinical research studies, e.g. estimation of the degree of arterial stenosis from peak velocity obtained from

the colour Doppler image rather than the spectral Doppler waveform.

Quantitative analysis of flow patterns

Blood flow patterns are known to alter considerably in disease, as described in Chapter 8. However, there is little attempt to use the colour Doppler images to provide quantitative information on the colour flow patterns in disease, as there are not yet methods of quantification which have been shown to be clinically useful.

Volume flow

Calculation of volumetric flow requires estimation of the vessel cross-sectional area and the mean velocity. Ideally, the mean velocity and cross-sectional area should be estimated throughout the cardiac cycle to account for the expansion of the arteries which occurs during the cardiac cycle. If colour flow images are obtained with the vessel imaged in the longitudinal plane, the velocity profile of the vessel can be obtained from the colour flow image, and the diameter obtained from the B-mode image. Estimation of the mean velocity requires an assumption that flow is symmetric within the vessel (i.e. all points at the same radius have the same velocity). Cross-sectional area is obtained from measured diameter assuming that the vessel has a circular cross section. Multiplication of the measured area and mean velocity give the volume flow.

This technique makes a number of assumptions, such as circular vessel, symmetric flow patterns, which limit its use to normal or relatively undiseased vessels. In practice, there is little call for volumetric flow measurement, and this method is not widely used.

TIME-DOMAIN SYSTEMS

The essential features of this technique are described in Chapter 7, where it is noted that the change in target depth between consecutive echoes is estimated. Target velocity is then calculated by dividing the change in depth by the pulse repetition interval. The time-domain approach was initially described by Bonnefous and Pesque (1986). It has been used in commercial colour flow systems, and is used for tissue Doppler on some systems.

The time delay is calculated by comparing the echo pattern of consecutive transmission pulses by a mathematical technique called 'cross-correlation'. This process is carried out by sliding one line of echoes past the other in a series of steps or time shifts, comparing the lines at each step. The time shift which gives the closest correlation for each part of the line gives a measure of how the corresponding target has moved between pulses. This method relies on the detailed echo pattern shifting as a whole, but not changing in its overall shape. However, as a region of tissue or blood moves, the echo pattern will change in overall shape as it moves farther away from its original site. This is associated with changes in the relative position of the scatterers as the blood or tissue moves, and also changes in the location of the red cells with respect to the transducer. The change in echo shape is called 'decorrelation', and after a certain distance the echo pattern has changed so much from the original echo pattern that the ultrasound machine cannot measure the time difference between the current and previous echo.

The time-domain method has a number of features which lead to differences in performance compared to the phase-domain method:

- *Aliasing*: As the time delay between pulses is measured, the technique does not suffer from aliasing. The upper limit of velocity detection is associated with the decorrelation previously described.
- *Accuracy*: The time-domain method calculates velocity more accurately than the autocorrelator, for the same ensemble length. Consequently, for the same accuracy, the time-domain approach requires a smaller ensemble size, with gains to be made in either frame rate or line density. However, modern colour flow systems use '2D autocorrelation', and the accuracy for this technique is comparable with the time-domain approach, so that this advantage is no longer present.

The time-domain technique calculates the movement along the beam. In other words, only the component of velocity in the direction of the beam is calculated. The time-domain technique

is, therefore, dependent on the angle between the beam and the direction of motion.

QUESTIONS

Multiple Choice Questions

Q1. Colour flow appears within a box rather than over the whole field of view in order to:
   a. Reduce clutter filter values
   b. Increase frame rate
   c. Make the image look nicer
   d. Increase B-mode penetration depth
   e. Decrease frame rate

Q2. Which ones do not lead to increase in frame rate?
   a. Reduce the depth of the bottom of the colour box
   b. Reduce the width of the colour box
   c. Reduce the colour-line density
   d. Reduce the vertical length of the colour box by lowering the top of the colour box
   e. Reduce the number of colour pulses per line

Q3. Concerning the mean Doppler frequency in colour flow:
   a. The maximum rather than the mean frequency is estimated
   b. All the frequency components are estimated then averaged to obtain the mean frequency
   c. The mean frequency is estimated directly
   d. The mean frequency is estimated using a Fourier transform method or a variant
   e. The mean frequency is estimated using an autocorrelator or a variant

Q4. Which are components of the colour flow processor?
   a. Transmit beam-former
   b. Blood-tissue discriminator
   c. Optical coherence estimator
   d. Clutter filter
   e. Demodulator

Q5. The blood-tissue discriminator:
   a. Tries to put colour flow in areas of true blood flow
   b. Chooses whether to display colour flow data or B-mode data at each pixel in the final image

c. Tries to put B-mode data in areas of true blood flow

d. Tries to put colour flow in areas where there is no blood flow

e. Tries to put B-mode data in areas where there is no blood flow

Q6. The outputs of the autocorrelator are:
   a. Mean frequency
   b. Variance
   c. Maximum frequency
   d. Power
   e. Elastic modulus

Q7. Which of the Doppler modes suffers from aliasing:
   a. Colour Doppler
   b. B-mode
   c. Power Doppler
   d. Spectral Doppler from a continuous-wave system
   e. Spectral Doppler from a pulsed-wave system

Q8. Concerning power Doppler:
   a. The colour image is of mean Doppler frequency
   b. The change in colour at the end of a vessel is due to the partial volume effect
   c. The colour image is of maximum Doppler frequency
   d. The displayed colour is highly dependent on angle between beam and direction of motion
   e. It does not suffer from aliasing

Q9. Concerning colour Doppler:
   a. The colour image is of mean Doppler frequency
   b. The change in colour at the end of a vessel is due to the partial volume effect
   c. The colour image is of maximum Doppler frequency
   d. The displayed colour is not dependent on angle between beam and direction of motion except near 90°
   e. It does not suffer from aliasing.

Q10. Clutter breakthrough:
   a. Gives rise to increase in Doppler frequency within vessels
   b. May be reduced by increasing the level of the clutter filter
   c. Gives rise to false colouration in tissues

d. Is not a problem of colour Doppler systems

e. Is only a problem for power Doppler and not for colour Doppler

Short-Answer Questions

Q1. Describe three different colour flow modes and state what quantities are displayed in each case.

Q2. In colour flow what does the 'colour flow processor' do and what are its five component parts?

Q3. Explain why colour flow is often displayed in a colour box rather than filling the whole field of view (FOV).

Q4. What is the purpose of the blood-tissue discriminator?

Q5. What is the effect of increase in colour-box depth and width on colour frame rate, and why?

Q6. Describe angle dependence (i.e. dependence of displayed colour on the angle between beam and direction of motion) and aliasing characteristics for colour Doppler and power Doppler.

Q7. What is 'clutter breakthrough' and when might this occur?

Q8. Name three types of noise which might be seen on colour flow images.

REFERENCES

Arning C, Eckert B. 2004. The diagnostic relevance of colour Doppler artefacts in carotid artery examinations. *European Journal of Radiology*, 51, 246–251.

Bonnefous O, Pesque P. 1986. Time domain formulation of pulse-Doppler ultrasound and blood velocity estimation by cross-correlation. *Ultrasonic Imaging*, 8, 73–85.

Campbell SC, Cullinan JA, Rubens DJ. 2004. Slow flow or no flow? Color and power Doppler US pitfalls in the abdomen and pelvis. *RadioGraphics*, 24, 497–506.

Cobbold RSC. 2007. *Foundations of Biomedical Ultrasound*. Oxford: Oxford University Press.

Evans DH. 2010. Colour flow and motion imaging. *Journal of Engineering in Medicine*, 224, 241–253.

Evans DH, Jensen JA, Nielsen MB. 2011. Ultrasonic colour Doppler imaging. *Interface Focus*, 1, 490–502.

Evans DH, McDicken WN. 2000. *Doppler Ultrasound: Physics, Instrumentation and Signal Processing*. Chichester: Wiley.

Hoskins PR, McDicken WN. 1997. Colour ultrasound imaging of blood flow and tissue motion. *British Journal of Radiology*, 70, 878–890.

Kamaya A, Tuthill T, Rubin JM. 2003. Twinkling artefact on color Doppler sonography: Dependence on machine parameters and underlying cause. *American Journal of Radiology*, 180, 215–222.

Kasai C, Namekawa K, Koyano A, Omoto R. 1985. Real time two-dimensional blood flow imaging using an autocorrelation technique. *IEEE Transactions on Sonics and Ultrasonics*, 32, 458–464.

Loupas T, Peterson RB, Gill RW. 1995a. Experimental evaluation of velocity and power estimation for ultrasound blood flow imaging by means of a two-dimensional autocorrelation approach. *IEEE Transactions on Ultrasonics, Ferroelectrics, and Frequency Control*, 42, 689–699.

Loupas T, Power JT, Gill RW. 1995b. An axial velocity estimator for ultrasound blood flow imaging, based on a full evaluation of the Doppler equation, by means of a two-dimensional autocorrelation approach. *IEEE Transactions on Ultrasonics, Ferroelectrics, and Frequency Control*, 42, 672–688.

Namekawa K, Kasai C, Tsukamoto M, Koyano A. 1982. Real-time blood-flow imaging system utilizing autocorrelation techniques. In Lerski RA, Morley P (Eds.), *Ultrasound' 82*. New York: Pergamon Press. pp. 203–208.

Nilsoon A. 2001. Artefacts in sonography and Doppler. *European Radiology*, 11, 1308–1315.

Rahmouni A, Bargoin R, Herment A, Bargoin N, Vasile N. 1996. Color Doppler twinkling artefact in hyperechoic regions. *Radiology*, 199, 269–271.

Rubens DJ, Bhatt S, Nedelka S, Cullinan J. 2006. Doppler artefacts and pitfalls (reprinted from Ultrasound Clinics, vol 1, 2006). *Radiologic Clinics of North America*, 44, 805–835.

Rubin JM, Bude RO, Carson PL et al. 1994. Power Doppler US: A potentially useful alternative to mean frequency based color Doppler US. *Radiology*, 190, 853–856.

Wells PNT. 1994. Ultrasonic colour flow imaging. *Physics in Medicine and Biology*, 39, 2113–2145.

11

Advanced techniques for imaging flow and tissue motion

PETER R HOSKINS AND ALINE CRITON

INTRODUCTION

It has been noted in previous chapters that the majority of clinical ultrasound practice and clinical ultrasound instrumentation is based on B-mode imaging, colour flow and spectral Doppler. For Doppler techniques Chapters 7, 9 and 10 have described 'classic' techniques used for measurement and imaging of blood flow. These classic techniques (i) use separate transmit and receive beam forming, (ii) are based on the Doppler effect and (iii) are single-beam techniques for which the estimated velocity is dependent on the angle.

This chapter describes a number of techniques available on commercial ultrasound scanners all concerned with imaging and measurement of blood flow and tissue motion, which in one or more ways are different to the classic techniques described in previous chapters. The chapter is broadly divided into flow techniques and tissue-motion techniques. Table 11.1 provides a summary of these techniques.

B-FLOW

As discussed in Chapter 7, the backscatter signal from blood is at a very low amplitude, hence appears dark on B-mode imaging. However, there are echoes present within blood vessels, and the echo pattern moves as a result of blood flow. If the B-mode gain is increased then these patterns can sometimes be seen. Analysis of the moving echoes has been used to obtain information on blood velocities in research studies (Trahey et al. 1987; Bohs and Trahey 1991).

In B-flow imaging the echoes from moving tissues are enhanced. Non-moving regions of blood or tissue are displayed as dark, and regions where the velocity is high are displayed bright (Figure 11.1a). When viewed in real time, or in video format, an impression of flow is obtained. The B-flow method is based on a pair of coded pulses, one pulse the complement of the other. Addition of the echoes from stationary tissues results in cancellation. However, any slight difference in the signal arising

Table 11.1 Angle dependence for various modalities concerned with measurement and visualisation of blood flow and tissue motion

	Modality	Angle dependence
B-flow	Blood flow	No
Vector flow imaging	Blood flow	No
High frame rate colour flow	Blood flow	Yes
High frame rate spectral Doppler	Blood flow	Yes
Small vessel imaging	Blood flow	Yes
Doppler tissue imaging	Tissue motion	Yes
Wall motion measurement	Tissue motion	Yes

from movement will result in a signal which can be displayed. The technique is dependent on the velocity, where higher velocities are displayed brighter, since higher velocities will be associated with greater difference between the echo pair. The B-flow image may be displayed alone (Figure 11.1a), or if desired the B-mode image can be mixed in to produce a more uniform display in which tissue and blood have similar brightness levels (Figure 11.1b). Greyscale B-flow imaging is not quantitative in that it does not measure blood velocity. From a static B-flow image the blood may be hard to distinguish from the tissue, and this is best done by video playback, where movement allows a much clearer distinction of the vessel. B-flow is claimed to have a number of advantages over colour flow, especially that the display is not angle dependent, it does not suffer from aliasing and the frame rate is higher. Several studies have compared B-flow with existing ultrasound modalities in the assessment of degree of stenosis (Bucek et al. 2002; Clevert et al. 2007).

It is also possible to modify the B-flow technique slightly to obtain colour B-flow images. Typically two pairs of pulses are used to generate two B-flow signals which are then processed using the colour flow processor to provide a rough indication of velocity and direction. The estimated velocity is not as accurate as would be obtained using 10–12 pulses typical of true colour flow. The velocity and direction information is then used to colour code the B-flow data (Figure 11.1c). This method is designed to retain the high-resolution, high-frame-rate capability of B-flow, while also allowing easier distinction of the vessels from the tissues by the use of colour.

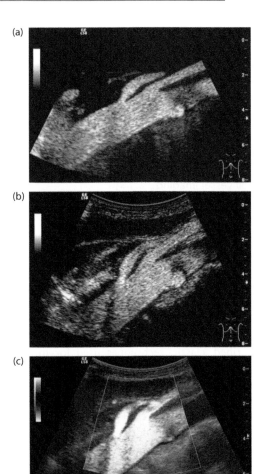

Figure 11.1 B-flow image of the coeliac trunk: (a) background off, (b) background on, (c) colour B-flow with background on. (Images provided courtesy of Dr. Seeger, University Hospital Kiel, Germany.)

VECTOR FLOW IMAGING

Conventional spectral Doppler and colour Doppler systems all suffer from angle dependence, in that the detected Doppler frequency shift is dependent on the angle between the beam and the direction of motion. Measurement of blood velocity in spectral Doppler requires the operator to tell the machine what the angle is, by aligning the angle cursor, allowing the ultrasound machine to provide an estimate of the blood velocity. Often the angle cursor is aligned with the vessel wall, which implicitly assumes that the blood flows parallel to the wall. Vector flow systems allow estimation of blood velocity without the need for manual input of the angle.

Key to understanding vector flow imaging is that blood velocity is a vector. A vector is a quantity which has both direction and magnitude, as opposed to a scalar quantity which has only magnitude. Examples of scalar quantities are mass, volume and temperature, while examples of vector quantities are velocity, acceleration and force. In ultrasound the quantities 'speed' and 'velocity' are often used interchangeably; however strictly speaking speed is the magnitude of velocity, so speed is a scalar quantity and velocity is a vector quantity. A vector quantity has three components at 90° to each other; in an x, y, z coordinate system there are components along each direction. A velocity V therefore has three components; V_x, V_y and V_z. Figure 11.2 shows a blood velocity vector with a component aligned along the Doppler beam, a component at 90° to the beam but in the scan plane and a third component at right angles to the scan plane. Conventional Doppler systems estimate only the component of velocity along the direction of the beam, which leads to the angle dependence as previously discussed.

The simplest way to obtain two velocity components is to use two Doppler beams as illustrated in Figure 11.3. This methodology is commonly referred to as 'vector Doppler'. Each Doppler beam measures the component of velocity along the beam. The true velocity in the scan plane is obtained by mathematically compounding the two components. These techniques have been described for spectral Doppler (Fox 1978) and colour Doppler (Fei et al. 1994; Hoskins et al.

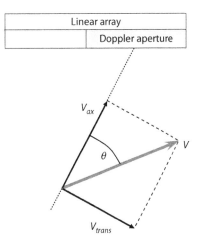

Figure 11.2 Blood velocity vector V decomposed into components along the beam (V_{ax}) and transverse to the beam (V_{trans}).

1994) and are reviewed in Dunmire and Beach (2000) and Jensen et al. (2016a). In principle vector Doppler techniques could be applied to array-based systems, by using a single transmit aperture then two receive apertures (Figure 11.4). While prototype versions of such a system were produced by at least one manufacturer (Steel et al. 2004), this approach has not been adopted commercially.

Figure 11.5 shows examples of vector flow images in different vessels. Commercial implementation of vector flow techniques has been

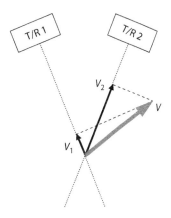

Figure 11.3 Vector Doppler system with two transmit/receive beams; each Doppler system measures a separate velocity component V_1 and V_2.

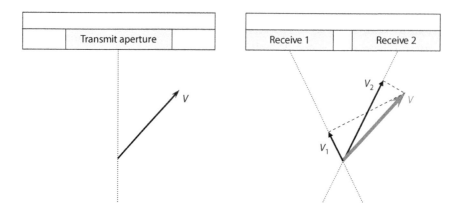

Figure 11.4 Possible implementation of vector Doppler using a linear array; there is a single transmit beam followed by two simultaneous receive beams, each beam providing a separate velocity component V_1 and V_2.

achieved using a technique called 'transverse oscillation' (TO) (Jensen and Munk 1998; Jensen 2000; Jensen et al. 2016a). Consider velocity estimation by the Doppler effect; this relies on creation of an oscillation at the transmit frequency in the direction of the ultrasound beam. There is no oscillation

transverse to the beam hence the velocity component transverse to the beam cannot be estimated. The TO technique involves creation of an oscillation in the transverse direction from which the transverse velocity component can be estimated. Creation of the TO is performed in reception

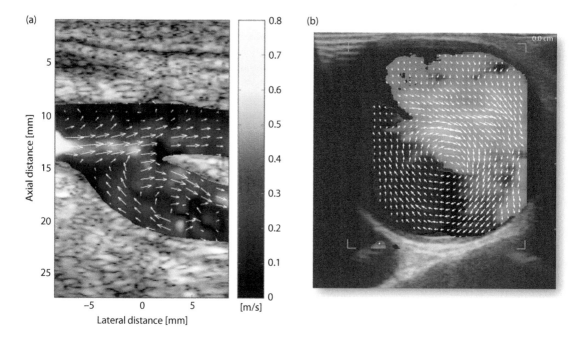

Figure 11.5 Examples of vector flow imaging. (a) Carotid bifurcation. (b) Spiral flow in an abdominal aortic aneurysm. ([a] Reprinted from Udesen J et al. 2007. *Ultrasound in Medicine and Biology*, 33[4], 541–548, Copyright 2007, with permission from the World Federation for Ultrasound in Medicine and Biology; [b] kindly provided by Prof. Jørgen Jensen, Technical University of Denmark, Lyngby, Denmark.)

which produces two beams from which the transverse velocity component is estimated. The transmit beam used in the TO technique is unchanged from conventional Doppler. This means that TO has advantages over vector Doppler in that the field of view and frame rate are larger.

A number of new possibilities arise from vector flow imaging including improved quantification of volumetric flow, and new quantities related to flow patterns which may be useful in diagnosing and quantifying arterial disease.

HIGH FRAME RATE DOPPLER

One of the main issues for conventional colour flow systems is achieving real-time frame rates. Chapter 10 described that for each pixel of the colour flow image 2–20 (typically 10) ultrasound pulses are required to estimate mean Doppler frequency. Compared to B-mode imaging this causes a fall in frame rate typically by a factor of 10. Restriction of the colour field of view (the 'colour box') helps improve frame rate; for superficial vessels such as the carotid and femoral real-time frame rates can generally be achieved. However achieving real-time visualisation is more difficult for deeper arteries, for example in the abdomen.

Achieving very high frame rates requires the use of plane wave imaging techniques, which were introduced in Chapter 3. For basic plane wave imaging technique there is no focusing on transmission, just on reception. This allows generation of very high frame rates of 10,000–20,000 s^{-1}.

A review of high frame rate techniques for colour flow is provided by Jensen et al. (2016b). This section concentrates on the technique described by Bercoff et al. (2011), called 'ultrafast compound Doppler' (UCD). There are two steps in the high frame rate colour flow process:

- *Construct a sequence of N_C compound images:* A series of N_A plane waves at different angles is transmitted and received and a series of low resolution images is formed. A higher resolution compound image is formed from the set of low resolution images. This process is repeated N_C times.
- *Colour flow processing:* Each pixel of the image contains N_C data points. These form the input to the colour flow processor with steps (Chapter 10) of demodulation, clutter filter, mean frequency estimation, etc. The processing for each pixel is performed in parallel in the computer. The value of N_C is typically 100+, comparable to the number of pulses used in spectral Doppler, and much higher than the 2–20 used in conventional colour flow imaging.

The number of angles N_A used to form each compound image is a key determinant of image quality for colour flow. Bercoff et al. (2011) showed that similar image quality to conventional colour flow was obtained when nine angles ($N_A = 9$) were used, but with a seven times higher frame rate. For example, if the frame rate for conventional colour flow is 30 s^{-1}, then for comparable image quality a frame rate of 210 s^{-1} may be achieved using UCD.

The availability of such a large amount of data in UCD means that choices can be made about which aspects of image quality to improve (Table 11.2). Figure 11.6 shows reduction in colour flow noise when frame rate is increased. The use of a greater number of angles leads to increase in signal to noise (and improvement in accuracy of velocity estimation) at the expense of reduction in frame rate.

Table 11.2 summarises typical improvements in UCD over conventional colour flow for clinical applications, with a 10 times improvement in

Table 11.2 Typical improvement in performance for ultrafast compound Doppler over conventional colour flow imaging for selected scenarios in clinical imaging

	Conventional colour flow	Equivalent UCD
Frame rate	20–40 s^{-1}	200–400 s^{-1}
Sensitivity	40 times higher than UCD	–
Minimum detectable velocity	5 mm·s^{-1}	0.5 mm·s^{-1}
Minimum vessel diameter visualised	500 μm	50 μm

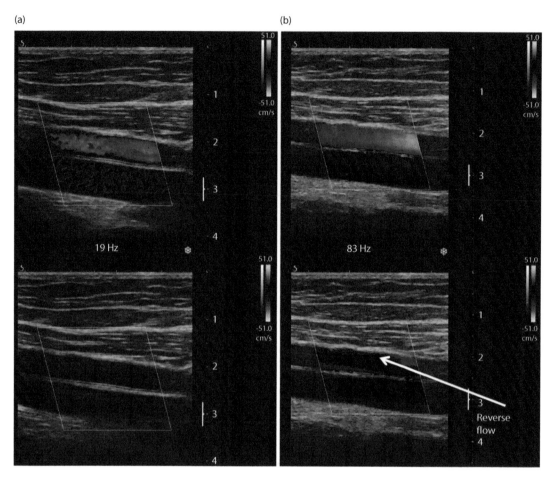

Figure 11.6 Comparison of **(a)** conventional colour flow and **(b)** ultrafast colour flow. There is reduced noise and improved sensitivity for ultrafast colour flow. (Images kindly provided by SuperSonic Imagine, Aix-en-Provence, France.)

frame rate, minimum detectable velocity and minimum vessel diameter visualised, but noting that not all can be achieved at the same time.

It was noted that there are a very large number of pulses for each pixel of the UCD image. The pulse repetition frequency for each pixel is several kilohertz. This is sufficient to allow the production of high-quality spectral data from each pixel; this information is available in the computer and may be accessed retrospectively. In other words the operator chooses locations for display of spectral Doppler data after acquiring the data, not before as is the practice in conventional spectral Doppler. The operator can obtain spectral data from several locations in the same UCD image (Figure 11.7).

Various measurements can be made from the spectral data, such as resistance index or pulsatility index (see Chapter 9). As spectral data are available in every pixel of the UCD image resistance index or pulsatility index may be estimated automatically for every pixel of the image and displayed as a colour image (Demené et al. 2014).

SMALL VESSEL IMAGING

Using conventional colour flow ultrasound the smallest diameter vessel which can be visualised is around 0.5 mm. A number of techniques have been developed which enable visualisation of vessels with diameters down to around 50 micron, a 10 times

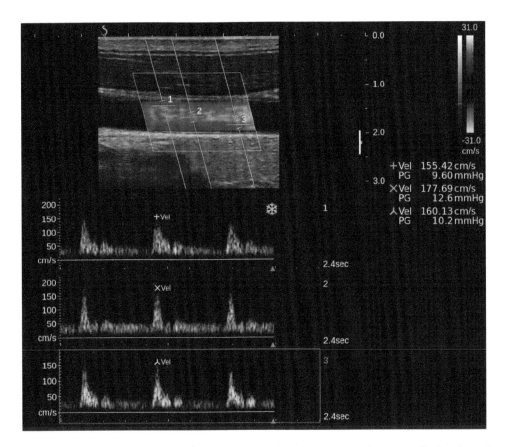

Figure 11.7 Spectral Doppler from high frame rate Doppler showing simultaneous display from three sites. (Images kindly provided by SuperSonic Imagine, Aix-en-Provence, France.)

improvement over conventional colour flow. This section briefly describes small vessel colour flow imaging, noting that manufacturers are loath to provide much in the way of technical details on this.

In conventional colour flow the clutter filter operates to suppress the large clutter component. In its simplest form the filter operates on the basis of Doppler frequency; frequencies below a threshold value are suppressed while those above the threshold are left unaltered. (In practice the clutter filter does not have a step-function response but there is a roll-off response centred around the threshold.) The result is that the large-amplitude clutter signal is suppressed. However the low-velocity signals are also suppressed. Blood velocities fall for smaller vessels; for vessels with diameters below 0.5 mm the mean velocity is only 0.5–1 cm.s^{-1}, hence the difficulty in visualising flow in small vessels in conventional colour flow.

Small vessel imaging therefore relies on improvements in the ability to distinguish clutter from flow at low velocities. Commercial methods appear to fall into two main areas; one is tinkering with conventional colour flow and the other is high frame rate Doppler.

Tinkering with the conventional colour flow beam-former and processor could involve continuous low frame rate power Doppler to provide information on tissue motion. The tissue motion data could then be used in real time as part of an adaptive clutter filter, though as noted previously the exact details of this remain unclear.

The previous section discussed high frame rate Doppler and the huge increase in the number of pulses compared to conventional Doppler. Simultaneously a large number of Doppler samples are acquired over the two-dimensional (2D) image and at a high frame rate. This allows the design of improved clutter filters based on temporal filtering and spatial filtering (Demené et al. 2015). This reduces the minimum velocity which can be detected from around $5 \text{ mm} \cdot \text{s}^{-1}$ (conventional colour flow) to around $0.5 \text{ mm} \cdot \text{s}^{-1}$ (Table 11.2) and hence allows visualisation of smaller vessels down to around 50 micron. The ability to visualise small vessels in the brain has led to the development of 'functional ultrasound' (mirroring 'functional magnetic resonance imaging') concerned with measuring changes in brain activity associated with changes in blood flow (Mace et al. 2011; Demené et al. 2018).

DOPPLER TISSUE IMAGING

The motion of tissues is traditionally observed using B-mode and M-mode imaging. These methods have been most commonly used to assess the motion of heart valves and the heart wall. It is also possible to obtain information on motion by using the Doppler effect, where it is known as Doppler tissue imaging (DTI) (McDicken et al. 1992). The DTI mode is a modification of the colour Doppler mode previously described for blood flow. The pulse repetition frequency and hence the velocity scale are lowered in order to account for the much lower velocities present within the tissues compared to the blood. Lowering the sensitivity results in recording of signals from the moving tissues rather than from blood, which appears as black on the DTI image. The clutter filter is deactivated as there is no longer the necessity to reject the clutter signal (as there is when blood flow signals are detected). The ensemble length (number of pulses) is also lowered. An example of a DTI image from the heart is seen in Figure 11.8. On modern ultrasound machines it is not necessary to adjust the colour flow controls in order to activate DTI. Instead the operator activates DTI directly, and sensitivity, velocity scale and other machine settings are automatically updated.

In addition to real-time 2D imaging, it is possible to acquire DTI M-mode data. This is analogous to the greyscale M-mode trace. In DTI M-mode, the ultrasound beam is fixed in a direction of interest through the tissues and the velocities at different depths are presented on the screen on a vertical

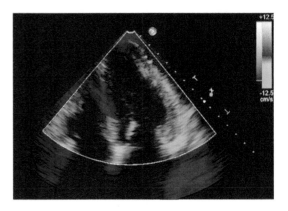

Figure 11.8 DTI of the heart.

line, which is swept across the screen (Figure 11.9) (Fleming et al. 1996). Having obtained the velocity information at each pixel in the image, it is possible to estimate how much the tissue has been stretched or compressed (strain) and how fast the tissue is stretched or compressed (strain rate) (Fleming et al. 1996; Abraham et al. 2007). In clinical use the DTI technique is mostly used in cardiac applications (Abraham et al. 2007), though it is possible to use DTI for the study of arterial motion.

Although DTI has been available on commercial ultrasound systems for over two decades it is not widely used in clinical practice. A large amount of velocity information is presented when real-time DTI is acquired, and it is often necessary to replay this slowly to appreciate the wall motion. The DTI mode also suffers from some of the same limitations as pulsed-wave Doppler for blood flow, in that it is possible to generate aliasing, and there is an angle dependence of the displayed data.

WALL MOTION MEASUREMENT

Arteries expand and contract as a result of the change in blood pressure from systole to diastole. Typically arteries change diameter by about 10% during the cardiac cycle, so for a carotid artery of 5 mm diameter the change in diameter is about 0.5 mm or 500 μm. While this motion can easily be seen on the B-mode or M-mode display, accurate measurement is not possible as the spatial resolution is close to the change in diameter.

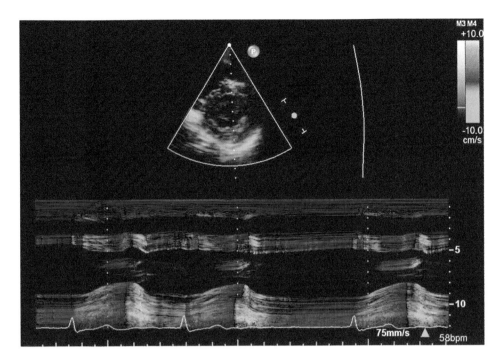

Figure 11.9 DTI M-mode scan of the heart.

Measurement of wall motion has a long history with the first reports dating back to 1968 involving measurement of A-line data on an oscilloscope (Arndt et al. 1968). To obtain high accuracy measurements of wall motion it is necessary to use the RF data. Early techniques involved tracking of the RF data (Hokanson et al. 1970), hence the term 'wall tracking'. Modern techniques involve a comparison of consecutive A-line data using correlation techniques, where accuracies of 1–10 μm can be achieved (Rabben et al. 2002). Figure 11.10 shows wall motion measured in the carotid artery.

Wall motion has been used to estimate the stiffness of arteries. There are two indices which may be estimated; the Young's modulus E (see Chapter 14 for further details of 'stiffness') and the pressure strain elastic modulus E_p (Equations 11.1 and 11.2). Typically pressure P is estimated from an arm cuff measurement while diameter d, change in diameter and wall thickness h are measured from ultrasound. Studies have shown changes in stiffness as a result of ageing and disease such as diabetes or atherosclerosis (reviewed in Hoskins and Bradbury 2012):

$$E = \frac{d_d}{2h} \cdot \frac{(P_s - P_d)}{(d_s - d_d)/d_d} \quad (11.1)$$

where subscripts s and d are 'systole' and 'diastole'.

$$E_p = \frac{P_s - P_d}{(d_s - d_d)/d_d} \quad (11.2)$$

Measurement of wall motion is rarely used in clinical practice; however a number of commercial systems incorporate this functionality (e.g. Figure 11.10).

QUESTIONS

Multiple Choice Questions

Q1. Which of the following techniques display information related to blood velocity:
   a. Wall motion display
   b. B-flow
   c. Vector Doppler
   d. Doppler tissue imaging
   e. Colour Doppler

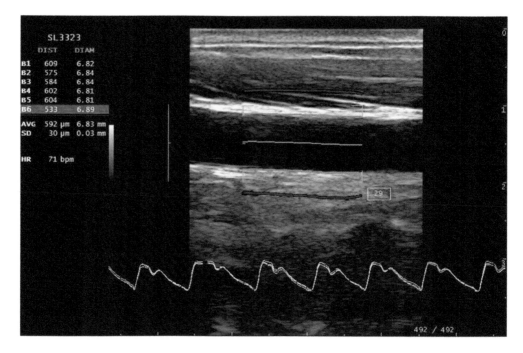

Figure 11.10 Wall motion from the carotid artery. The orange lines show the position of the arterial walls; the blue lines show the motion (magnified). The lower trace shows motion with time for each wall. (Images kindly provided by Dr. Peter Brands, Esaote Group, Genoa, Italy.)

Q2. Which of the following has a cosine dependence for the displayed quantity:
    a. Vector Doppler
    b. Colour Doppler
    c. B-flow
    d. Power Doppler
    e. Doppler tissue imaging

Q3. Vector Doppler techniques implemented using a linear array allow measurement of:
    a. Both axial and transverse blood velocity components
    b. Only axial velocity component
    c. Only transverse component
    d. All three velocity components
    e. No velocity components

Q4. Comparing high frame rate colour flow and conventional colour flow; for comparable image quality frame rate is:
    a. Higher by a factor of 1,000
    b. Higher by a factor of 100
    c. Higher by a factor of 10

    d. The same
    e. Lower by a factor of 10

Q5. For visualisation of small vessels the minimum vessel diameter which can be visualised is:
    a. 0.5 mm for ultrafast Doppler
    b. 0.5 mm for conventional colour flow
    c. 50 micron for ultrafast Doppler
    d. 50 micron for conventional colour flow
    e. 5 micron for ultrafast Doppler

Q6. Concerning ultrafast Doppler:
    a. Involves the use of plane waves on transmission
    b. Involves stepping of the transmit beam along the array on transmission
    c. Involves typically 1,000+ Doppler data points per pixel
    d. Allows retrospective display of spectral Doppler at any point in the 2D image
    e. Requires display of colour in a box superimposed on the 2D image to increase frame rate

Q7. Compared to colour Doppler, Doppler tissue imaging has:
   a. Lower velocity scales as tissue has lower velocities than blood
   b. Display of blood flow and tissue motion simultaneously
   c. Reduced sensitivity as the Doppler signal from tissue is much higher than from blood
   d. The same sensitivity as the Doppler signal from tissue and blood are comparable
   e. High-velocity scales as tissue has higher velocities than blood

Q8. Measurement of arterial wall motion typically requires accuracy of motion detection of:
   a. 0.1–1 micron
   b. 1–10 micron
   c. 10–100 micron
   d. 100–1000 micron
   e. 1–10 mm

Q9. Estimation of elastic modulus of the wall of the carotid artery requires measurement of:
   a. Diameter and pressure only
   b. Blood velocity and diameter only
   c. Flow rate and blood velocity only
   d. Change in diameter and pressure only
   e. Diameter, change in diameter and pressure only

Short-Answer Questions

Q1. Briefly describe how a vector Doppler system estimates true blood velocity and direction of motion.

Q2. Briefly describe how a B-flow image is produced.

Q3. Briefly describe how a high frame rate colour flow image can be produced using plane wave imaging.

Q4. Briefly describe how high frame rate Doppler can be used to visualise small vessels in the microcirculation.

Q5. Briefly describe how a Doppler tissue imaging system has been optimised compared to colour Doppler.

Q6. Briefly describe how elastic modulus of a carotid artery can be measured using ultrasound.

REFERENCES

Abraham TP, Dimaano VL, Liang HY. 2007. Role of tissue Doppler and strain echocardiography in current clinical practice. *Circulation*, 116, 2597–2609.

Arndt JO, Klauske J, Mersch F. 1968. The diameter of the intact carotid artery in man and its change with pulse pressure. *Pfluegers Archiv*, 301, 230–240.

Bercoff J, Montaldo G, Loupas T, Savery D, Mézière F, Fink M, Tanter M. 2011. Ultrafast compound Doppler imaging: Providing full blood flow characterization. *IEEE Transactions on Ultrasonics, Ferroelectrics, and Frequency Control*, 58, 134–147.

Bohs LN, Trahey GE. 1991. A novel method for angle independent ultrasonic-imaging of blood-flow and tissue motion. *IEEE Transactions on Biomedical Engineering*, 38, 280–286.

Bucek RA, Reiter M, Koppensteiner I et al. 2002. B-flow evaluation of carotid arterial stenosis: Initial experience. *Radiology*, 225, 295–299.

Clevert DA, Johnson T, Jung EM et al. 2007. Color Doppler, power Doppler and B-flow ultrasound in the assessment of ICA stenosis: Comparison with 64-MD-CT angiography. *European Radiology*, 17, 2149–2159.

Demené C, Deffieux T, Pernot M et al. 2015. Spatiotemporal clutter filtering of ultrafast ultrasound data highly increases Doppler and ultrasound sensitivity. *IEEE Transactions on Medical Imaging*, 34, 2271–2285.

Demené C, Mairesse J, Baranger J, Tanter M, Baud O. 2018 April 10. Ultrafast Doppler for neonatal brain imaging. *NeuroImage*. Available at: https://doi.org/10.1016/j.neuroimage.2018.04.016.

Demené C, Pernot M, Biran V, Alison M, Fink M, Baud O, Tanter M. 2014. Ultrafast Doppler reveals the mapping of cerebral vascular resistivity in neonates. *Journal of Cerebral Blood Flow and Metabolism*, 34, 1009–1017.

Dunmire B, Beach KW. 2000. Cross-beam vector Doppler ultrasound for angle-independent velocity measurements. *Ultrasound in Medicine and Biology*, 26, 1213–1235.

Fei DY, Fu CT, Brewer WH, Kraft KA. 1994. Angle independent Doppler color imaging: Determination of accuracy and a method of display. *Ultrasound in Medicine and Biology*, 20, 147–155.

Fleming AD, Palka P, McDicken WN, Fenn LN, Sutherland GR. 1996. Verification of cardiac Doppler tissue images using grey-scale M-mode images. *Ultrasound in Medicine and Biology*, 22, 573–581.

Fox MD. 1978. Multiple crossed-beam ultrasound Doppler velocimetry. *IEEE Transactions on Sonics and Ultrasonics*, SU-25, 281–286.

Hokanson DE, Strandness DE, Miller CW. 1970. An echotracking system for recording arterial wall motion. *IEEE Transactions on Sonics and Ultrasonics*, SU-17, 130–132.

Hoskins PR, Bradbury AW. 2012. Wall motion analysis. In: Nicolaides A, Beach KW, Kyriakou E, Pattichis CS (Eds.), *Ultrasound and Carotid Bifurcation Atherosclerosis*. New York: Springer. pp. 325–339.

Hoskins PR, Fleming A, Stonebridge P, Allan PL, Cameron DC. 1994. Scan-plane vector maps and secondary flow motions. *European Journal of Ultrasound*, 1, 159–169.

Jensen JA. 2000. Algorithms for estimating blood velocities using ultrasound. *Ultrasonics*, 38, 358–362.

Jensen JA, Munk P. 1998. A new method for estimation of velocity vectors. *IEEE Transactions on Ultrasonics Ferroelectrics and Frequency Control*, 45, 837–851.

Jensen JA, Nikolov SI, Yu AC, Garcia D. 2016a. Ultrasound vector flow imaging—Part I: Sequential systems. *IEEE Transactions on Ultrasonics, Ferroelectrics, and Frequency Control*, 63, 1704–1721.

Jensen JA, Nikolov SI, Yu AC, Garcia D. 2016b. Ultrasound vector flow imaging—Part II: Parallel systems. *IEEE Transactions on Ultrasonics, Ferroelectrics, and Frequency Control*, 63, 1722–1732.

Mace E, Montaldo G, Cohen I, Baulac M, Fink M, Tanter M. 2011. Functional ultrasound imaging of the brain. *Nature Methods*, 8, 662–664.

McDicken WN, Sutherland GR, Moran CM, Gordon LN. 1992. Colour Doppler velocity imaging of the myocardium. *Ultrasound in Medicine and Biology*, 18, 651–654.

Rabben SI, Bjaerum S, Sorhus V et al. 2002. Ultrasound-based vessel wall tracking: An auto-correlation technique with RF center frequency estimation. *Ultrasound in Medicine and Biology*, 28, 507–517.

Steel R, Ramnarine KV, Criton A, Davidson F, Allan PL, Humphries N, Routh HF, Fish PJ, Hoskins PR. 2004. Angle-dependence and reproducibility of dual-beam vector Doppler ultrasound in the common carotid arteries of normal volunteers. *Ultrasound in Medicine and Biology*, 30, 271–276.

Trahey GE, Allison JW, von-Ramm OT. 1987. Angle independent ultrasonic-detection of blood-flow. *IEEE Transactions on Biomedical Engineering*, 34, 965–967.

Three-dimensional ultrasound

PETER R HOSKINS AND TOM MACGILLIVRAY

INTRODUCTION

Three-dimensional (3D) imaging techniques such as computed tomography (CT), magnetic resonance imaging (MRI) and positron emission tomography (PET), will be familiar to the modern imaging specialist. The strength of ultrasound imaging lies in its real-time ability and, as this book has discussed to this point, this has been based on two-dimensional (2D) imaging. The operator moves the 2D transducer around and, where necessary, builds up a 3D map in his or her own head of the 3D structures in the body. Many modern ultrasound systems now come with a 3D scanning option which is available for the operator to use. Clinical uses for these systems are becoming established, mainly in obstetrics and cardiology as discussed later. The purpose of this chapter is to describe the technology of 3D ultrasound, including examples of clinically useful applications and measurements. Further reading, on the history and technology of 3D ultrasound, may be found in review articles by Fenster et al. (2001) and Prager et al. (2009).

Terminology

The terms '1D', '2D', '3D' and '4D' are used. The 'D' in every case refers to 'dimension'. '1D' is one spatial dimension, in other words a line; for example a 1D transducer consists of a line of elements. '2D' is two spatial dimensions, which is an area. A 2D transducer consists of a matrix of elements. '3D' is three spatial dimensions, in other words a volume. 3D scanning refers to the collection of 3D volume data. '4D' is three spatial dimensions and one time dimension. 4D scanning refers to the collection of several 3D volumes over a period of time. Examples include the visualisation of the heart during the cardiac cycle, or of the fetus during fetal movement. Table 12.1 summarizes this terminology.

In addition to the spatial and time dimensions, there is the issue of velocity measurement through Doppler ultrasound. Blood flow is complex and the velocity of blood at any particular

Table 12.1 Dimension terminology

Term	Description
1D	Line
2D	Area
3D	Volume
4D	Data set consisting of several volumes collected over a period of time

point in the vessel generally requires knowledge of all three components of velocity (along *x*, *y* and *z*). Conventional clinical ultrasound systems provide only one component of velocity, in the direction of motion of the transducer, as discussed in Chapters 7, 9 and 10. This limitation is also true for 3D ultrasound in that it too only obtains one component of velocity, in the direction of the ultrasound beam.

3D/4D ULTRASOUND SYSTEMS

These systems can be divided into two types which depend on the ultrasound transducer used.

- *2D-array transducers:* The operator positions the transducer and the ultrasound beam is electronically swept through a 3D volume.
- *Conventional ultrasound transducers:* These steer the beam within a 2D plane; hence collection of 3D data is achieved by movement of the transducer. The 3D volume is then built up of many 2D scan planes. A common requirement for these systems is that there must be knowledge of where the data are collected from, in order that the 2D slices are put in the correct location within the 3D volume. There are several solutions to this problem depending on the system used.

The following sections describe the systems which are commonly used for collection of 3D/4D data.

Freehand systems

In freehand 3D scanning a conventional ultrasound transducer is swept across the patient's skin (Figure 12.1a). In this system knowledge of the transducer position in space is obtained using a tracking device. The result is a 3D data set consisting of 2D scan planes whose position is correctly allocated in the 3D space (Figure 12.2).

The basic components of the tracking system are a target attached to the transducer, and a remote sensor or receiver several feet away which is able to track the position and orientation of the target in space. Note that it is insufficient just to track the position. At a single location in space the transducer may be pointed in many different directions

so that knowledge of the transducer orientation is also needed. Two commonly used tracking systems are optical and electromagnetic:

- *Electromagnetic systems:* A small transmit device placed near the patient generates a magnetic field. A sensor coil is attached to the ultrasound transducer, and is connected to a receiver. Movement of the coil through the magnetic field induces voltage in the coil from which the receiver estimates the position and orientation in space. These systems do not require line of sight (i.e. a clear uninterrupted path between coil and receiver); however, they may suffer distortions due to the presence of other hospital equipment.
- *Optical systems:* Typically several infrared sources such as LEDs (light-emitting diodes) are attached to the ultrasound transducer (Figure 12.3), and the positions of the LEDs are recorded using two cameras. A processing unit combines the images of the LEDs from the two cameras and provides information on the position and orientation of the LEDs. These systems require clear lines of sight between the transducer and the tracking device. Optical tracking is more widely used for freehand scanning than electromagnetic tracking due to the effect of hospital equipment on electromagnetic systems, as noted previously.

For freehand systems it is the position and orientation of the target (LEDs or coil) which are tracked, not the transducer. As the target and the transducer are rigidly connected, the computer system used in freehand scanning is able to estimate the position and orientation of the 2D scan plane from that of the target. In order to do this the tracking system must be calibrated; i.e. the computer system must be provided with sufficient information to convert target position and orientation coordinates into scan-plane position and orientation coordinates. The simplest calibration method consists of scanning a steel plate, moving the transducer through a series of translations and rotations. More complex calibration methods require dedicated phantoms (Fenster et al. 2001; Mercier et al. 2005).

Freehand systems allow virtually any ultrasound system to be converted at a later stage into a 3D

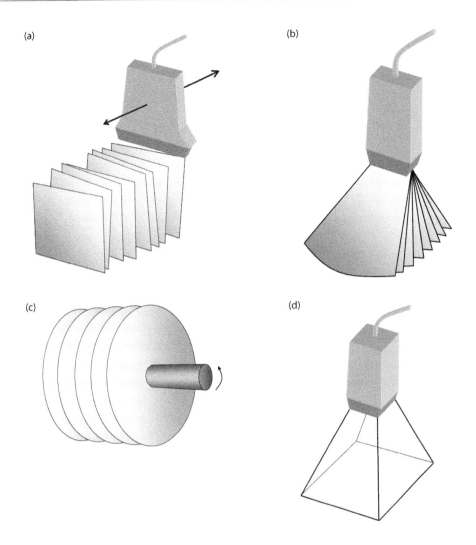

Figure 12.1 Different 3D scanning modes. (a) Freehand 3D scanning where the transducer is swept over the skin by the operator and the 3D volume is built up as a series of 2D scan planes. (b) Mechanically steered array real-time 3D, where the 2D scan planes are swept back and forth through a mechanical motion of the linear array within its housing to produce a series of 3D volumes. (c) Endoprobe 3D ultrasound, where retraction of the ultrasound transducer builds up the 3D data set as a series of 2D scan planes which are assumed to be parallel to each other. (d) Matrix-array 3D ultrasound where the 3D volume is built up by electronic steering of the beam throughout the 3D volume to produce a pyramid-shaped data set.

system. There is no restriction on the type of transducer which may be used, so that systems based on linear arrays, phased arrays and mechanically driven transducers are all suitable for conversion into 3D ultrasound systems using the freehand approach.

The time for acquisition of a 3D volume is typically a few seconds. If the operator moves the transducer too quickly then the distance between adjacent 2D slices may be too great and parts of the 3D image will not be visualised. The use of colour flow slows the frame rate, requiring slower scanning speeds. In arterial applications the vessel wall moves during the cardiac cycle and it may be necessary to cardiac-gate the received data, so

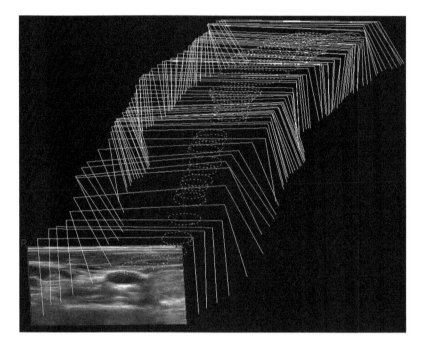

Figure 12.2 A series of 2D images is shown after positioning using an optical tracking system. Also shown are the lumen boundaries of the artery which were obtained by the operator.

that only one 2D frame is acquired every cardiac cycle. This requires very slow scanning speeds of 0.5–1 mm per second, with total scan times of typically 1–2 minutes.

The requirements for the setting up of a tracking system, transferring image data to an external computer and 3D acquisition and analysis mean that medical physics support is needed. The support needed may be lessened by the availability of freely downloadable software for acquisition and processing, such as that provided by Cambridge University in the United Kingdom (Prager et al. 2002).

Mechanically steered array systems

In these systems a linear array is contained within a fluid-filled housing. The transducer is held against the patient, and the linear array is swept back and forth within its housing in a controlled manner using a mechanical driver. The 2D scan plane is swept through the tissues, producing a pyramid-shaped 3D volume (Figure 12.1b). There is still the requirement that the system knows the location of the 2D scan planes within the 3D volume, and this is looked after by the machine itself with no need for

the operator to get involved in calibration. Each time the transducer sweeps back and forth a new volume is produced. This is useful in applications where there is movement, such as in obstetrics. Movement of fetal limbs and facial movements, such as mouth opening, can be observed in real time. Typically 5–10 volumes per second can be obtained provided that the 2D sector angle and the 3D sweep angle are reduced. Larger sector angle and sweep angle will result in a larger 3D volume being interrogated, but at the expense of a reduction in the number of volumes which can be obtained per second. These systems are commonly used in obstetrics and general radiology.

Endoprobe 3D ultrasound

Three-dimensional imaging may be achieved by controlled retraction of the transducer by the operator or using a mechanical pull-back system. This is typically used in trans-rectal scanning and in intravascular ultrasound. In the 3D reconstruction it is assumed that the 2D slices are aligned parallel with each other (Figure 12.1c). Though this is a reasonable assumption in, for example, trans-rectal imaging, it

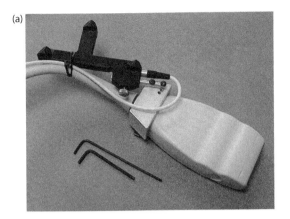

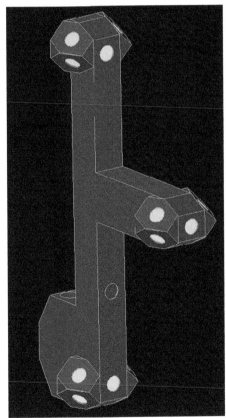

Figure 12.3 **(a)** A linear array is shown to which is attached a tool with a series of infrared light sources. The light sources are detected by the two cameras of the tracking unit. **(b)** Close-up of the tool with LEDs attached.

is unlikely to be true in intravascular ultrasound (IVUS) where vessels are curved, and where the operator has no control over the orientation of the

catheter and transducer within the vessel. Typically a 3D volume can be acquired in 5–10 seconds.

2D phased-array probes

These matrix transducers are designed to electronically steer the beam within a 3D volume (Figure 12.1d). The operator positions the transducer on the patient's skin, and a series of 3D volumes are automatically collected. The small footprint of the transducer makes it ideally suited for cardiac applications. Typically 10–20 volumes per second are collected for adult applications, and 20–40 for paediatric applications. Matrix-array probes are not widely adopted commercially, possibly because of their cost and complexity. Clinical applications have concentrated on cardiac applications, however array probes suitable for vascular and abdominal scanning have also been introduced (at the time of writing).

VISUALISATION

A problem with all types of 3D data sets, whether they are produced by MRI, CT or ultrasound, is visualisation. The possible types of visualisation which have been developed for MRI and CT are also applicable to ultrasound, and are described in the following sections.

2D display

Display of all the 3D raw data at the same time would be very confusing. The 3D data sets are actually most commonly viewed as a series of 2D slices. Figure 12.4 shows a 3D fetal study in which three 2D orthogonal (at 90° with respect to each other) views are provided simultaneously, along with a shaded surface view (see later) of the 3D structures. Figure 12.5 shows a 3D rectal study in which three 2D orthogonal views are shown, along with a composite image of the orthogonal planes. The position of each of the 2D slices within the 3D volume can be controlled by the operator.

Other variants of this display method allow the operator to choose the position and orientation of the 2D plane. One advantage of the 3D technique is that 2D planes can be visualised which would be impossible to obtain using conventional 2D imaging.

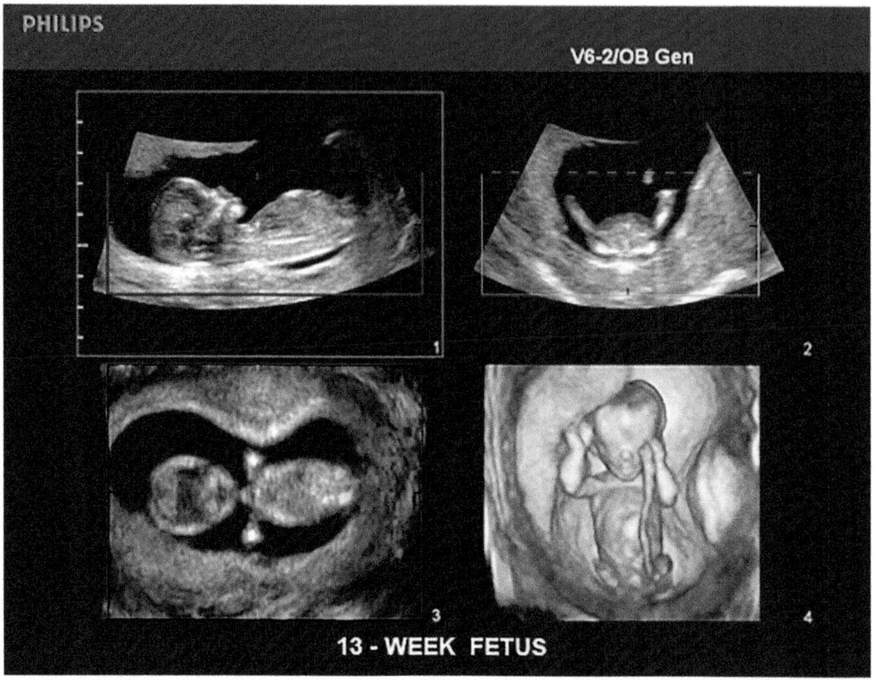

Figure 12.4 Display of the 3D data set using three orthogonal 2D slices of a 13-week fetus. The fourth image shows a surface-shaded view of the fetus.

Shaded surface display

In this method the surface of an object is highlighted and displayed. One common example is highlighting of the fetal face and body, to look at facial features and real-time actions such as mouth opening (Figure 12.6). The boundary or surface of the structure of interest is ideally identified using fully automated image-processing methods as these are rapid and require no user intervention. Automated boundary identification is possible when the boundary is sufficiently clear. When automated surface identification fails then a semi-automated method involving some user input of key surface points may be tried, or even a fully manual method involving specification of many points on each 2D slice of the 3D image. Boundaries are clearest when the structure of interest is adjacent to fluid; for example the fetus, which is surrounded by liquor, and the heart chambers, which contain blood. The intensity of the echoes received from the tissues of the fetus or heart is much higher than the intensity of echoes received from the liquor or blood, making the boundary of the heart or fetus easy to identify. For other features of interest such as tumours, the echo brightness may be similar to the surrounding tissue, making it more difficult for automated methods to identify the tumour boundary. For these reasons most 3D ultrasound applications of surface shading involve the heart and fetus.

Once the surface is identified the displayed grey level for each pixel of the surface is adjusted to make the displayed object look realistic. One regime might be that the grey level depends on the orientation of the surface; this is a simple technique that does not need operator intervention. Other more complex techniques involve illumination from a virtual light source. This technique may require operator intervention to position the virtual light source. For viewing, it is common practice to rotate the shaded surface, though static images are used for recording in patients' notes or reports.

Surface shading is an excellent method for allowing the operator to visualise the 3D shape of a structure. The method is intuitive as, in our everyday lives, we all perceive objects as solid structures

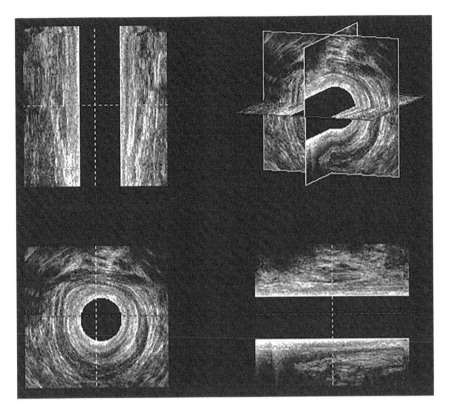

Figure 12.5 Display of the 3D data set using three orthogonal 2D views in a rectal study. The upper right image is a composite showing all three 2D images. (Images courtesy of BK Medical, Herlev, Denmark.)

which we can walk around to gain an impression of their shape. The surface-shading method is replicating this process within the computer.

The mathematical details of the surface-shading method may be complex, and are the subject of considerable research. Interested readers could start with the text by Udapa and Herman (1999). The key step is boundary identification. Where the contrast between different tissue types is high and noise is low, automated methods have reasonably high success in identifying the correct structures. In other 3D imaging techniques, especially CT, these methods are now routinely used to provide shaded images showing different organs. In ultrasound scanning, as previously noted, the surface-shading method is most easily suited to fetal and cardiac applications.

Stereoscopic viewing

The display methods described involve projecting a 3D object onto a 2D plane and so are often best suited to relatively simple structures. Most anatomy has some curvature and contains complex structures that overlap or partially obscure each other (Fronheiser et al. 2007). It can be difficult to appreciate all of this detail with conventional display methods and so this presents a greater challenge. Stereoscopic viewing offers an alternative method of visualising objects in a 3D ultrasound scan by creating the sensation of depth. This can enhance perception of a scan and also complement other display methods.

Stereoscopic viewing takes advantage of the human visual system. Human eyes are separated by a small horizontal distance and so the left and right eyes receive slightly different perspectives of the same scenery. The brain processes these differences to create a stereo image enabling the perception of depth. So by providing each eye with an image that views an object in an ultrasound scan from a slightly different perspective it is possible to improve viewing.

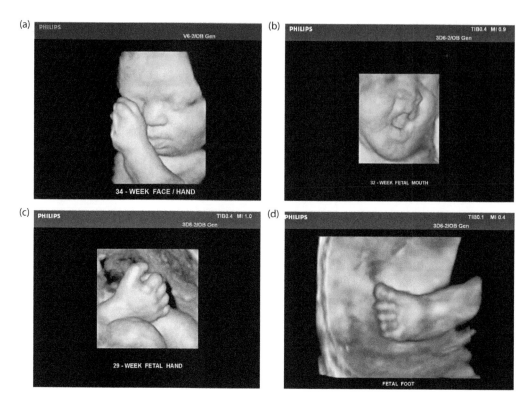

Figure 12.6 Surface-shaded view of a fetus. **(a)** Face and hands at 34 weeks, **(b)** face at 32 weeks with open mouth, **(c)** hand at 29 weeks, **(d)** foot.

There are various means of producing and displaying stereographic images. These include the use of polarizing filters to view separate right and left image pairs, complementary colour image pairs viewed through red/green glasses (similar to those used in the entertainment industry for viewing 3D films), liquid crystal light glasses that display right and left image pairs to each eye (similar to those used for virtual reality) and modern computer screens such as lenticular displays and liquid crystal displays.

For stereoscopic viewing to work with ultrasound there must be sufficient contrast between the objects of interest and surrounding tissue. Some anatomical structures such as cysts, ventricles and abscesses as well as lesions surrounded by low echogenic tissue, i.e. some tumours and haematomas, provide good contrast in ultrasound images and are therefore well suited. Breast tumours and cardiac structures such as the mitral valve are also suitable for stereo display. In obstetrics stereoscopic viewing improves visibility of primary features such as fetal skull and spine (Nelson et al. 2008). Augmented stereoscopic vision has applications in surgery as it allows the combination of radiographic data from different sources (e.g. ultrasound, CT and MRI) with the surgeon's vision (Gronningsaeter et al. 2000).

APPLICATIONS

Fetal

Pictures of the baby's face generated by surface shading are now widely used to help parental bonding with the baby (Figure 12.6). The 3D imaging also aids in the systematic examination of organ structure and the detection of abnormalities, commonly referred to as sonoembryology (Benoit et al. 2002).

Examination of the fetal heart in real time in 3D is difficult due to the high heart rate (two to three beats per second). However, a technique has been developed which involves reconstruction of the 3D volume in the computer (Yagel et al. 2007). Typically a slow sweep through the fetal heart is used in which 2D

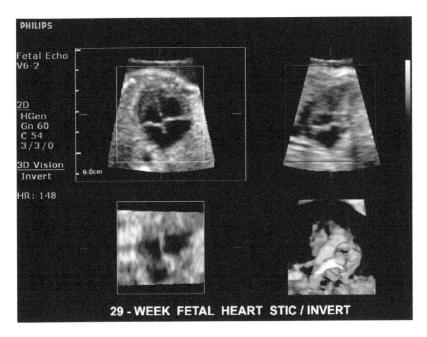

Figure 12.7 Fetal cardiac images taken using the STIC technique. (Image kindly provided by Philips Medical Systems, Eindhoven, Netherlands.)

frames are collected for 10–30 seconds. Image processing is used to detect those frames corresponding to the systolic peak, and this then allows the timing of each frame within its cardiac cycle to be estimated. A series of 3D volumes are then generated from the 2D slices, each volume corresponding to a different time point in the cardiac cycle. This technique is referred to as spatio-temporal image correlation (STIC). Both 3D B-mode and 3D colour flow data can be acquired using this approach (Figure 12.7). Clinical applications of 3D fetal heart ultrasound concern the identification of congenital abnormalities.

Cardiac applications in adults

Three-dimensional cardiac ultrasound systems have been used for three decades (Salustri and Roelandt 1995). Commercial systems have been based on acquisition of 2D data using both trans-cutaneous and trans-oesophageal approaches. Clinical applications of 3D cardiac ultrasound have only become routinely used since the introduction of 2D matrix transducers. The difference in echo brightness between cardiac tissues and blood allows the generation of surface-shaded images which are useful for real-time visualisation of the

3D structures of the heart. Figure 12.8 shows an apical four-chamber view. Common clinical uses of 3D ultrasound involve the measurement of chamber volumes, the assessment of congenital abnormalities and valvular motion and flow dynamics (Lang et al. 2006; Morbach et al. 2014).

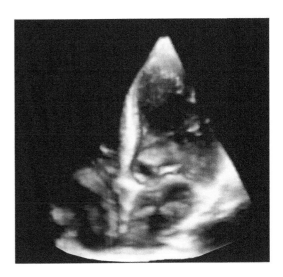

Figure 12.8 Surface-shaded apical four-chamber view of an adult heart.

Trans-rectal

The use of 2D ultrasound is well established for examination of rectal tissues, for applications such as tumour diagnosis and staging and detection of rectal tears and fistulae. The use of rotating 3D transducers may have advantages over 2D due to the ability to make 3D measurements of volume, and to examine 3D structures and anatomy (Figure 12.9).

Vascular ultrasound

Three-dimensional IVUS scanning has been mainly used in research labs, where measurements of plaque volume and structure are of interest over time, or following intervention (Kawasaki et al. 2005; Cervinka et al. 2006). The 3D data may also be acquired trancutaneously using matrix-array probes.

Other applications

Three-dimensional ultrasound techniques have been used in many other applications, but mainly on a research basis. Examples include abdominal (liver, spleen, kidney), oesophageal, arterial and muscular.

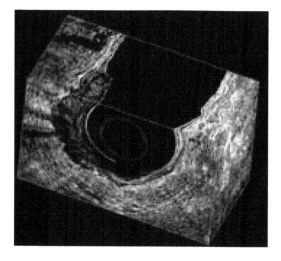

Figure 12.9 Box display for a 3D rectal study showing a rectal tumour. (Image courtesy of BK Medical, Herlev, Denmark.)

MEASUREMENTS

It is often desirable to extract measurements from medical images in order to reveal and diagnose disease, examine a patient's response to treatment or in obstetrics to monitor the progression of pregnancy. In 3D ultrasound a wider range of measurement possibilities exists compared to conventional 2D ultrasound, similar to CT and MRI.

Distances

As a volume of data is collected with 3D ultrasound it is possible to measure distances through any direction within the scan. Measurements are not just restricted to straight lines but can take the form of any path through the scan volume. This involves operator interaction to define the measurement path by selecting points by hand on a series of 2D slices. Examples of such measurements include tumour length and fetal body measurements which reflect the gestational age of the fetus.

2D shapes

As previously mentioned with 3D ultrasound it is possible for the operator to choose the position and orientation of a 2D plane through the scan volume revealing images of anatomy which would be impossible to obtain using conventional 2D imaging. This allows the quantification of shapes and areas of structures or lesions. In IVUS or endoscopic ultrasound the scan is often viewed lengthwise so that the operator can identify areas of plaque along a vessel or tumour shapes down the oesophagus that are sometimes missed by conventional 2D viewing of such data.

3D and 4D volumes

One of the greatest strengths of 3D ultrasound is the capacity for measuring volume. Three-dimensional ultrasound does not require assumptions of any specific geometry as is the case with conventional 2D ultrasound. The operator can identify the boundary of an object of interest in a series of slices through the scan and then obtain its volume. One example is the measurement of the volume of atherosclerotic plaque in diseased arteries (Delcker and Diener 1994; Ainsworth et al. 2005; Makris et al. 2011). The vessel wall may be segmented by hand, however automated systems may also be used (Figure 12.10). Change in plaque volume can be used to evaluate the effect of plaque-reducing therapies such as statins or atorvastatin (Makris et al. 2011; Lindenmaier et al. 2013). Further details of 3D volume measurement are provided in Chapter 6.

Using a mechanically steered array or a 2D array it is possible to record a sequence of 3D images over time. In cardiology measurements of volume at different points throughout the cardiac cycle provide valuable information (Figure 12.11). Left ventricular volume is determined by semi-automatic segmentation, i.e. the operator marks points on the object boundary while computational processing performs a more detailed search

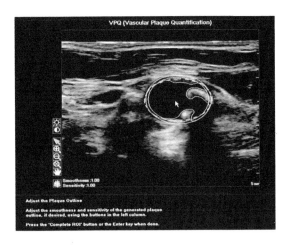

Figure 12.10 Automated segmentation of atherosclerotic plaque used for calculating plaque volume. (Images provided courtesy of Philips Electronics UK Ltd., Guildford, United Kingdom.)

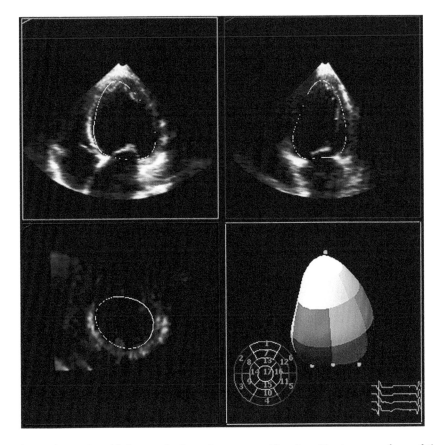

Figure 12.11 Three-dimensional left ventricular volume quantification. The inner surface of the left ventricle is obtained using a semi-automated method allowing the left ventricular volume to be obtained at each time point in the cardiac cycle.

to complete the segmentation (Pedrosa et al. 2016). Repeating this process at regular time intervals not only returns a series of volume measurements but also enables the analysis of additional parameters such as ejection fraction (i.e. the fraction of blood pumped out of a ventricle with each heartbeat), wall thickening and global and regional left ventricular rotation.

QUESTIONS

Multiple Choice Questions

Q1. Which of the following transducers can be used for 3D ultrasound?
   a. Linear array transducer with collection of 2D frames placed in a 3D volume by electromagnetic tracking
   b. Mechanically steered array transducer
   c. 2D array transducer
   d. Rotating transducer with mechanical pullback
   e. Phased-array transducer with collection of 2D frames placed in a 3D volume by optical tracking

Q2. Freehand 3D ultrasound:
   a. Cannot be used with linear array transducers
   b. In principle can be retro-fitted to any conventional 2D imaging ultrasound system
   c. Can only be used for phased-array transducers
   d. Involves a system for tracking the position of the transducer in space
   e. Is widely used in clinical practice

Q3. A mechanically steered array transducer:
   a. Gives real-time (several frames per second) 3D imaging
   b. Uses a 2D array
   c. Sweeps the 2D scan plane to and fro to build up the 3D volume using a mechanical mechanism
   d. Cannot be used in obstetrics
   e. Involves use of an optical tracking system

Q4. Which of the following do not use any mechanical means to sweep the beam to create a 3D volume?
   a. Freehand 3D ultrasound based on an array transducer
   b. Freehand 3D ultrasound based on a mechanically scanned transducer
   c. Endoprobe scanning with mechanical pullback
   d. Mechanically steered array transducer
   e. 2D array-based 3D ultrasound

Q5. Which of the following use only electronic beam steering to build up the 3D volume?
   a. Freehand 3D ultrasound based on an array transducer
   b. Freehand ultrasound based on a mechanically scanned transducer
   c. End-probe scanning with mechanical pullback
   d. Mechanically steered array transducer
   e. 2D array-based 3D ultrasound

Q6. Orthogonal plane display:
   a. Is only used in CT and MRI, not in 3D ultrasound
   b. May be used in visualisation of 3D ultrasound images of the liver
   c. Involves simultaneous display of three planes at 90° through the 3D volume
   d. Involves simultaneous display of two planes at 90° through the 3D volume
   e. Involves display of one plane though the 3D display

Q7. Surface shading:
   a. Involves highlighting particular surfaces of organs and of the fetus
   b. Involves simultaneous display of orthogonal 2D images
   c. Works best when contrast between tissues is poor
   d. Only works for images from 2D array transducers
   e. Is widely used in 3D obstetric studies

Q8. STIC:
   a. Stands for sono-transthoracic inverse collection
   b. Is used in 4D imaging of the fetal heart
   c. Is used in 4D imaging of the adult heart
   d. Stands for spatio-temporal image correlation
   e. Involves construction of a single composite cardiac cycle from 50 to 60 cardiac cycles

Short-Answer Questions

Q1. Describe the basic principles of operation of freehand 3D scanning using optical tracking.

Q2. Describe how a 2D array can be used for 3D imaging and describe advantages this approach may have over 3D data collection using a mechanically steered array transducer.

Q3. Describe how a 3D data set may be acquired using endoprobe transducers and give two examples of their use.

Q4. Describe surface shading display of 3D images and give examples of clinical use.

Q5. Describe one way by which the 3D volume of an organ may be measured.

REFERENCES

Ainsworth CD, Blake CC, Tamayo A et al. 2005. 3D Ultrasound measurement of change in carotid plaque volume: A tool for rapid evaluation of new therapies. *Stroke*, 26, 1904–1909.

Benoit B, Hafner T, Kurjak A et al. 2002. Three-dimensional sonoembryology. *Journal of Perinatal Medicine*, 30, 63–73.

Cervinka P, Costa MA, Angiolillo DJ et al. 2006. Head-to-head comparison between sirolimus-eluting and paclitaxel-eluting stents in patients with complex coronary artery disease: An intravascular ultrasound study. *Catheterization and Cardiovascular Interventions*, 67, 846–851.

Delcker A, Diener HC. 1994. Quantification of atherosclerotic plaques in carotid arteries by 3-dimensional ultrasound. *British Journal of Radiology*, 67, 672–678.

Fenster A, Downey DB, Cardinal HN. 2001. Three-dimensional ultrasound imaging. *Physics in Medicine and Biology*, 46, R67–R99.

Fronheiser PMP, Noble JR, Light E, Smith SW. 2007. Real time stereo 3D ultrasound. *IEEE Ultrasonics Symposium*, 2239–2242.

Gronningsaeter A, Lie T, Kleven et al. 2000. Initial experience with stereoscopic visualization of three-dimensional ultrasound data in surgery. *Surgical Endoscopy*, 14, 1074–1078.

Kawasaki M, Sano K, Okubo M et al. 2005. Volumetric quantitative analysis of tissue characteristics of coronary plaques after statin therapy using three-dimensional integrated backscatter intravascular ultrasound. *Journal of the American College of Cardiology*, 45, 1946–1953.

Lang RM, Mor-Avi V, Sugeng L, Nieman PS, Sahn DJ. 2006. Three-dimensional echocardiography. The benefits of the additional dimension. *Journal of the American College of Cardiology*, 48, 2053–2069.

Lindenmaier TJ, Buchanan DN, Pike D et al. 2013. One-, two- and three-dimensional ultrasound measurements of carotid atherosclerosis before and after cardiac rehabilitation: Preliminary results of a randomized controlled trial. *Cardiovascular Ultrasound*, 11, 39.

Makris GC, Lavida A, Griffin M, Geroulakos G, Nicolaides AN. 2011. Three-dimensional ultrasound imaging for the evaluation of carotid atherosclerosis. *Atherosclerosis*, 219, 377–383.

Mercier L, Lango T, Lindseth F, Collins DL. 2005. A review of calibration techniques for freehand 3-D ultrasound systems. *Ultrasound in Medicine and Biology*, 31, 449–471.

Morbach C, Lin BA, Sugeng L. 2014. Clinical application of three-dimensional echocardiography. *Progress in Cardiovascular Diseases*, 57, 19–31.

Nelson TR, Ji EK, Lee JH, Bailey MJ, Pretorius DH. 2008. Stereoscopic evaluation of fetal bony structures. *Journal of Ultrasound in Medicine*, 27, 15–24.

Pedrosa J, Barbosa D, Almeida N, Bernard O, Bosch J, D'hooge J. 2016. Cardiac chamber volumetric assessment using 3D ultrasound – A review. *Current Pharmaceutical Design*, 22, 105–121.

Prager R, Gee A, Treece G, Berman L. 2002. Freehand 3D ultrasound without voxels: Volume measurement and visualisation using the Stradx system. *Ultrasonics*, 40, 109–115.

Prager RW, Ijaz UZ, Gee AH, Treece GM. 2009. Three-dimensional ultrasound imaging. *Journal of Engineering Medicine*, Available at: https://doi.org/10.1243/09544119JEIM586.

Salustri A, Roelandt JRTC. 1995. Ultrasonic 3-dimensional reconstruction of the heart. *Ultrasound in Medicine and Biology*, 21, 281–293.

Udapa JK, Herman GT. 1999. *3D Imaging in Medicine*. Boca Raton, FL: CRC Press.

Yagel S, Cohen SM, Shapiro I, Valsky DV. 2007. 3D and 4D ultrasound in fetal cardiac scanning: A new look at the fetal heart. *Ultrasound in Obstetrics and Gynecology*, 29, 81–95.

Contrast agents

CARMEL MORAN AND MAIRÉAD BUTLER

INTRODUCTION

Contrast agents are used in all imaging modalities to increase the sensitivity of the imaging technique by altering the image contrast between different structures. Ultrasonic contrast agents are composed of a solution of gas-filled microbubbles. It is the dramatic increase in ultrasound scattering which can be achieved with these microbubbles which has made them useful both in difficult-to-image patients and in vessels and organs for which the ultrasound signal to noise is inadequate to make a clinical diagnosis. Injection of the contrast microbubbles and observation of their real-time dynamic passage in an ultrasound image sequence can significantly improve the sensitivity and diagnostic power of ultrasonic imaging.

Ultrasonic contrast agents were initially used by a cardiologist in the 1960s, who used physiological saline as an intra-cardiac ultrasound contrast medium in the anatomic identification of mitral valve echoes. Microbubbles, formed in the saline after shaking, resulted in an increase in the ultrasound signal from the blood and hence the contrast in the images. Intravenously injected agitated saline is still used today in cardiac studies in the assessment of patent foramen ovale. However, it was not until the mid-1980s that clinical interest resulted in development of a wide range of potential ultrasound contrast agents, some of which were the early precursors of current commercially available contrast agents. A history of the early development of ultrasonic contrast agents can be found in Ophir and Parker (1989).

CONTRAST MICROBUBBLES

Commercially available ultrasonic contrast agents are microbubbles filled with a gas and surrounded by a thin outer layer or shell (Figure 13.1). Gas-filled microbubbles scatter ultrasound much more effectively than liquid or solid-filled microbubbles of comparable size and as a result when injected intravenously, increase the magnitude of the ultrasound signal from vascular spaces. Free gas microbubbles, that is microbubbles with no shell, dissolve rapidly

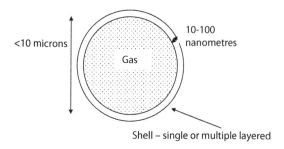

Figure 13.1 A contrast microbubble illustrating the gaseous interior surrounded by a thin outer shell.

in blood. The addition of a shell ensures that the microbubbles survive passage through the lungs and to the organ of interest. In commercial preparations of microbubbles, the shells of the microbubbles must be biocompatible and tend to be either fat (lipid) or protein (albumin) based (Figure 13.2, Table 13.1). Different shell compositions can create microbubbles with different properties, e.g. lifetime and response to ultrasound signals.

Early commercially available contrast microbubbles, such as Levovist and Albunex, are often referred to as first-generation agents. These agents were air filled but proved to be relatively unstable in comparison to more recent microbubble preparations in which air has been replaced by high

Figure 13.2 Optical microscopic image of contrast microbubbles (Definity/Luminity).

molecular weight gases such as fluorocarbons (or SF_6). Such gases dissolve less readily in blood and thus the gas can be visualised for longer within the vasculature.

The mean diameter of commercially produced ultrasonic contrast microbubbles lies in the range of 2–5 μm, comparable in size to red blood cells. Prior to the commercial manufacture of ultrasonic contrast agents, agitated saline or dyes were used for some clinical applications. However, microbubbles formed in this manner tended to be large (>10 μm) and so were unable to pass through the lung capillary bed.

Table 13.1, shows the most widely used commercial contrast agents. This is not an exhaustive list but demonstrates the range of compositions, size of microbubbles, recommended clinical applications and dosages of the agents which are used in European countries.

The behaviour of the microbubble within the ultrasound beam is complex and depends on many factors including size, shell composition, transmit frequency and transmit pressure. This behaviour leads to a number of possibilities for imaging over and above a simple increase in received echo strength. These issues are addressed in more detail in the sections that follow.

Commercially available ultrasound contrast agents

Ultrasound contrast microbubbles have a finite lifetime as microbubbles, and as such activation of the agents is recommended just prior to clinical usage. Activation tends to include either mechanical or manual agitation or, be by injection of saline followed by manual agitation. Figures 13.3–13.6 show the four contrast agents currently available and the kit required for activation of the agents.

INTERACTION OF MICROBUBBLES AND ULTRASOUND

When the ultrasound pulse insonates a microbubble the compressibility of the gas within the microbubble and the difference in acoustic impedance between the gas-filled microbubble and the surrounding tissue will cause the ultrasound to be strongly scattered.

Table 13.1 Examples of commercially available ultrasonic contrast agents

Contrast agent	Manufacturer	Outer material	Gas	Diameter and concentration	Charge	Licensing and use	Supplied and reconstitution	Dose
Definity (United States/Canada) Luminity (Europe)	Lantheus Medical Imaging Inc.	Lipids: DPPA, DPPC MPEG5000, DPPE	Octafluoro-propane	Mean range 1.1–2.5 μm 98% <10 μm 6.4 × 10⁹ microspheres per millilitre (150 μL gas per millilitre)	Negative	Europe and United States approved for suboptimal echocardiograms, for opacification of the left ventricle and to improve delineation of the endocardial border	Single vial containing lipid solution with gas in the vial headspace; is rapidly shaken for 45 seconds using a Vialmix. Needs to be refrigerated	Bolus: 10 μL/kg with saline flush Infusion: 1.3 mL in 50 mL saline, initial rate 4 mL per minute Max: Two bolus or one infusion dose
Sonazoid	GE Healthcare	Lipid (HEPS)	Perfluorobu-tane	Median 2.6 μm 99.9% <7 μm 8 μL gas per millilitre	Negative	Japan and Korea approved Focal liver lesions, focal breast lesions	Reconstituted with 2 mL sterile water and shaken for 1 minute	0.12 μL/kg bolus with saline flush
SonoVue	Bracco Diagnostics Inc.	Lipids Macrogol 4000, DSPC, DPPG, Palmitic acid	Sulphur hexafluo-ride	99% <11 μm 8 μL microspheres per millilitre	Negative	European Union approved Imaging chambers in the heart and Doppler of macro- and micro- (liver) vasculature	Supplied in vial of 25 mg dry powder and glass syringe of 5 mL saline. Saline is added to the powder and the mixture hand shaken for 20 seconds	2–2.4 mL bolus with saline flush Intravesical paediatric: 1 mL
Optison	GE Healthcare	Human albumin	Octafluoro-propane	Mean range 2.5–4.5 μm 95% <10 μm 5–8 × 10⁸ microspheres per millilitre (0.19 mg gas per millilitre)	Slight nega-tive	United States, Australia and Europe approved To opacify cardiac chambers and to improve the delineation of the left ventricular endocardial borders	Supplied as a single vial of clear solution with a white liquid layer the gas is in the vial headspace. To re-suspend the microbubbles the vial is inverted and gently rotated	0.5 mL at no greater than 1 mL per second Max 5 mL in 10 minutes, 8.7 mL per patient study

Abbreviations: DPPA, dipalmitoyl glycerophosphate; DPPC, dipalmitoyl glycerophosphocholine; DPPE, dipalmitoyl glycerophosphoethanolamine; DPPG, dipalmitoyl glycerophosphoglycerol; DSPC, distearoyl glycerophosphocholine; HEPS, hydrogenated egg phosphatidyl serine.

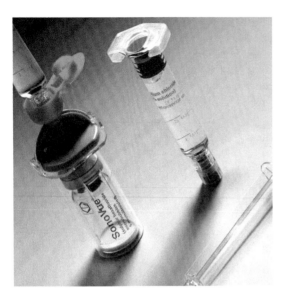

Figure 13.3 SonoVue contrast agent. Once the mini-spike is inserted into the vial, the saline is injected into the vial and the solution is gently agitated.

Microbubble oscillations

When the ultrasound pulse insonates a gaseous microbubble it forces the microbubble to undergo oscillation – contracting during the compression (positive) part of the cycle and expanding during the rarefaction (negative) part of the cycle (Figure 13.7).

Microbubble resonant frequency

The degree of scattering (and oscillation) is maximal if the ultrasonic wave is at the resonant frequency of the microbubbles. The resonant frequency is the frequency at which the microbubbles oscillate most easily. This is similar to a wine glass 'singing' at a specific frequency on rubbing the rim. The resonant frequency depends on both the shell characteristics and the size of the microbubble. The resonant frequency of a microbubble, of a similar size to a red blood cell, is such that it lies within the clinical diagnostic frequency range of

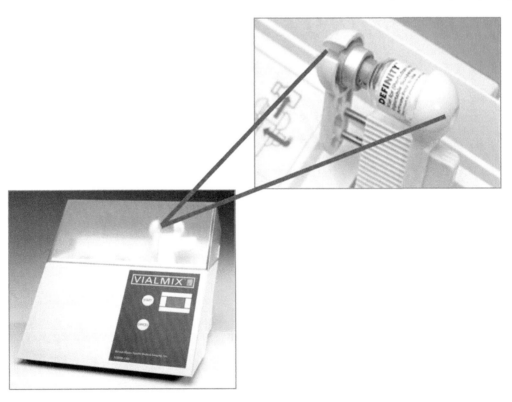

Figure 13.4 Vialmix with a vial of Definity. The vial is rapidly agitated for 45 seconds in the mixer prior to withdrawal from the vial. The rapid shaking forces the head gas into the lipid solution.

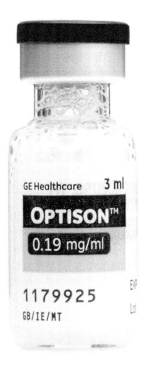

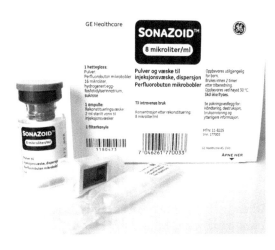

Figure 13.6 Sonazoid is activated by injection of sterile water and gentle manual rotation of the vial.

Figure 13.5 Vial of Optison. Optison is activated by gentle inversion and rotation of the vial to re-suspend the microbubbles, yielding a milky-white solution.

ultrasound. It must be remembered that although average diameters are often quoted by commercial contrast agent manufacturers, the range in size may vary from sub-micron up to 10 μm diameter (see Table 13.1); thus there is likely to be a broad range of diagnostic frequencies over which resonant behaviour may be observed.

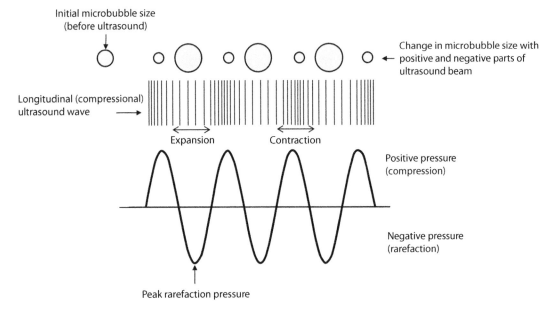

Figure 13.7 A microbubble contracting during the positive part of the ultrasound wave and expanding during the negative part of the wave.

Microbubble and acoustic pressure

The reaction of the contrast microbubbles with varying acoustic pressure is also of importance when imaging contrast agents *in vivo*. On most ultrasound scanners the acoustic pressure is not directly displayed but both the thermal and mechanical safety indices are displayed. The mechanical index (MI) is defined as:

$$\text{MI} = \frac{p_r}{\sqrt{f}} \qquad (13.1)$$

where p_r is the peak rarefaction pressure of the ultrasound wave *in situ* (also known as the peak negative pressure) and f the insonation frequency. It describes the likelihood of onset of inertial cavitation assuming the presence of cavitation nuclei in a homogeneous tissue. This topic is addressed more fully in Chapter 16. At low acoustic pressure ($p_r < 0.1$ MPa), and low MI the majority of microbubbles oscillate symmetrically in the ultrasound beam. The scattered ultrasound signal from the microbubble will therefore be of the same frequency as the incident wave. This is known as linear behaviour (Figure 13.8).

As the acoustic pressure is increased ($p_r > 0.1$ MPa), the bubble cannot contract as much as it can expand; this is due to the presence of the gas within the microbubbles. As a result the bubble begins to oscillate asymmetrically or unevenly, with the microbubble expanding more than it contracts. This is referred to as non-linear behaviour. Because of the asymmetric oscillations, the scattered signal will be different in both magnitude and frequency content to the scattered wave from linear oscillations which are obtained at lower acoustic pressures. The non-linear scattered wave will contain the frequency of the incident beam as well as frequencies known as harmonics (see Chapter 2). The second harmonic is twice the incident frequency. This is seen in music: for the bugle, the lowest note is the first (fundamental) frequency (equivalent to the incident ultrasound beam). On changing lip shape while blowing, the note played is changed, and the next note heard is twice the frequency of the first (second harmonic). The next highest note is three times the fundamental frequency (third harmonic). For detection of microbubbles, usually it is the second-harmonic signal from non-linear (asymmetric) oscillations produced from higher-pressure incident ultrasound beams, which is utilized to form the contrast image.

At even higher acoustic pressures ($p_r > 0.4$ MPa, MI > 0.2 for a 4 MHz pulse), the shell of the microbubbles can be forced to crack or rupture, releasing the gas from the microbubble (see Figure 13.8). The resulting free gas bubbles will strongly scatter ultrasound. However, they will dissolve rapidly

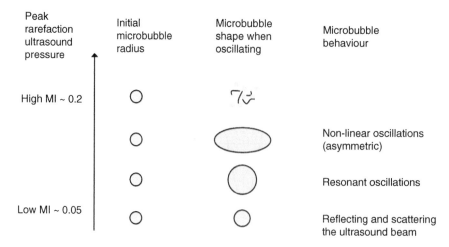

Figure 13.8 Interaction of microbubbles with increasing acoustic pressure. As the acoustic pressure increases, the microbubbles oscillate non-linearly and eventually collapse, releasing free gas bubbles.

into the bloodstream and therefore can only be visualised in one or two frames of ultrasound.

This violent interaction between the microbubbles and the high-pressure ultrasound pulse will result in a scattered wave which has a wide range of frequency components.

The specific acoustic pressure at which each of these phenomena takes place has been shown to be dependent upon many parameters, including the composition of the encapsulating shell, the size of the microbubbles and the proximity of the microbubbles to other structures such as cell and vessel walls. It must also be emphasized that although an indication of the maximum acoustic pressure generated in the scan plane is given by the displayed MI value on the screen, there is a wide variation of the acoustic pressure both laterally across the scan plane and at depth which is likely to cause a range of different interactions within each scan plane. By modifying the acoustic output of the scanner the predominant acoustic interaction can be changed but that is not to say that the other interactions do not occur.

This short section summarizes a vast range of research literature on the interaction between microbubbles and ultrasound waves (Kollmann 2007).

CONTRAST-SPECIFIC IMAGING TECHNIQUES

As previously described, the scattered ultrasound beam from the microbubbles can contain different frequencies to the incident ultrasound beam. Although the scattered signal from tissue can also contain different frequencies, these are generated by other mechanisms and are more likely to occur at depth and at higher pressures (see Chapter 2). Current techniques for imaging contrast microbubbles rely predominantly on separating the different frequency components in echoes from tissue and microbubbles.

For simplicity, contrast-specific techniques can be viewed as either low-MI imaging techniques where the acoustic pressure is such that the majority of the contrast microbubbles are not disrupted by the ultrasound wave or high-MI techniques where the acoustic pressure is such that microbubble destruction is maximized in the scan plane.

Before reviewing these imaging techniques it is worthwhile reflecting upon both fundamental and second-harmonic imaging techniques which were initially used to image contrast agents clinically.

Fundamental imaging

Prior to the development of contrast-specific imaging techniques, contrast microbubbles were imaged in a manner similar to that used for routine B-mode imaging such that the bandwidth over which the scattered ultrasound wave was received matched that of the transmitted ultrasound wave. In addition, doses of contrast agents tended to be of the order of several millilitres compared to doses now, which tend to be smaller. Use of injections of several millilitres resulted in significant, and in some instances long-lasting, attenuation distal to, or in some instances within, the vessel or organ of interest. This was particularly true for cardiac studies. However, the use of contrast agents proved to be of great importance in colour Doppler studies specifically in small vessels, such as capillaries, where without contrast the blood signal level was too low to be detected. In such vessels, injection of contrast agents could enhance the strength of the backscatter signal from the blood, thus raising the signal above the level of the clutter signal and increasing the diagnostic potential of the study.

Second-harmonic imaging

As discussed previously, at relatively modest acoustic output pressures (0.1–0.3 MPa) and at diagnostic ultrasound frequencies, the wave that is scattered from contrast microbubbles is nonlinear and incorporates harmonics of the incident fundamental frequency (including sub-harmonics, which are harmonics generated at fractions of the fundamental frequency). Second-harmonic imaging isolates the received ultrasound wave centred at twice the transmit frequency, and uses this to form the B-mode image. By using this technique at modest acoustic pressures the ratio of the backscatter signal from the contrast agent to the backscatter signal from the tissue, known as the contrast-to-tissue signal ratio, can be increased significantly. However, the technique has limitations, because even at these acoustic pressures

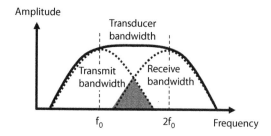

Figure 13.9 The overlap (shaded region) in the transmitted and receive bandwidths in second harmonic imaging.

non-linear propagation of ultrasound through soft tissue results in generation of tissue harmonics (see Chapter 2). Generation of tissue harmonics is more pronounced at depth and becomes increasingly important as the acoustic pressure is increased. Second-harmonic imaging necessitates the use of broad-bandwidth transducers to ensure that the received second harmonic signal can be separated from the fundamental frequency. Consequently, the transmit frequency bandwidth is narrowed to reduce overlap with the receive bandwidth (Figure 13.9). This limitation in transmit and receive bandwidths can result in a reduction in spatial resolution (see Chapter 4).

Low mechanical index techniques

Low-MI techniques have two distinct characteristics. First, using these techniques the contrast microbubbles are not destroyed and yet generate a significant non-linear, harmonic component and, second, at low-MI, tissue harmonics are not generated to the same level as microbubble harmonics, resulting in improved contrast-to-tissue signal ratio, i.e. there is more signal detected from contrast microbubbles than from the tissue. Low-MI techniques rely upon the processing of the scattered signal from multiple (two or more) transmitted pulses. These pulses can vary in amplitude or phase or in both amplitude and phase. Because of the requirement for multiple transmitted pulses these low-MI techniques have been shown to work best in vessels in which the microbubbles are moving slowly and thus there is minimal movement of the bubbles between consecutive insonations. Moreover there is a reduction in frame rate due to

the multiple transmitted pulses required. However, because this is a low-MI technique (MI 0.05–0.1) the majority of the contrast microbubbles are not destroyed between consecutive insonations. These low-MI techniques are routinely used in abdominal contrast imaging and for cardiac stress-echo studies. Various low-MI techniques are discussed in the following sections.

Low mechanical index: Pulse inversion (harmonic imaging)/phase inversion

The bandwidth limitations of second-harmonic imaging are overcome using pulse-inversion/phase-inversion techniques where the echoes from consecutive transmitted pulses received over the full transducer bandwidth are combined to remove the linear echoes. In these techniques, two consecutive ultrasound pulses are transmitted – the second pulse being inverted with respect to the first (180° out of phase). Upon reception, when the two received signals are added together the linear signals summate to zero (cancel); that is, they respond equally to positive and negative pressures. As discussed, microbubbles respond differently to positive and negative pressures, hence the received signals do not summate to zero and, since contrast microbubbles are the principal source of non-linearity, these images will have a high contrast-to-tissue ratio (Figure 13.10).

Low mechanical index: Amplitude modulation/power modulation

In a similar manner to that employed in pulse-inversion imaging, in amplitude (AM) or power modulation (PM), the amplitude of the pulse is changed rather than the phase of the signal. Two, or in some instances three, consecutive pulses of varying amplitude (often in a half-amplitude, full-amplitude, half-amplitude sequence) are emitted from the transducer in succession. Low-amplitude pulses generate fewer harmonics than higher-amplitude pulses and subtracting the low-amplitude response from the high-amplitude response removes the linear contribution to the scattered signal, with the remaining signal being from the non-linear scatterers (Figure 13.11).

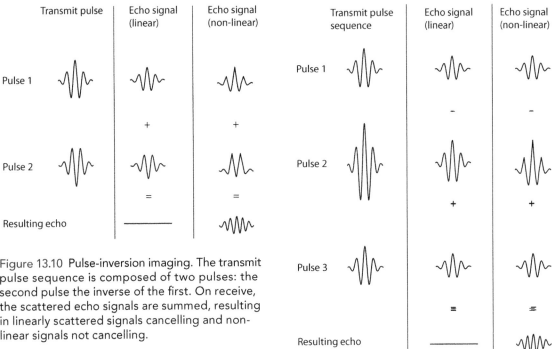

Figure 13.10 Pulse-inversion imaging. The transmit pulse sequence is composed of two pulses: the second pulse the inverse of the first. On receive, the scattered echo signals are summed, resulting in linearly scattered signals cancelling and non-linear signals not cancelling.

Low mechanical index: Pulse-inversion amplitude modulation

Pulse-inversion amplitude modulation (PIAM) comprises a combination of the previous two techniques with consecutive insonating pulses varying in both amplitude and phase. Although requiring more processing to form an image, this method of detecting signals from microbubbles has been found to be very sensitive.

High mechanical index techniques: Stimulated acoustic emission, flash imaging, intermittent imaging

In the majority of abdominal imaging clinical cases, low-MI techniques are used to observe the wash-in and wash-out dynamics of blood within vessels and potential lesions. In these instances the contrast microbubbles are not destroyed and can remain in the vasculature for extended periods. Higher-MI techniques are known to destroy the contrast microbubbles. By following a strict protocol, high-MI techniques can be used to give an indication of the perfusion of different organs. This technique was used previously to most effect

Figure 13.11 Pulse amplitude/power modulation. The incident pulse sequence is composed of three pulses – one half-amplitude pulse followed by a full pulse followed by a second half-mplitude pulse. On reception since the half-amplitude scattered pulses contain mainly linear signal and the scattered full-amplitude pulse contains linear and non-linear components, the subsequent subtraction process removes the linear component from the full-amplitude pulse.

in studying the perfusion of the myocardium. The technique involves initially visualising the microbubbles entering a blood vessel of interest using a low-MI technique. By emitting one, or several, high-MI pulses (flash pulse), the contrast microbubbles are forced to collapse, releasing free gas which gives a much enhanced backscattered signal (stimulated acoustic emission) for a short time while also destroying the microbubbles in the scan field. By observing and quantifying the subsequent re-filling dynamics of the vessel at low-MI values, an indication of perfusion can be measured. The high-pressure MI pulse was triggered directly from the electrocardiogram or from a pre-defined time sequence in the contrast set-up, thus generating the name 'intermittent imaging'. Although this technique was developed initially for the assessment of myocardial perfusion, there was never widespread adoption of the technique. However, newer applications in assessing cerebral blood flow using contrast agents are now being explored (Bilotta et al. 2016).

PERFORMING A CONTRAST SCAN

A baseline scan is performed using non-contrast software initially to determine whether the patient requires contrast to aid diagnosis. If a contrast study is necessary, the ultrasound scanner is set up in a contrast protocol and, assuming no contraindications, the contrast is prepared/activated and the patient set up for an IV injection into the femoral vein. Usually, a three-way tap is used so that a saline flush can be injected immediately after the contrast. The bolus is drawn from the vial and injected at a rate of no more than 1 mL s^{-1} over a period of 5–10 seconds and is followed immediately by a saline flush at the same speed of injection. The time of the injection is noted and the patient is scanned throughout. When a low-MI study is being undertaken, the output power of the scanner is kept very low (MI <0.1) to ensure that the contrast microbubbles do not collapse during the scanning session. In the majority of instances, switching to a low-MI contrast-specific imaging mode will result in the image becoming very dark and in some instances very little anatomical information may be visualised on the screen. These images are formed from the non-linear signals and hence, without contrast present, there is limited generation of non-linearities from tissue at these low output powers. As a result the technique is reliant on the expertise of the sonographer to maintain the position and orientation of the ultrasound beam when scanning. In most high-end commercial scanners, when contrast mode is selected the image on the screen sub-divides into two – one showing the normal B-mode image formed over the whole bandwidth of the transducer, at low-MI values, and on the other the image formed from the non-linear signals. The ability to be able to simultaneously visualise the structural information while observing the filling dynamics of vessels is of great benefit and ensures that the correct

scanning position is maintained throughout a scanning session.

Once the contrast has been injected, it is generally visualised initially in a feeding vessel to the organ of interest. Initially strong attenuation of the ultrasonic beam may occur causing shadowing of distal structures. As the bolus of contrast distributes throughout the organ, the more distal structures may then become visible again. Study of the filling dynamics and distribution of the agent around lesions, specifically in the liver, has been shown to have high diagnostic power (see European Federation of Societies for Ultrasound in Medicine and Biology [EFSUMB] guidelines, Claudon et al. 2013). Acquisition of images and cine loops during contrast enables a review of the acquired images. Dependent on the quantity of agent used and the manufacturer's recommendations a second and possibly third dose of contrast agent may be given to provide further diagnostic information.

For Sonazoid, which is licenced solely for the detection of focal hepatic lesions, the liver can be scanned in the vascular phase to give a differential diagnosis as well as in the post-vascular phase when the contrast is taken up by the Kupffer cells, 10–15 minutes after injection.

Infusion injections are often used in cardiac stress-echo studies to maintain a steady flow of contrast. The contrast agent is diluted with saline and placed in an infusion pump and run continuously throughout the study. Infusion studies generally take longer than bolus studies. In many instances the contrast has to be gently agitated within the infusion pump to ensure that the bubbles are evenly distributed in the syringe.

Machine settings

During ultrasound contrast studies, it is important that none of the settings of the scanner are adjusted while the study is being undertaken. When the contrast-specific imaging mode is selected, the gain is set at a relatively low level to ensure that enhancement is visualised as the contrast arrives in the scan plane. This gain setting as in normal B-mode imaging will not affect the beam transmitted from the probe but only the image that is displayed on the screen. The majority of high-end ultrasound scanners have contrast-specific imaging protocols offering either a pulse-inversion or amplitude-modulation imaging technique. Once selected these packages switch to a low-MI value (low power output), which should not be adjusted by the operator throughout the scan. Varying the power output will affect the amplitude of the transmitted beam and, as discussed previously, can determine the microbubble response to the beam. In addition, as with B-mode scanning, higher frequencies will increase resolution and lower frequencies will increase penetration.

Quantification of dynamic contrast enhancement

Quantification of microbubble enhancement is challenging due to the difficulty in controlling the number of bubbles in a region of interest; therefore, comparisons between different scans or scans acquired using different contrast agents are difficult even when machine settings are kept the same. However, since microbubble contrast agents are blood-pool agents (vascular) they can be useful as indicators of physiological parameters such as vascular flow volume, blood flow velocity and perfusion rate, although accuracy is an issue because of the reasons previously described. A method of looking at contrast agent enhancement in a region of interest is the time-intensity curve, where backscattered intensity from microbubbles is plotted against time. This curve shows how the concentration of microbubbles in a region of interest changes as the blood flows through. These curves can be used to assess the change of echo intensity with time, with their shape depending on the organ of interest and whether the contrast is administered as a bolus or infusion. A bolus will provide enhancement for up to 2 minutes as the microbubbles wash in and wash out of the region of interest as they circulate, while an infusion can provide enhancement for longer periods as the microbubbles are administered at a steady rate over time (see Table 13.1 for rates of infusion and bolus administration methods). Figure 13.12 shows typical time-intensity curves for a bolus injection and an infusion of contrast agent. The data can be collected from regions of interest within an organ or vessel. The intensity of the backscatter signal is measured over the duration of the bolus injection

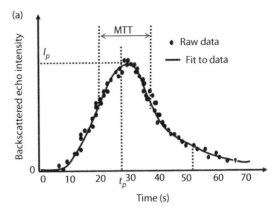

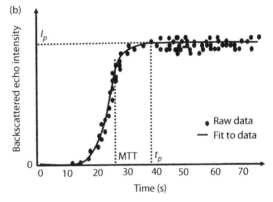

Figure 13.12 Typical time-intensity curves after a bolus **(a)** and infusion **(b)** injection of contrast agents showing key parameters which can be measured including peak intensity (I_p), time to peak intensity (t_p) and mean transit time (MTT).

or infusion of contrast. From these curves, parameters such as peak intensity, mean transit time and time to peak can be measured, and these parameters can provide valuable information on tissue or lesion perfusion (Dietrich et al. 2012a).

CLINICAL APPLICATIONS OF CONTRAST AGENTS

Ultrasonic contrast agents are becoming increasingly accepted in diagnostic clinics when used in combination with contrast-specific imaging techniques as previously described. A comprehensive review of the clinical applications in which the use of contrast agents is considered to improve diagnostic potential has been published by the EFSUMB, in which guidelines and recommendations for good practice are outlined. Although commercially available contrast agents are licenced principally for use in the liver, breast or for cardiac endocardial border definition, guidelines are also provided for assisting in studies which are not currently licenced (off-licence) including renal, spleen and pancreatic, transcranial, urological and cardiac perfusion studies (Piscaglia et al. 2012). Low-MI techniques are recommended for all of these organs. Moreover, in 2017 SonoVue (see Table 13.1) was licenced for some limited paediatric applications and EFSUMB has published a position paper upon the wider use of contrast agents for paediatric applications (Sidhu et al. 2017).

Liver

The use of contrast agents to aid in the detection and diagnosis of focal liver lesions is now well established and after cardiac applications, the liver is the organ most widely studied using contrast. After a bolus contrast injection there are three distinct phases of enhancement within the liver. These are the arterial, portal-venous and late phase, occurring 10–20, 30–45 and more than 120 seconds after injection, respectively. By studying the enhancement patterns of focal liver lesions during these three phases, detection and classification of the lesions are possible (Figure 13.13). More detailed information can be found in the EFSUMB/World Federation for Ultrasound in Medicine and Biology (WFUMB) guidelines (Claudon et al. 2013).

Kidney

Although some contrast agents are licenced for imaging large vessels (SonoVue) there is no agent which is licenced for imaging the kidney; therefore, this represents an off-licence application. After a bolus injection, enhancement is visualised first in the cortex, 10–15 seconds after bolus injection. Medullary enhancement then follows. Shadowing can occur due to the high vascularity of the cortex attenuating the signal from the medulla. Currently differentiation between malignant and benign renal lesions within the kidney is not possible (Figure 13.14).

Spleen and pancreas

Ultrasonic contrast agents can be used off-licence to study both the spleen (Figure 13.15) and pancreas (Figure 13.16). The pancreas can be studied using contrast-enhanced endoscopic ultrasound, to improve delineation, sizing and diagnosis of pancreatic lesions or for studying blunt trauma injury (Dietrich et al. 2012b).

Transcranial

SonoVue is licenced for use in transcranial studies. Colour Doppler or duplex sonography can be used, after a bolus injection, to aid differentiation of transcranial vessels which are occluded or which have low flow. It is specifically useful in those patients who have a poor signal-to-noise ratio in unenhanced ultrasound cerebral studies as it enhances

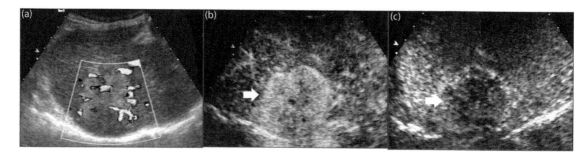

Figure 13.13 **(a)** Colour Doppler image of liver metastasis. **(b)** Image acquired using contrast-specific imaging in arterial phase after 2.4 mL injection of SonoVue – note the homogeneous enhancement in the liver metastasis indicated by the arrow. **(c)** Subsequent wash-out in portal venous phase. (Reproduced with permission from Dr. Emilio Quaia, Edinburgh Royal Infirmary.)

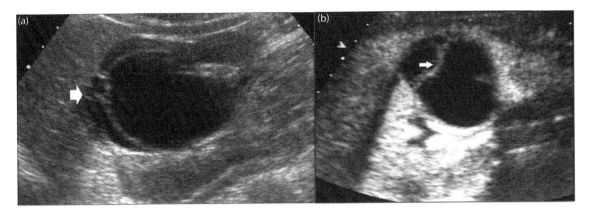

Figure 13.14 **(a)** Image suggests a complex renal cyst in fundamental B-mode imaging. **(b)** Acquired after a 1.2 mL bolus injection of SonoVue using a low-MI contrast-specific imaging technique. Arrow indicates an enhancing septa within the cystic renal lesion. (Reproduced with permission from Dr. Emilio Quaia, Edinburgh Royal Infirmary, Edinburgh.)

simultaneously in several vessels, showing brain anatomy and areas of blood flow.

Cardiac applications

Almost all ultrasonic contrast agents have been licenced for specific cardiac applications such as endocardial border definition and left ventricular opacification. Identification and enhancement of cardiac chamber borders is of specific benefit during stress-echo studies where the patient is either

physically (by exercise) or pharmacologically stressed. During such studies the heart is ultrasonically scanned at specific time points, generally at rest, low stress, peak stress and recovery, and the movements of individual cardiac chamber walls are continuously assessed using ultrasound. In difficult-to-image patients, sections of the heart walls may be indistinct. Injection of contrast enhances the endocardial borders, enabling a global assessment of cardiac wall movement. Such cardiac studies are performed in real time, using low-MI

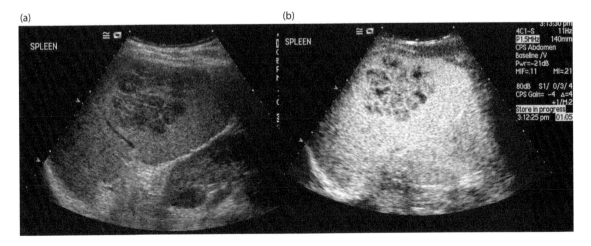

Figure 13.15 **(a)** Image suggests a possible splenic cyst in fundamental imaging. **(b)** Acquired after a bolus injection of SonoVue and using a low-MI technique to image the spleen. Honeycomb pattern suggestive of a spleen abscess. (Reproduced with permission from Dr. Paul Sidhu, King's College Hospital, London.)

(a) (b)

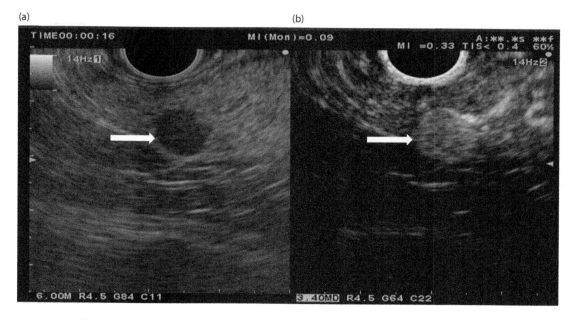

Figure 13.16 Endoscopy image of the pancreas with an insulinoma. **(a)** A 7 mm hypoechoic mass in the pancreatic body. **(b)** After injection of 2.4 mL of contrast, the lesion, highlighted by an arrow, is hyperenhancing with prompt homogeneous uptake of contrast in the arterial phase. (Reproduced with permission from Dr. Ian Penman, Edinburgh Royal Infirmary.)

contrast-specific imaging techniques. Contrast can be injected as a bolus but more routinely as an infusion (Figure 13.17). In addition, contrast agents are often used in cardiac for detection of mural thrombus (Figure 13.18).

MOLECULARLY TARGETED CONTRAST AGENTS

Targeting specific biological markers can be achieved using an antibody-antigen link on the surface of the microbubble along with a targeting ligand specific to a cell marker. For example the targeting ligand can be specific to a marker for inflammation. Targeted microbubbles flowing through the vascular system come into contact with the cell marker of interest, bind to the marker and provide site-specific imaging of regions of inflammation. Novel contrast agents targeted to specific biological molecular markers have been developed and several clinical applications are currently under investigation in first-in-human studies. One molecule expressed in neoangiogenesis (vascular endothelial growth factor receptor type 2 – VEGFR2) has been targeted in a first-in-human

prostate study (Smeenge et al. 2017), while kinase insert domain receptor (KDR) targeted contrast microbubbles were used in the first-in-human study with patients with breast and ovarian lesions (Willmann et al. 2016). Both studies have demonstrated the potential of using microbubbles targeted to specific molecules.

SAFETY OF CONTRAST AGENTS

The main risks associated with the use of ultrasound contrast agents are embolic risk, allergic reaction, toxicity and biological effects due to acoustic cavitation. The safety of two commercially available contrast agents (Definity/Luminity and Optison) was addressed retrospectively in a large multi-centre study which indicated that the incidence of severe adverse reactions was less than that associated with contrast agents used for other imaging modalities. However, it is the likelihood of acoustic cavitation occurring under routine clinical application which is of most concern. Gas bodies do not exist under normal conditions within the blood or soft tissue. When contrast microbubbles are injected into the body, they provide a potential source of cavitation

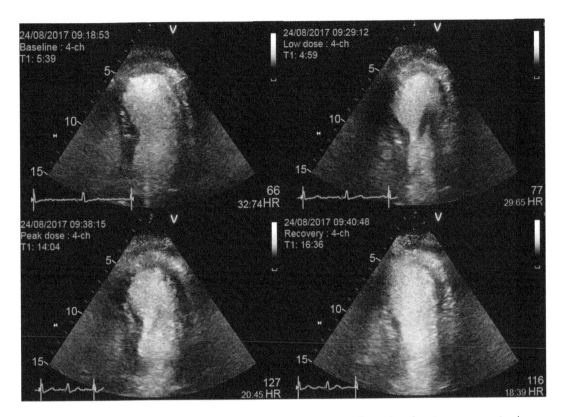

Figure 13.17 Cardiac stress-echo exam with SonoVue. Parasternal four-chamber images acquired at baseline (top left), low dose (top right) and high dose (bottom left) dobutamine and recovery (bottom right). (Reproduced with permission from Dr. Patrick Gibson, Consultant Cardiologist, Royal Infirmary of Edinburgh.)

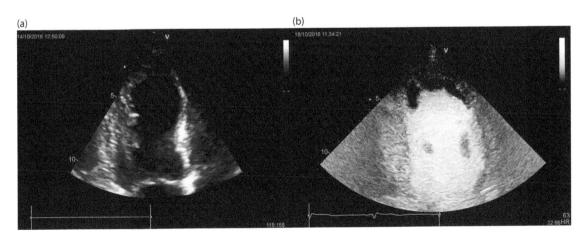

Figure 13.18 **(a)** Heart apex is akinetic in two-chamber view. **(b)** After injection of contrast (SonoVue), mural thrombus is confirmed at apex. (Reproduced with permission from Dr. Patrick Gibson, Consultant Cardiologist, Royal Infirmary of Edinburgh.)

nuclei. Cavitation has previously been shown to result in high-speed fluid jets, localised high temperatures (>1000°C) and sonochemical reactions. Whether intact microbubbles are capable of causing cavitation effects within the body is still under investigation. The potential bioeffects from these processes have been reviewed by the WFUMB and current recommendations for the safe use of ultrasonic contrast agents include (1) scanning at low MI, (2) scanning at higher frequencies, (3) reducing total acoustic exposure time, (4) reducing contrast dose and (5) adjusting the timing of cardiac triggering during cardiac contrast studies (Barnett et al. 2007). In addition, due to increasing use of contrast agents for off-label applications, specific care should be taken in tissues and organs (e.g. eye, neonate and brain) for which damage to the microvasculature could result in significant clinical implications (Piscaglia et al. 2012). Recently EFSUMB has also published a position paper on the use of ultrasound in paediatrics (Sidhu et al. 2017) while SonoVue has recently been approved for assessment of liver lesions in paediatrics. The safety of contrast agents is also addressed in Chapter 16.

ARTEFACTS IN CONTRAST IMAGING

In Chapter 5, the sources of the most commonly observed artefacts in clinical ultrasound imaging were discussed. In this section, we address those artefacts which are particularly relevant to contrast-enhanced clinical scanning.

Propagation artefacts

In ultrasound imaging, it is assumed that attenuation within soft tissue is constant and this is corrected on the ultrasound image by the use of time-gain compensation sliders. Of particular diagnostic interest are localized regions of increased attenuation which create shadows (reduced brightness) beyond the region of increased attenuation. A common example of the diagnostic potential of this is shadowing due to calcified plaque in vessels. Ultrasound contrast agents, as well as being strong scatterers of ultrasound, are also strong attenuators. As the contrast agent arrives in an organ via a feeding vessel, it is usual that the distal part of

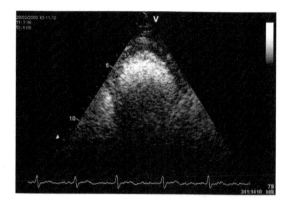

Figure 13.19 Shadowing caused by bolus injection of contrast (SonoVue). In this two-chamber view of the heart, as the contrast bolus enters the left ventricle, it strongly attenuates the ultrasound beam, and distal regions of the heart are difficult to visualise. After several cardiac cycles, the contrast will begin to clear and the whole ventricle will be visualised. (Image courtesy of Dr. Stephen Glen, Stirling Royal Infirmary, Scotland.)

the organ temporarily disappears from the screen due to the high attenuation of the microbubbles. After a period dependent upon contrast agent dose, rate of injection, scanner set-up and organ dynamics, the contrast will begin to clear from the organ and the distal part of the image will begin to reappear. This is particularly evident in bolus injections where a large number of scatterers are injected over a period of 1–2 seconds. Although shadow artefacts are useful in diagnosis in clinical scanning, due to their transient nature in contrast-enhanced studies they are currently not considered useful (Figure 13.19).

Multiple scattering

Within an imaging system, it is assumed that the transmitted ultrasound pulse travels only to targets on the beam axis and back to the transducer. When contrast agents are injected into this path, a range of different phenomena can occur. First, due to the number of scatterers present, multiple scattering is likely to occur, resulting in a delay in the return path of the ultrasound beam, and hence echoes from beyond the contrast-enhanced region will be displayed at greater depth than the true position of the source.

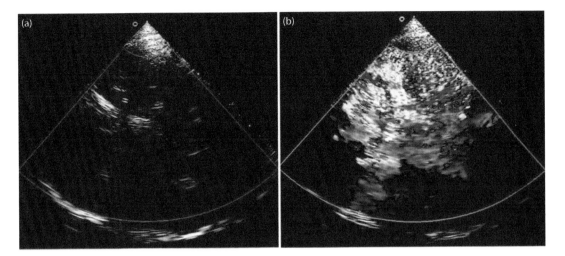

Figure 13.20 Blooming effect during ultrasound contrast application in a patient with inadequate acoustic cranial bone window: **(a)** unenhanced examination; **(b)** blooming phase after ultrasound contrast bolus injection. (Reprinted from *Seminars in Cerebrovascular Diseases and Stroke*, 5, Stolz EP and Kaps M, 111–131, Copyright 2005, with permission from Elsevier.)

Colour blooming

In colour Doppler imaging, after the injection of contrast an artefact known as colour blooming may occur. As contrast enters a vessel the magnitude of the backscattered signal increases, giving a corresponding large increase in Doppler signal. This increase in signal can expand beyond the width of the vessel, causing 'blooming', and can be corrected by reducing the Doppler gain (Figure 13.20).

SUMMARY

Contrast agents for use with ultrasound are in the form of microbubbles – gas bubbles encapsulated in a thin outer shell. These microbubbles expand and contract in the ultrasound beam, generating linear and non-linear signals. There are contrast-specific scanning modes available on ultrasound scanners which make use of the non-linear scattering from contrast microbubbles, allowing the signals from microbubbles to be distinguished from tissue. Ultrasound contrast agents can be used clinically for scanning vascular organs such as the heart and liver, and used appropriately, ultrasound contrast agents can improve the diagnostic ability of ultrasound scans.

ACKNOWLEDGEMENT

The authors would like to acknowledge that grant funding from the British Heart Foundation supported Dr. Mairéad Butler.

QUESTIONS

Multiple Choice Questions

Q1. Concerning ultrasound contrast agents:
    a. They can help in assessing difficult-to-image patients
    b. They are gas-filled microbubbles
    c. They are blood-pool agents
    d. They contain ionising radiation
    e. Their mean diameter is less than 6 microns

Q2. Ultrasound contrast agents, when subjected to an ultrasound pressure wave can:
    a. Reflect the ultrasound because of the change in acoustic impedance
    b. Oscillate symmetrically in the ultrasound beam
    c. Oscillate unsymmetrically in the ultrasound beam
    d. Be ruptured by the ultrasound beam
    e. Provide information about vessel wall dynamics

Q3. The second harmonic frequency is useful for imaging ultrasound contrast agents and is:
   a. Half the fundamental frequency
   b. The same as the fundamental frequency
   c. Twice the fundamental frequency
   d. Not related to the fundamental frequency
   e. One-and-a-half times the fundamental frequency

Q4. The behaviour of a contrast agent in an ultrasound pressure wave depends on the:
   a. Microbubble size
   b. Microbubble shell composition
   c. Transmit frequency
   d. Transmit pressure
   e. Gain setting on the scanner

Q5. Considering the safe use of contrast agents, the recommendations are for:
   a. Scanning at higher frequencies
   b. Reducing total acoustic exposure time
   c. Using the maximum possible contrast dose
   d. Scanning at low MI
   e. Scanning at lower frequencies

Q6. Pulse-inversion amplitude modulation imaging:
   a. Uses multiple pulses in a sequence
   b. Uses different frequency pulses
   c. Displays non-linear echoes
   d. Uses pulses of different acoustic pressures
   e. Displays linear echoes

Short-Answer Questions

Q1. How does the resonant frequency of a microbubble change with microbubble size?

Q2. How does the response of a microbubble to ultrasound change with acoustic pressure?

Q3. How does harmonic imaging work?

Q4. Why does harmonic imaging work well with contrast agents?

Q5. What are the main risks and hazards associated with ultrasound contrast agents?

Q6. Summarize some clinical areas where contrast agents can be used.

REFERENCES

Barnett SB, Duck F, Ziskin M. 2007. Recommendations on the safe use of ultrasound contrast agents. *Ultrasound in Medicine and Biology*, 33, 173–174.

Bilotta F, Robba C, Santoro A, Delfini R, Rosa G, Agati L. 2016. Contrast-enhanced ultrasound imaging in detection of changes in cerebral perfusion. *Ultrasound in Medicine and Biology*, 42, 2708–2716.

Claudon M, Dietrich CF, Choi BI et al. 2013. Guidelines and good clinical practice recommendations for contrast enhanced ultrasound (CEUS) in the liver – Update 2012: A WFUMB-EFSUMB initiative in cooperation with representatives of AFSUMB, AIUM, ASUM, FLAUS and ICUS. *Ultrasound in Medicine and Biology*, 39, 187–210.

Dietrich CF, Averkiou MA, Correas JM, Lassau N, Leen E, Piscaglia F. 2012a. An EFSUMB introduction into dynamic contrast-enhanced ultrasound (DEC-US) for quantification of tumour perfusion. *Ultraschall in Der Medizin*, 33, 344–351.

Dietrich CF, Sharma M, Hocke M. 2012b. Contrast-enhanced endoscopic ultrasound. *Endoscopic Ultrasound*, 1(3), 130–136.

Kollmann C. 2007. New sonographic techniques for harmonic imaging – Underlying physical principles. *European Journal of Radiology*, 64, 164–172.

Ophir J, Parker KJ. 1989. Contrast agents in diagnostic ultrasound. *Ultrasound in Medicine and Biology*, 15, 319–333.

Piscaglia F, Nolsoe C, Dietrich CF et al. 2012. The EFSUMB guidelines and recommendations on the clinical practice of contrast enhanced ultrasound (CEUS): Update 2011 on non-hepatic applications. *Ultraschall in Der Medizin*, 33, 33–59.

Sidhu P, Cantisani V, Degannello A et al. 2017. Role of contrast-enhanced ultrasound (CEUS) in paediatric practice: An EFSUMB position statement. *Ultraschall in Der Medizin*, 38, 33–43.

Smeenge M, Tranquart F, Mannaerts CK, de Reijke TM, van de Vijver MJ, Laguna MP, Pochon S, de la Rosette JJMCH, Wijkstra H. 2017. First-in-human ultrasound molecular imaging with a VEGFR2-specific ultrasound molecular contrast agent (BR55) in prostate cancer: A safety and feasibility pilot study. *Investigative Radiology*, 52(7), 419–427.

Willmann JK, Bonomo L, Testa AC, Rinaldi P, Rindi G, Vallura KS, Petrone G, Martini M, Lutz AM, Gambhir SS. 2016. Ultrasound molecular imaging with BR55 in patients with breast and ovarian lesions: First-in-human results. *European Journal of Radiology*, 84(9), 1685–1693.

Elastography

PETER R HOSKINS

INTRODUCTION

The term 'elastography' is used to describe techniques which provide information related to the stiffness of tissues. It has long been known that diseased tissues such as tumours are stiffer than the surrounding normal tissue. Indeed one of the oldest diagnostic methods is the assessment of the stiffness of tissues by palpation; if a stiff lump is found then the lump may well be diseased. Ultrasound techniques generally work by replicating this process; the tissue is squeezed and the response to squeezing is measured using the ultrasound system.

Ultrasound elastographic techniques in this chapter are classified in two ways based on the underlying measurement principle:

- *Strain techniques*: These rely on the compression of the tissues, and the measurement of the resulting tissue deformation and strain using ultrasound. These may be referred to as 'static' methods.

- *Shear-wave techniques*: These rely on the generation of shear waves, and the measurement of shear-wave velocity within the tissues using ultrasound, from which elastic modulus may be estimated. These may be referred to as 'dynamic' methods.

The first of these techniques provides information mainly on strain; the estimation of elastic modulus is challenging and not much performed outside research labs. The second technique does provide information on elastic modulus, but is more technically challenging, especially for two-dimensional (2D) image formation.

This chapter concentrates on elastography methods which have been adopted commercially, in order to offer understanding to sonographers and other users of the techniques available in routine clinical practice. Other elastographic techniques have been described which, though promising, have yet to find their way into commercial systems. Details of these can be found in review articles such as those by Ophir et al. (1999)

and Greenleaf et al. (2003). Other general reviews of elastography are by Wells and Liang (2011), Palmeri and Nightingale (2011), Hoskins (2012) and Dewall (2013).

ELASTICITY

When a material is subject to a force it is stretched or compressed. The general terminology and principles used to describe this process can be illustrated using a simple material such as a rubber band or a wire (Figure 14.1). If a weight is attached to a strip of the material, the strip stretches. If the weight is increased the strip stretches further. Eventually the strip will break when the weight becomes too much.

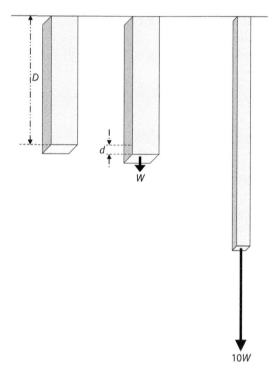

Figure 14.1 Stretching of a simple elastic material by a force W. The unstretched material has a length D. A weight W stretches the material by an amount d. In the case of soft tissue the material is incompressible (density does not change) so that the stretch will result in reduction of the cross-sectional area. This is small for small stretches, but when a large force is applied the material will stretch by a large amount and there is a clear reduction in cross-sectional area.

In Figure 14.1 a weight W is attached to a strip of material. The stretch from the original length is d. For a simple material, such as rubber, the stretch is proportional to the added weight. A simple equation describing this behaviour is:

$$\text{Stress/Strain} = \text{Constant} \quad (14.1)$$

For Figure 14.1 the stress is the force W divided by the cross-sectional area A. This has units of Nm^{-2}, the same units as pressure:

$$\text{Stress} = \text{Force/Area} \quad (14.2)$$

For the example of Figure 14.1 this gives:

$$\text{Stress} = W/A \quad (14.3)$$

The strain is the change in length d divided by the original (non-distended) length D. This has no units:

$$\text{Strain} = \frac{\text{Length (after)} - \text{Length (before)}}{\text{Length (before)}} \quad (14.4)$$

For the example of Figure 14.1 this gives:

$$\text{Strain} = d/D$$

The constant in Equation 14.1 is commonly referred to as Young's modulus, E. Equation 14.1 then becomes:

$$E = \text{Stress/Strain} \quad (14.5)$$

For many materials, including human soft tissues, the density of the material is virtually unchanged by stretching or compression. In Figure 14.1 a large stretch will therefore result in reduction of the cross-sectional area of the strip; in other words, extension in one direction is compensated by contraction in the other directions.

Young's modulus may be measured in a tensile testing system where narrow strips of the material are subject to a known force and the extension is measured. The force is gradually increased and the material is stretched until it breaks. Young's modulus is measured from the early part of the graph as the ratio of stress divided by strain.

Table 14.1 Young's modulus values for harder materials

Material	E (GPa)
Non-human materials	
Diamond	1220
Steel	200
Wood (oak)	11
Nylon	1–7
Vulcanized rubber	0.010–0.100
Human tissues	
Tooth enamel	20–84
Femur	11–20

Table 14.2 Young's modulus values for softer materials

Material	E (kPa)
Non-human materials	
Silicone rubber	500–5000
PVA cryogel tissue mimic	35–500
Agar/gelatine tissue mimic	10–70
Human tissues	
Artery	700–3000
Cartilage	790
Tendon	800
Healthy soft tissues[a]	0.5–70
Cancer in soft tissues[a]	20–560

[a] Breast, kidney, liver, prostate.

Typical values of Young's modulus are given for hard materials in Table 14.1 and soft materials including soft tissues in Table 14.2. Note that there is a factor of a million between the values in the two tables. A compilation of published work on Young's modulus of biological tissues can be found in Duck (2012) and Sarvazyan (2001). Tissue mimics suitable for construction of elastography phantoms continue to be developed. Candidate materials have included polyvinyl alcohol cryogel (PVAc) (Dineley et al. 2006), agar and gelatin-based tissue mimics (Madsen et al. 2005; Manickam et al. 2014) and oil-based gel tissue mimics (Cabrelli et al. 2017).

General principles of strain elastography

The steps needed for estimation of the elastic modulus using lab-based or imaging-based methods are illustrated in Figure 14.2.

- *Apply known force*: The force or stress compressing or stretching the tissues must be known.
- *Measure change in dimensions*: The change in dimensions of the tissues is measured, and the strain calculated from Equation 14.4.

- *Estimate Young's modulus*: Young's modulus is estimated from the equation.

This apparently simple set of steps is difficult to achieve except for the most simple of circumstances, such as the isolated thin strip of Figure 14.1. In the body there are a number of complexities in the estimation of Young's modulus which are explored in more detail later in this chapter. The main problem which is discussed at this stage is the lack of knowledge of the force or stress to which the tissue is subject. Difficulty in estimating the true tissue stress means that strain elastography concentrates on the estimation of tissue strain, not on the estimation of Young's modulus.

Strain and strain ratio

Figure 14.3 shows an idealized uniform tissue in which are embedded two lesions, one stiff and one soft. The block of tissue is subject to a uniform force which causes compression of the block in the vertical direction, and also expansion of the block in the horizontal direction. For materials such as soft tissue whose density does not change when they

Figure 14.2 Steps required for estimation of elastic modulus.

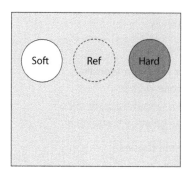
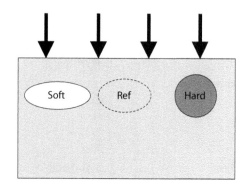

Figure 14.3 Effect of compression on a block of material in which are embedded two lesions, one soft and one hard. There is also a third reference region. Application of a uniform force compresses the tissue. As there is no change in density compression the block of tissue expands in the direction perpendicular to the stretch. There is little change in the dimensions of the hard lesion, but the soft lesion alters significantly.

are compressed this is the usual result – compression in one direction leads to stretching in another direction. In the compressed block, the soft lesion is compressed more than the surrounding tissue, which in turn is compressed more than the hard lesion. The change in dimensions which arise as a result of compression is the strain (Equation 14.4). The strain in the direction of the applied force (vertically in the case of Figure 14.3) is shown, as this is the measurement usually made using ultrasound. As stiff lesions are resistant to compression they are usually associated with low strain values. A strain value of zero would imply that the lesion was very hard. Figure 14.4 shows the two lesions and the reference region from Figure 14.3, pre- and post-compression. The reference region is usually a part of the image which is known or assumed to be

normal, to which the lesion can be compared. The strain ratio is:

$$\text{Strain-ratio} = \frac{\text{Strain (reference region)}}{\text{Strain (lesion)}} \quad (14.6)$$

In clinical practice the strain ratio is sometimes used as a surrogate index for stiffness in the absence of a true index of stiffness such as Young's modulus.

Clinical and research strain elastography systems

In the research literature there are several methods for elastography based on measuring the change in tissue dimensions (the strain) arising from an

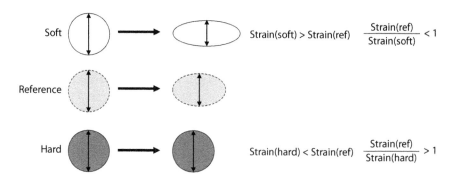

$$\text{Strain(soft)} > \text{Strain(ref)} \quad \frac{\text{Strain(ref)}}{\text{Strain(soft)}} < 1$$

$$\text{Strain(hard)} < \text{Strain(ref)} \quad \frac{\text{Strain(ref)}}{\text{Strain(hard)}} > 1$$

Figure 14.4 The dependence of strain on stiffness for the two lesions and the reference region of Figure 14.3.

applied force. An external compression device may be used, in order to control the rate and frequency at which the tissues are compressed (reviewed in Ophir et al. 1999; Greenleaf et al. 2003). Commercial systems rely on the use of the transducer or the acoustic radiation force to compress the tissue. These methods have the advantage of being clinically acceptable as the operator can both image and compress the tissues with one hand, without the involvement of separate compression devices which would require a second person to operate. The strain-based methods described in the next sections are based on the technology available on current commercial systems.

STRAIN ELASTOGRAPHY USING AN EXTERNALLY APPLIED FORCE

Commercial ultrasound systems generally use the transducer itself to compress the tissues. Older systems and also some current ultrasound systems require the operator to physically press the transducer in and out of the tissues while the ultrasound system records images. Increased ability to track tissue motion (see later in the text) means that very small natural movements are sufficient, such as those produced by the natural movement of the transducer, or caused within the patient as a result of breathing or pulsation from the heart.

This improved sensitivity means that the operator does not need to physically press the transducer in and out. For endoluminal elastography controlled movement of the transducer is difficult, and compression is achieved by the inflation of a plastic sheath, by the injection of water into the sheath using a syringe which the operator controls.

There are two main methods for estimation of tissue movement, one using the amplitude of the received echoes extracted from the radiofrequency (RF) data, and one using the Doppler tissue imaging (DTI) data.

Strain estimation from A-lines

Figure 14.5 shows a uniform tissue in which there is a stiff lesion. A-scan lines are shown passing through the centre of the lesion pre- and post-compression. The received echoes as a function of depth are shown in Figure 14.6b. The ultrasound system estimates the amount by which the tissue has moved at each point as a result of the compression (Figure 14.6c), and from this estimates the strain (Figure 14.6d).

This method for estimation of tissue displacement relies on the use of signal-processing methods which compare the A-line pre- and post-compression. Essentially, these are search algorithms which identify any change in position or stretching of image features. In its most simple

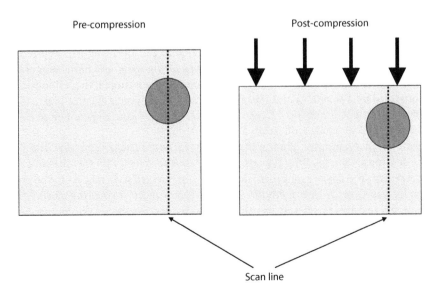

Pre-compression

Post-compression

Scan line

Figure 14.5 Location of an A-scan line pre- and post-compression (see Figure 14.6).

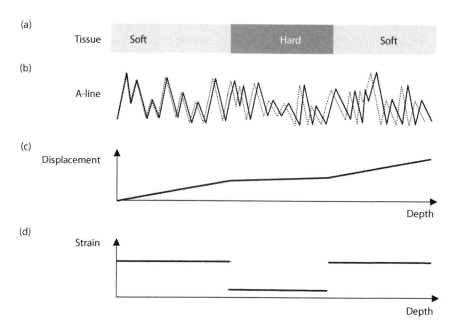

Figure 14.6 Steps in the estimation of strain from A-scan lines pre- and post-compression. **(a)** Idealized tissue consisting of a hard lesion surrounded by softer tissue; **(b)** A-lines from Figure 14.5 are shown with the post-compression line (dotted) displaced; **(c)** displacement estimated from (b); **(d)** strain estimated from **(c)**.

form the method assumes that the tissue only moves in the direction of the A-line. However, it was previously noted that tissue spreads sideways when compressed. To account for this, some systems incorporate search features which explore up and down an individual A-line (axially), and also between adjacent A-lines (transversely). This helps improve the estimation of strain and the appearance of the strain image. Further details of correlation-based search algorithms used for strain estimation are provided by Yamakawa and Shiina (2001) and Shiina et al. (2002).

Figure 14.7 shows typical images of strain in hard and soft lesions in phantoms. Figure 14.8 shows examples of clinical images. The displayed strain may be presented in colour or in greyscale. Strain elastography based on A-line methods is especially suitable for real-time use, and has the advantage over off-line methods of immediate visualisation of strain. A further advantage is that the degree of compression tends to be less than that required for DTI methods (below), which reduces artefacts associated with lesions moving out of the beam during compression.

Strain estimation from Doppler tissue imaging

Methods for estimation of strain were described in Chapter 11, in the context of cardiac imaging using DTI. The same methods may be used to estimate strain in elastography. The steps of the process are given in the following list. Figure 14.9 shows each step for imaging of a hard lesion in a breast phantom. In the example, the lesion is clearly visible on the B-mode image (Figure 14.9a), though the lesion in clinical studies may not be so clearly visible.

- *Estimate tissue velocities*: The estimation of tissue velocity by the ultrasound system is performed with respect to the position of the transducer. As the transducer is pushed into the tissues, the tissues are compressed, and in the ultrasound image this appears as a movement of the tissues towards the transducer. Similarly, when the transducer is withdrawn, the appearance in the ultrasound image is of the tissue moving away from the transducer. These movements give rise to a DTI image of velocity

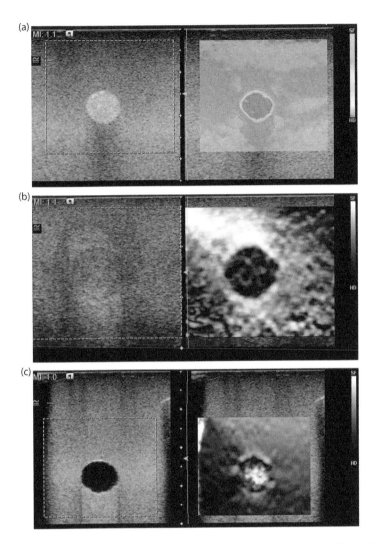

Figure 14.7 Strain images for lesions in breast phantoms. Each figure shows the B-mode image on the left and the strain image on the right. **(a)** Stiff echogenic lesion on B-mode, with low strain (red); **(b)** stiff lesion in which the lesion has the same echogenicity as the surrounding tissue, with low strain (black); **(c)** fluid-filled structure shown as an echo-free region on the B-mode, with high strain (white).

(Figure 14.9b). A sequence of DTI images is recorded as the transducer is pressed in and out of the tissue.

- *Estimate velocity gradient*: The change in velocity with distance along each image line is estimated from the DTI data (Figure 14.9c).
- *Estimate strain*: The strain is estimated from the velocity gradient image (Figure 14.9d). The strain values within the hard inclusion are much reduced compared to the adjacent tissue. Note also the red halo at the top and bottom

of the inclusion on the strain image. This is a commonly seen artefact associated with the strain estimation process.

In order to generate a sufficiently large velocity, the tissue must be pressed in several millimetres, more than is required for A-line based methods. This approach results in two potential artefacts: one is the displacement of lesions out of the imaging plane, and the other is the halo artifact as shown in Figure 14.9d.

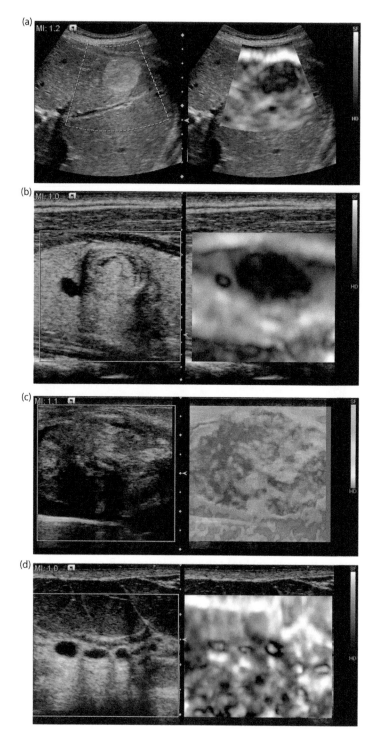

Figure 14.8 Strain images of lesions in patients. Each figure shows the B-mode image on the left and the strain image on the right. **(a)** Liver haemangioma with decreased strain compared to the surrounding tissue; **(b)** thyroid mass with decreased strain; **(c)** muscle mass with decreased strain. **(d)** breast cysts showing high strain within the cysts.

Figure 14.9 DTI estimation of the strain. **(a)** B-mode image of a stiff lesion in a breast phantom; **(b)** DTI velocity image; **(c)** velocity gradient image; **(d)** strain image showing reduced strain in the lesion, and also regions of increased stiffness above and below the lesion which are artefacts.

Quantification of elasticity from strain images

As previously noted the estimation of Young's modulus requires knowledge of the true stress applied to the tissue. In general this is difficult to obtain, and the strain ratio may be used instead as an index of elastic modulus. The estimation of strain ratio involves comparison of the strain in the lesion with the strain in an adjacent 'reference' region (Equation 14.6). In Figure 14.10 regions

of interest are shown placed in the hard lesion of an elastography phantom, and an adjacent 'reference' region at the same depth. The strain values as a function of time during the compression of the tissues are also shown for both regions. The strain reaches a maximum at the point of maximum compression, and the strain-ratio measurement may be taken from this time point. In the example of Figure 14.10 the strain ratio has a value of 10, indicating that the lesion is some 10 times as stiff as the surrounding tissue. Figure 14.11 shows

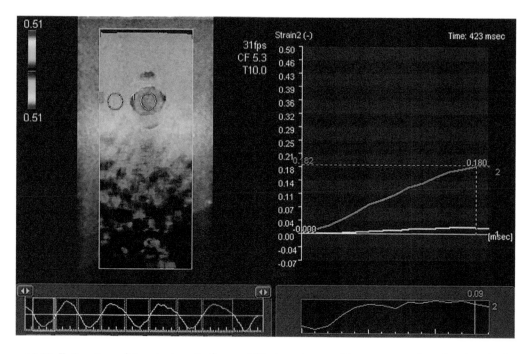

Figure 14.10 Estimation of the strain ratio for a stiff lesion. Areas of interest are placed within the lesion and in an adjacent region at the same depth. A graph of strain versus time is shown during the compression phase. The maximum compression values are 0.180 and 0.018, giving a strain-ratio value of 10.

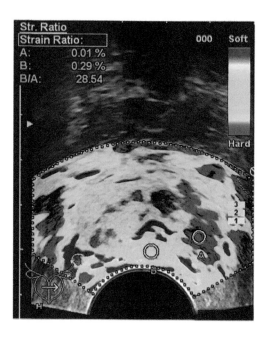

Figure 14.11 Estimation of strain ratio in a prostate lesion. Areas of interest are placed within the lesion and in an adjacent region. The strain ratio has a value of 28.

the strain ratio calculated in a strain image from a diseased prostate; in this case the strain ratio has a value of 28, indicating that the lesion is very stiff.

Change in strain with depth

When the tissue is compressed externally with the transducer, the force of compression is not uniform throughout the image. The force of compression decreases with depth. This means that a lesion, of similar hardness and size, will compress more the nearer it is to the transducer (Figure 14.12). The decrease in strain with depth within the tissue of an elastography phantom can also be seen in Figure 14.9d. This has two important consequences. The first consequence is that elastography by compression using the transducer is most effective for lesions near the surface, in practice less than about 5 cm depth. The second consequence is that, when estimating the strain ratio, the reference region should be placed at the same depth as the lesion. The use of a reference region at a different depth will lead to a different strain ratio.

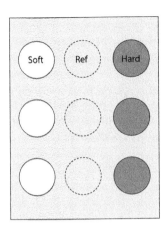
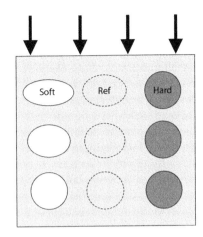

Figure 14.12 Effect of depth on strain. For a uniform force applied to the block of tissue, the degree of compression reduces with depth.

STRAIN ELASTOGRAPHY USING ACOUSTIC RADIATION FORCE

An ultrasound beam generates a radiation force which is directed in the direction of wave propagation. The magnitude of the force is highest in the focal zone, and increases with overall output power.

Acoustic radiation force imaging (ARFI) provides a 2D strain image and involves the use of high-output beams to displace a region of tissue, and imaging beams to monitor the displacement (Nightingale et al. 2001, 2002, 2006). The minimum sequence of beams used in each line of the ARFI image is illustrated in Figure 14.13. The ARFI technique sends a

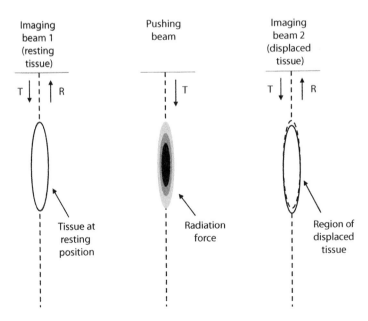

Figure 14.13 Sequence of ultrasound beams used for ARFI elastography. (left) An imaging beam records the position of the tissue in its resting position; (middle) a pushing beam produces a radiation force which displaces tissue in the focal region; (right) a second imaging beam sent immediately after the pushing beam records the position of the displaced tissue.

high-output ultrasound pulse to produce displacement of the tissue in the focal region. This is usually called the 'pushing beam' or 'pushing pulse' as its job is to produce a small movement of the tissues. The displacement produced is typically 1–20 μm, sufficient to be detected using an ultrasound system. The displacement reaches a peak after 1 ms and the tissue restores to its normal position within 5 ms. The pushing pulse is not used for imaging purposes; this requires separate imaging pulses. At least two ultrasound pulses are used in a conventional transmit/receive imaging manner, one sent before the pushing pulse to monitor the position of tissue prior to displacement, and a second immediately after the pushing pulse to monitor the position of the tissue while it is displaced. In practice more complicated sequences may be used involving several pushing pulses interleaved with imaging pulses, followed by a series of imaging pulses to monitor the return of the tissue to its undisturbed position (Nightingale et al. 2002). Comparison between the received A-lines for each imaging pulse is performed to estimate the movement of the tissues. Each line of the ARFI image therefore requires a minimum of three ultrasound pulses: a pushing pulse and at least two imaging pulses. Figure 14.14 shows an image of a lesion in the liver obtained using the ARFI

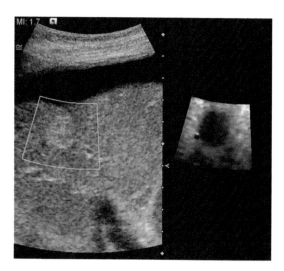

Figure 14.14 ARFI elastography in the liver. The B-mode image shows an indistinct lesion. The strain image shows that the lesion has increased hardness with much improved image contrast.

technique. It was previously noted that compression of the tissues using an external force may be ineffective for deeper tissues. In these situations ARFI may be used to overcome this limitation.

ESTIMATION OF ELASTIC MODULUS USING STRAIN ELASTOGRAPHY

The estimation of elastic modulus using strain elastography relies on methods for estimation of the applied stress at each point within the tissue. A method is needed to estimate the force applied to the tissue from the ultrasound transducer or the radiation force, and then a method to determine how that force is distributed through the tissue. Measurement of the applied force could be achieved using a force transducer attached to the ultrasound transducer, or through calculation of the radiation force from the pushing pulse. The force distribution in tissue is not uniform; this is shown in Figures 14.9d and 14.12 where the resulting strain decreases with depth. In general the force distribution in patients is affected by the shape of the transducer, the shape and mechanical properties of the underlying tissues and the size and duration of the applied force. Current commercial ultrasound systems do not include the technology to estimate the stresses within the tissues, and hence do not provide a means for estimating elastic modulus from strain elastography.

ELASTIC MODULI AND WAVE GENERATION

This section discusses elastic moduli and wave generation, in order to understand how shear waves can be used to estimate stiffness (as discussed in the next section). In general for waves such as those used in medical ultrasound systems, the speed of the wave as it travels through the tissue is determined by the density of the tissue and the appropriate elastic modulus.

Bulk modulus B and pressure waves

When a material is subject to increased pressure the material will be compressed. The bulk modulus describes the change in volume of a material occurring as a result of compression. Figure 14.15a–c shows a cube of material of volume V subject to a pressure P on all sides. As a result of a change in pressure to $P + \delta P$ the cube compresses, causing a reduction in volume by an amount δV. The bulk modulus B expresses these changes as an equation. The minus sign indicates that an increase in pressure results in a decrease in volume:

$$B = \frac{-\text{Change in pressure}}{\text{Fractional change in volume}} = \frac{-\delta P}{\delta V / V} \quad (14.7)$$

When a tissue is subject to a change in pressure, for example by being struck or compressed, pressure waves will be generated which propagate through the tissue. The waves generated by a diagnostic ultrasound transducer are called either 'compressional waves' or 'longitudinal waves'. The speed of propagation of the pressure wave is controlled by the bulk modulus, as shown in Equation 14.8:

$$c_l = \sqrt{\frac{B}{\rho}} \quad (14.8)$$

where c_l is the speed of the longitudinal wave, B is the bulk modulus and ρ is the average density.

Figure 14.16a illustrates the propagation of compressional waves. At any one point the material oscillates to and fro in the direction of motion of the wave. There are changes in density as the wave propagates, illustrated in Figure 14.16a, since changes occur in the area of each element at different locations.

Shear modulus G and shear waves

When an object is subject to a force parallel to one surface, that surface will be dragged in the direction of the force. This force is called a 'shear force'. This is illustrated in Figure 14.15d–f, where the shear force drags one surface of a cube in the direction of the force. The shear force is transmitted through the rest of the cube, causing the cube to be distorted, or sheared, in the direction of the force. The shear modulus describes the ability of the material to withstand a shear force F. The amount of shear is represented by the angle θ. The shear modulus expresses these changes as an equation:

$$G = \frac{\text{Shear stress}}{\text{Shear strain}} = \frac{F/A}{\tan(\theta)} \quad (14.9)$$

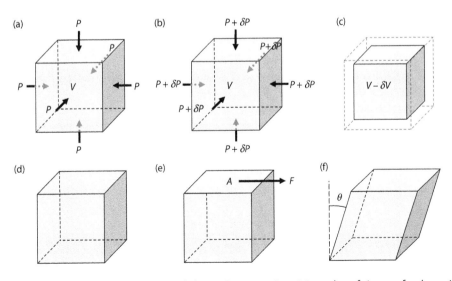

Figure 14.15 Effect of tissue compression and shear. Compression: **(a)** a cube of tissue of volume V is shown subject to pressure P on all sides; **(b)** an increase in pressure to $P + \delta P$ results in **(c)** compression of the cube by an amount δV. Shear: **(d)** a cube of material is shown; **(e)** this is subject to a force acting parallel to one of the sides resulting in **(f)** the cube being pulled over to one side, by an amount shown by the angle θ.

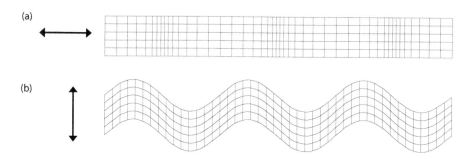

Figure 14.16 Propagation of waves. **(a)** In compressional waves the tissue moves to and fro in the same direction as the wave motion; **(b)** in shear waves the tissue moves transverse to the wave motion.

When a shear force is applied to a material, shear waves are generated which travel through the material. The propagation of the shear waves is controlled by the shear modulus as shown in Equation 14.10:

$$c_s = \sqrt{\frac{G}{\rho}} \qquad (14.10)$$

where c_s is the speed of the shear wave, G is the shear modulus and ρ is the density.

Figure 14.16b illustrates the propagation of shear waves. At any one point the material oscillates to and fro perpendicular to the direction of motion of the wave. For shear-wave propagation there is no change in density with time. This is illustrated in Figure 14.16b by the area of each element remaining the same at different locations. The passage of shear waves relies on the ability of adjacent elements of the tissue to remain connected while a shear force is applied. Though solids and soft tissues support shear waves, application of a shear force in a fluid results in gross motion of the fluid, resulting in disconnection of adjacent elements, and therefore fluids do not support shear waves. Table 14.3 briefly summarizes the characteristics of shear waves and compressional waves.

Provided that the tissue can be assumed to be incompressible (no change in density) and uniformly elastic the shear modulus G is related to Young's modulus E by the following equation:

$$E = 3G \qquad (14.11)$$

Combining Equations 14.10 and 14.11 gives:

$$E = 3\rho c_s^2 \qquad (14.12)$$

Equation 14.12 provides a means for determining tissue stiffness from the measured shear-wave velocity.

Values of elastic moduli in body tissues

Figure 14.17 illustrates the values for bulk modulus B, shear modulus G and Young's modulus E for different body tissues. The ranges of G and E are very large, by a factor of 10^7, or 10 million.

Table 14.3 Features of shear and compressional waves

	Shear waves	Compressional waves
Changes in local density	No	Yes
Elastic modulus	G (shear modulus)	B (bulk modulus)
Speed of waves in soft tissue	1–10 m s^{-1}	1400–1600 m s^{-1}
Can wave travel through fluid?	No (at low frequencies)	Yes

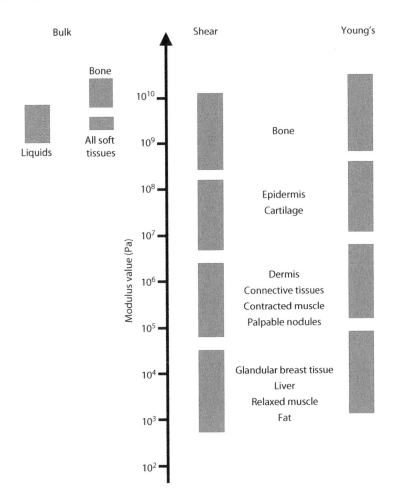

Figure 14.17 Values of bulk modulus B, shear modulus G and Young's modulus E for different tissues. The values for B in soft tissues occupy a narrow range which is similar to that for fluids. The values for G and E occupy over seven orders of magnitude (a factor >10,000,000). (Reprinted with permission from Sarvazyan AP et al. 1998. *Ultrasound in Medicine and Biology*, 24, 1419–1435.)

SHEAR-WAVE ELASTOGRAPHY

The steps for estimating tissue stiffness from shear wave velocity are described as follows and illustrated in Figure 14.18.

- *Induce shear waves in the tissue*: If the tissue is vibrated then shear waves will be produced

which will travel in all directions. These travel at a speed which is dependent on the local density and the local elastic modulus, according to Equation 14.10. Typical frequencies for which shear waves are generated are in the range 10–500 Hz. The speed of propagation of shear waves is typically 1–10 m s^{-1}. For 50-Hz shear waves in healthy liver the speed is about

Figure 14.18 Steps in the estimation of elastic modulus using shear waves.

1 m s^{-1} and the corresponding wavelength is 20 mm.

- *Measure the speed of propagation through the tissue of interest*: For the region of tissue which is of interest the speed of propagation of the shear waves is measured using the ultrasound system.
- *Estimate the stiffness*: Use of Equation 14.12 provides a value for the stiffness, calculated from the measured shear-wave velocity and the density of the tissue of interest. As density cannot be measured *in vivo* non-invasively, this requires an assumption to be made on the density of tissues. Manufacturers give little detail of this. Values for the density of various tissues are provided in Table 14.4 (from Duck 2012). The average density for a range of soft tissues (breast, prostate, liver, kidney) is $1047 \pm 5 \text{ kg m}^{-3}$.

The generation of shear waves relies on a method to induce motion within the tissues. Commercial methods use a handheld transducer to both generate and detect the shear waves. Methods may be divided into those that generate shear waves by the use of an external vibrator applied to the skin, and those which generate shear waves internally within the tissue by acoustic radiation force.

Table 14.4 Density of soft tissues

Tissue	Density (kg m^{-3}) mean (range)
Fat	928 (917–939)
Muscle – skeletal	1041 (1036–1056)
Liver	1050 (1050–1070)
Kidney	1050
Pancreas	1040–1050
Spleen	1054
Prostate	1045
Thyroid	1050 (1036–1066)
Testes	1040
Ovary	1048
Tendon (ox)	1165
Average soft tissues,[a] mean (S.D.)	1047 (5)

Note: Values are taken from Duck (2012). Values are reported as mean (range), or mean, or range.
[a] Excluding fat and tendon.

Elastic modulus estimation using shear waves generated from an external actuator

Methods described in the literature include external vibrators separate from the ultrasound transducer (reviewed in Ophir et al. 1999; Greenleaf et al. 2003). This section concentrates on methods adopted clinically in which the ultrasound transducer is used to both produce and detect the shear waves (Sandrin et al. 2002, 2003). This is referred to as 'transient elastography' as it produces a shear wave which is of short duration, as opposed to the continuous vibration produced in magnetic resonance imaging (MRI)–based dynamic elastography techniques (discussed in Greenleaf et al. 2003). The method described by Sandrin et al. (2002, 2003) is not based on the use of an imaging device, instead the operator must manually place the probe on the patient's skin in a region where the operator knows that the liver will be insonated. This method produces a single value of Young's modulus from one location.

The system described by Sandrin et al. (2002, 2003) and adopted commercially, uses a combined vibrator and ultrasound transducer (Figure 14.19a). The vibrator is a piston source which indents the tissue to produce shear waves and the ultrasound transducer measures the shear-wave velocity. This is an A-line–based technique; in other words a single line of ultrasound is acquired, and a single measurement is made, of the velocity of shear waves. No image is acquired. The steps involved are illustrated in Figure 14.19 and described as follows:

- *Production of shear waves*: The piston vibrator is pressed against the patient's skin by the operator, usually to insonate the liver. Activation of the vibrator indents the tissue at a low frequency of 50 Hz for 20 ms. The shear waves which are produced travel through the tissues (Figure 14.19b).
- *Ultrasound acquisition*: The ultrasound system is operated in pulse-echo mode (Figure 14.19c), acquiring ultrasound data along a single beam direction. The shear waves cause the tissue to be displaced in the direction of the beam, so that in each consecutive A-line the position of the shear wave appears slightly deeper.

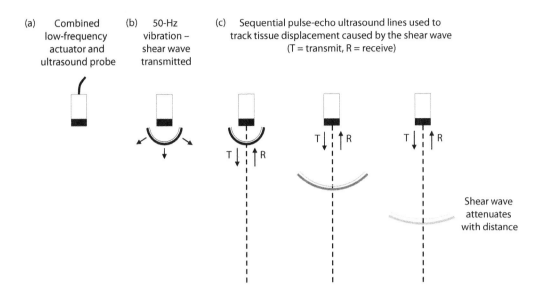

Figure 14.19 Production of shear waves and acquisition of ultrasound beams using a combined actuator-ultrasound system. **(a)** Schematic diagram of the combined transducer; **(b)** shear waves are produced by the actuator; **(c)** sequential A-line data are captured.

- *Ultrasound processing*: The A-line data are processed using cross-correlation algorithms to estimate the displacement of the tissue between consecutive A-lines, from which the resulting strain is calculated. The shear wave velocity is then calculated from the change in strain with time (Figure 14.20).

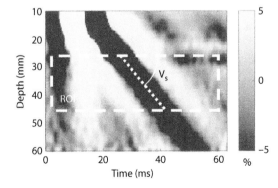

Figure 14.20 Strain values for sequential A-lines are displayed for a period of 60 ms following initiation of the shear wave. The wave velocity is the slope of the wave pattern. (Reprinted with permission from Sandrin L et al. 2003. *Ultrasound in Medicine and Biology*, 29, 1705–1713.)

- *Estimation of elastic modulus*: Young's modulus may be estimated from the shear-wave velocity using Equation 14.12. An assumed value for the tissue density is required.

Unlike pulse-echo techniques, the transmission and reception phases of the shear wave system described here overlap (Figure 14.21). The effect of vibrator movement on the received A-lines must therefore be removed before shear-wave velocity can be correctly estimated. Sandrin et al. (2002) assume that the shear waves attenuate quickly, so that the vibrator movement is deduced from the displacement estimated from deep echoes.

In order to measure the shear-wave velocity, a sufficient number of A-lines must be acquired while the shear wave passes from the vibrator to the bottom of the ultrasound field of view. The ultrasound system measures to a depth of about 6 cm. For a shear wave whose velocity is 3 m s^{-1}, the wave will travel 6 cm in 20 ms. For at least 20 A-lines to be acquired while the shear wave travels 6 cm requires a time between ultrasound pulses of at least 1 ms, or a pulse repetition frequency of 1000 Hz. This is easily achievable for a single beam; however, to produce a 2D image involving many image lines requires high-frame-rate techniques, as discussed later.

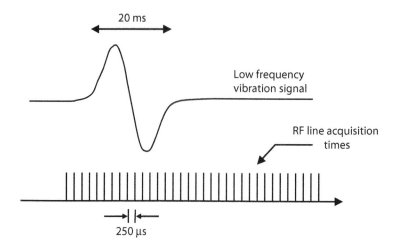

Figure 14.21 Demonstration that the 50-Hz vibration and ultrasound pulse-echo signals overlap. (Reprinted with permission from Sandrin L et al. 2003. *Ultrasound in Medicine and Biology*, 29, 1705–1713.)

Elastic modulus estimation from shear waves produced using radiation force

SINGLE-REGION METHODS

The use of radiation-force methods to produce distension of the tissue within the ultrasound beam focal region is previously described. The basic radiation-force method produces a single region, within the tissue, of high acoustic output. The resulting tissue distension within the region causes the production of shear waves which propagate through the tissue in three dimensions (3D). The ultrasound system may be used to monitor the propagation of the waves through the tissue (Figure 14.22). This technique, proposed by Sarvazyan et al. (1998), has been used in published studies on phantoms and in excised tissues (Nightingale et al. 2003), and *in vivo* in the liver of volunteers (Palmeri et al. 2008).

Commercial implementations of this method provide a single value of shear-wave velocity corresponding to a local region of tissue, usually the

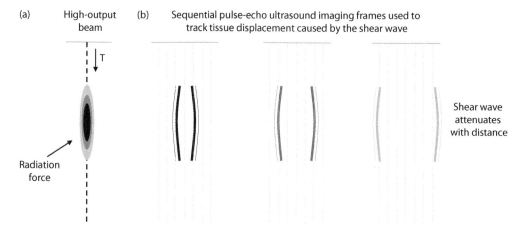

Figure 14.22 Shear-wave imaging using ARFI. (a) A high-output ultrasound beam produces a radiation force which displaces tissue in the focal region producing shear waves which propagate in 3D. (b) High-frame-rate imaging techniques are used to track the tissue displacement caused by the shear wave.

liver (Castera et al. 2005; Foucher et al. 2006). The location of the region, highlighted on the B-mode image, is controlled by the operator (Figure 14.23). Once the region is positioned, the operator actuates the high-output pulse which produces shear waves. These are generated just to one side of the region of interest. The shear waves travel from the high-output focal area through the region of interest. The progression of the shear wave through the region of interest is monitored using imaging beams, enabling estimation of the time of travel of the shear wave from its origin at the high-output focus to the region being sampled. The shear-wave velocity is displayed, from which the operator can estimate elastic modulus using Equation 14.12.

MULTI-REGION (SUPERSONIC) METHODS

A recognized problem with shear waves generated from radiation force is that the shear waves have low amplitude. This limits the distance over which the shear waves can be tracked using ultrasound imaging. Higher-amplitude shear waves may be generated by increasing the ultrasound output power; however,

this has patient safety and transducer-heating implications. The shear-wave amplitude may be increased by the use of multiple regions (Bercoff et al. 2004). In this technique high-output source regions are generated at increasing depths, one after the other (Figure 14.24a). This is equivalent to a high-output source moving rapidly through the tissues, where the speed of the source is greater than the speed of the shear waves which are generated. The shear waves from each source summate to produce a cone (referred to as a 'Mach cone') which travels through the tissue (Figure 14.24b). In general, when the source of waves moves faster than the wave speed, the source is said to be moving at 'supersonic' speed. This effect most commonly occurs when an aircraft travels faster than the speed of sound in air. The supersonic technique produces shear waves of increased amplitude, which therefore are easier to detect using ultrasound imaging. Tracking of the shear waves requires high-frame-rate imaging. The methods for production of a high frame rate, up to 20,000 frames s^{-1}, are described in Chapter 3. From the detected ultrasound images, the tissue displacement and strain

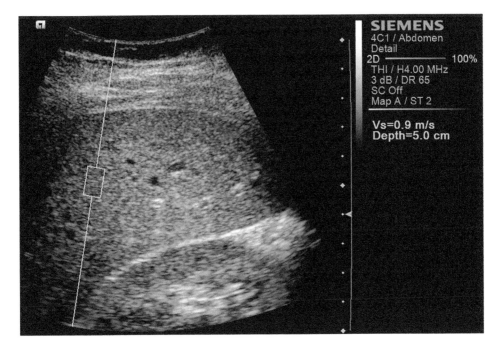

Figure 14.23 ARFI-based estimation of shear-wave velocity from a single region. The operator places the region (green box) at the desired location within the tissue, in this example the liver. The ARFI pulse is activated by the operator and the machine displays the shear-wave velocity, in this case 0.9 m s^{-1}.

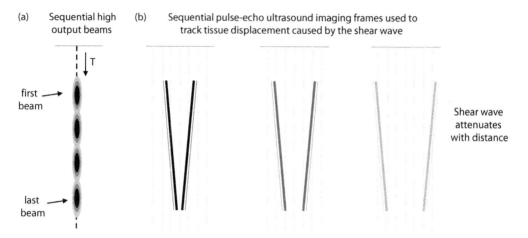

Figure 14.24 Shear-wave imaging using supersonic ARFI techniques. **(a)** Sequential high-output beams are generated with focal regions of increasing depth along the same line. **(b)** A shear wave cone is formed, and high-frame-rate imaging techniques are used to track the tissue displacement caused by the shear wave.

and the shear-wave velocity are estimated. Local tissue stiffness may then be calculated from Equation 14.11 and displayed in colour overlaid on the B-mode image. Clinical examples are shown in Figure 14.25. Stiffness measurements may vary due to a variety of factors and manufacturers have developed a quality display to help demonstrate when measurements are reliable (Figure 14.26).

Further considerations for shear-wave elastography

The explanations so far in this chapter represent the usual way of explaining the difference between compressional waves and shear waves; in one the oscillation is in the direction of wave motion, in the other it is transverse to the direction of wave motion. However, the examples in Figure 14.16 apply to plane waves; that is, waves generated by a very large source. In shear-wave elastography the source cannot be assumed to be very large. In some cases it is more appropriate to consider a point source, and in this case the propagation of waves is more complicated. Several papers discuss this in detail (Catheline et al. 1999b, 1999a; Sandrin et al. 2004; Carstensen et al. 2008; Giannoula and Cobbold 2008). There are three waves produced: a classic compressional wave which exhibits longitudinal motion, a classic shear wave which exhibits transverse motion,

and a third wave arising from the 'coupling term'. The paper by Sandrin et al. (2004) is especially enlightening, revealing that the third wave has the same velocity as that of the shear wave, yet has a longitudinal component. This is especially important in the technique previously described based on the work of Sandrin et al. (2002, 2003), which measures longitudinal displacement in order to measure shear wave velocity.

FURTHER DEVELOPMENTS

The approach taken in this chapter has been to introduce the essential components of tissue elastic behaviour by assuming that the tissue is uniform, incompressible and fully elastic in that once the stress is removed the tissue returns fully and immediately to its unstressed state. In practice real soft tissues are more complex than this. Different organs and components of organs are 3D, and their elastic properties may differ when examined from different directions. The material may not be fully elastic in that there may be a time delay before the tissue is restored to its unstressed state; in these circumstances the tissue is said to be 'viscoelastic'. For the idealized tissues considered in this chapter, a uniform, elastic, incompressible material needs only one constant to describe its behaviour when subject to stress, Young's modulus. The reader has seen the difficulty involved in obtaining just one

(a)

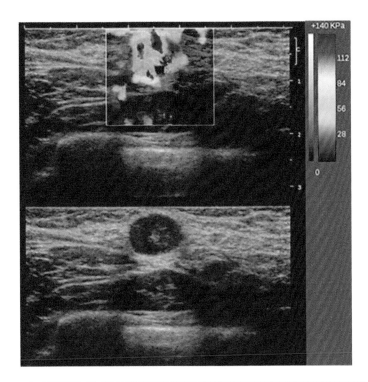

(b)

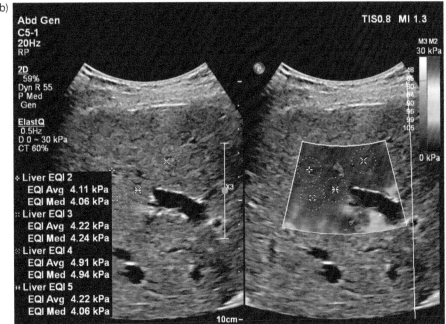

Figure 14.25 Clinical examples using supersonic shear-wave imaging. (a) Breast imaging – an echolucent region on the B-mode image has high stiffness, confirmed as a metastatic lymph node by pathology. (b) Liver imaging – stiffness measurements are displayed at multiple locations. (Image kindly provided by Philips Medical Systems, Eindhoven, The Netherlands.)

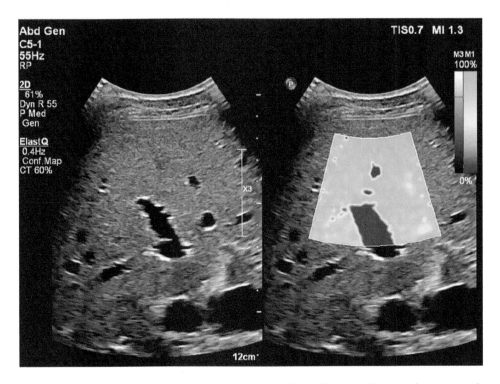

Figure 14.26 Display of a quality index indicating the level of confidence which can be assigned to the corresponding stiffness values. (Image kindly provided by Philips Medical Systems, Eindhoven, The Netherlands.)

constant, Young's modulus, from ultrasound imaging. More accurate descriptions of the mechanical behaviour of tissues require the estimation of a greater number of physical constants and in order to achieve this the complexity of the imaging procedures must increase. Recent years have seen the development of methods for estimation of both elastic modulus and viscosity and further reading is provided in Urban et al. (2012).

QUESTIONS

Multiple Choice Questions

Q1. Which of the following is true concerning strain?
   a. Strain has units of centimetres
   b. Strain is defined as change in length divided by original length
   c. Strain has no units
   d. Strain is defined as force divided by area

   e. Strain is not commonly measured in ultrasound elastography
Q2. Concerning Young's modulus:
   a. It has no units
   b. It is a measure of the volume of a material
   c. It is a measure of the change in shape of a material
   d. It is a measure of the stiffness of a material
   e. It has units of Pa (Pascals)
Q3. Concerning strain imaging:
   a. Strain in the tissues may be induced by natural movements such as breathing and cardiac pulsation
   b. The strain generally increases with increasing depth from the transducer
   c. The displayed image is of elastic modulus
   d. Soft tissues generally display low strain
   e. Hard tissues generally display low strain

Q4. Concerning strain ratio:
   a. This has units of Pa (Pascals)
   b. This may be measured using shear-wave elastography
   c. This is defined as strain in the reference region divided by strain in the lesion
   d. This may be measured in strain elastography
   e. This is defined as strain in the lesion divided by strain in the reference region

Q5. Concerning shear waves:
   a. In soft human tissue these typically travel at a speed of 1400–1600 m s^{-1}
   b. In soft human tissue these typically travel at a speed of 1–20 m s^{-1}
   c. These are concerned with particle motion in the same direction as the wave travels
   d. In soft human tissue these typically travel at a speed of 0.1–2 m s^{-1}
   e. These are concerned with particle motion transverse to the direction the wave travels

Q6. Which of the following quantities is concerned with the relationship between applied pressure and volume for a material:
   a. Pressure P
   b. Shear modulus G
   c. Poisson ratio ν
   d. Elastic modulus E
   e. Bulk modulus B

Q7. Which of the following quantities is concerned with the relationship between applied force and resulting deformation in the direction of the force:
   a. Pressure P
   b. Shear modulus G
   c. Poisson ratio ν
   d. Elastic modulus E
   e. Bulk modulus B

Q8. Which are the three main steps (in order) in shear-wave imaging:
   a. Induction of shear waves – track shear waves – estimate local shear velocity
   b. Apply ARFI pulse – apply external force to induce strain – estimate local shear velocity
   c. Apply ARFI pulse – track shear waves – apply ARFI pulse

   d. Cross correlation to estimate strain – apply force to induce strain – estimate shear modulus
   e. Track shear waves – estimate local shear velocity – induction of shear waves

Q9. In shear-wave ultrasound imaging shear waves may be induced using which of the following?
   a. Vibration of the transducer
   b. Manually moving the transducer backwards and forwards
   c. By the displacements produced at the focus of a high-output beam
   d. Naturally by the patient breathing
   e. A separate actuator applied on the patient's skin

Short-Answer Questions

Q1. Describe what happens when an elastic material is stretched in terms of force and extension; define stress and strain.
Q2. What is Young's modulus and how can it be measured?
Q3. Describe briefly the two types of elastography systems available commercially.
Q4. What is the measurement termed the 'strain ratio', and how should it be performed?
Q5. Describe briefly the principles of shear-wave elastography.
Q6. Describe the principles of strain elastography based on the use of A-line data.
Q7. Describe the minimum sequence of beams used in acoustic radiation force imaging (ARFI).
Q8. Describe two methods by which radiation-force techniques may be used to obtain elastography images.

REFERENCES

Bercoff J, Tanter M, Fink M. 2004. Supersonic shear imaging: A new technique for soft tissue elasticity mapping. *IEEE Transactions on Ultrasonics, Ferroelectrics, and Frequency Control*, 51, 396–409.

Cabrelli LC, Grillo FW, Sampaio DRT, Carneiro AAO, Pavan TZ. 2017. Acoustic and elastic properties of glycerol in oil-based gel phantoms. *Ultrasound in Medicine and Biology*, 43, 2086–2094.

Carstensen EL, Parker KJ, Lerner RM. 2008. Elastography in the management of liver disease. *Ultrasound in Medicine and Biology*, 34, 1535–1546.

Castera L, Vergniol J, Foucher J et al. 2005. Prospective comparison of transient elastography, fibrotest, APRI, and liver biopsy for the assessment of fibrosis in chronic hepatitis C. *Gastroenterology*, 128, 343–350.

Catheline S, Thomas JL, Wu F, Fink MA. 1999a. Diffraction field of a low frequency vibrator in soft tissues using transient elastography. *IEEE Transactions on Ultrasonics, Ferroelectrics, and Frequency Control*, 46, 1013–1019.

Catheline S, Wu F, Fink M. 1999b. A solution to diffraction biases in sonoelasticity: The acoustic impulse technique. *Journal of the Acoustical Society of America*, 105, 2941–2950.

Dewall RJ. 2013. Ultrasound elastography: Principles, techniques, and clinical applications. *Critical Reviews in Biomedical Engineering*, 41, 1–19.

Dineley J, Meagher S, Poepping TL, McDicken WN, Hoskins PR. 2006. Design and characterisation of a wall motion phantom. *Ultrasound in Medicine and Biology*, 32, 1349–1357.

Duck FA. 2012. *Physical Properties of Tissue*. York: Institute of Physical Sciences in Medicine.

Foucher J, Chanteloup E, Vergniol J et al. 2006. Diagnosis of cirrhosis by transient elastography (FibroScan): A prospective study. *Gut*, 55, 403–408.

Giannoula A, Cobbold RSC. 2008. Narrowband shear wave generation by a finite-amplitude radiation force: The fundamental component. *IEEE Transactions on Ultrasonics, Ferroelectrics, and Frequency Control*, 55, 343–358.

Greenleaf JF, Fatemi M, Insana M. 2003. Selected methods for imaging elastic properties of biological tissues. *Annual Review of Biomedical Engineering*, 5, 57–78.

Hoskins PR. 2012. Principles of ultrasound elastography. *Ultrasound*, 20, 8–15.

Madsen EL, Hobson MA, Shi HR, Varghese T, Frank GR. 2005. Tissue-mimicking agar/gelatin materials for use in heterogeneous elastography phantoms. *Physics in Medicine and Biology*, 50, 5597–5618.

Manickam K, Machireddy RR, Seshadri S. 2014. Characterization of biomechanical properties of agar based tissue mimicking phantoms for ultrasound stiffness imaging techniques. *Journal of the Mechanical Behaviour of Biomedical Materials*, 35, 132–143.

Nightingale K, McAleavey S, Trahey G. 2003. Shear-wave generation using acoustic radiation force: In vivo and ex vivo results. *Ultrasound in Medicine and Biology*, 29, 1715–1723.

Nightingale K, Palmeri M, Trahey G. 2006. Analysis of contrast in images generated with transient acoustic radiation force. *Ultrasound in Medicine and Biology*, 32, 61–72.

Nightingale K, Soo MS, Nightingale R, Trahey G. 2002. Acoustic radiation force impulse imaging: In vivo demonstration of clinical feasibility. *Ultrasound in Medicine and Biology*, 28, 227–235.

Nightingale KR, Palmeri ML, Nightingale RW, Trahey GE. 2001. On the feasibility of remote palpation using acoustic radiation force. *Journal of the Acoustical Society of America*, 110, 625–634.

Ophir J, Alam SK, Garra B et al. 1999. Elastography: Ultrasonic estimation and imaging of the elastic properties of tissues. *Journal of Engineering in Medicine*, 213, 203–233.

Palmeri ML, Nightingale KR. 2011. Acoustic radiation force-based elasticity imaging methods. *Interface Focus*, 1, 553–564.

Palmeri ML, Wang MH, Dahl JJ, Frinkley KD, Nightingale KR. 2008. Quantifying hepatic shear modulus in vivo using acoustic radiation force. *Ultrasound in Medicine and Biology*, 34, 546–558.

Sandrin L, Cassereau D, Fink M. 2004. The role of the coupling term in transient elastography. *Journal of the Acoustical Society of America*, 115, 73–83.

Sandrin L, Fourquet B, Hasquenoph JM et al. 2003. Transient elastography: A new noninvasive method for assessment of hepatic fibrosis. *Ultrasound in Medicine and Biology*, 29, 1705–1713.

Sandrin L, Tanter M, Gennisson JL, Catheline S, Fink M. 2002. Shear elasticity probe for soft tissues with 1-D transient elastography. *IEEE Transactions on Ultrasonics, Ferroelectrics, and Frequency Control*, 49, 436–446.

Sarvazyan AP. 2001. Elastic properties of soft tissue. In: Levy, M Bass, HE Stern, R (Eds.), *Handbook of Elastic Properties of Solids, Liquids and Gases, vol. 3. Elastic Properties of Solids: Biological and Organic Materials, Earth and Marine Sciences*. New York: Academic Press, pp. 107–127.

Sarvazyan AP, Rudenko OV, Swanson SD, Fowlkes JB, Emelianov SY. 1998. Shear wave elasticity imaging: A new ultrasonic technology of medical diagnostics. *Ultrasound in Medicine and Biology*, 24, 1419–1435.

Shiina T, Nitta N, Ueno E, Bamber JC. 2002. Real time tissue elasticity imaging using the combined autocorrelation method. *Journal of Medical Ultrasound*, 29, 119–128.

Urban MW, Chen S, Fatemi M. 2012. A review of shearwave dispersion ultrasound vibrometry (SDUV) and its applications. *Current Medical Imaging Reviews*, 8, 27–36.

Wells PNT, Liang HD. 2011. Medical ultrasound: Imaging of soft tissue strain and elasticity. *Journal of the Royal Society Interface*, 8, 1521–1549.

Yamakawa M, Shiina T. 2001. Strain estimation using the extended combined autocorrelation method. *Japanese Journal of Applied Physics*, 40, 3872–3876.

Quality assurance

NICK DUDLEY, TONY EVANS, AND PETER R HOSKINS

INTRODUCTION

The term 'quality assurance' (QA) has many definitions and is used in a variety of contexts. Most often it refers to schemes for maintaining the outcomes of some process or activity as measured against a required standard. In clinical ultrasound, the process outcome is usually the creation of a series of images which have a clinical utility and so a full QA programme would include not only the equipment but every element of the process from patient referral to final report. The clinical process is beyond the scope of this text and this chapter focuses on QA of the equipment.

There is controversy surrounding the reliability and value of QA techniques for comparing between scanners or against a purchase specification (Dudley et al. 2017). The emphasis in a QA programme should therefore be on the detection of a fault, including those present at commissioning, or a change in performance in an individual scanner at as early a stage as possible in order that technical help can be summoned as appropriate. Following commissioning, a QA programme should determine whether the condition or performance of the equipment has changed over time. The QA process also includes the actions to be taken when a change has been demonstrated.

Since the outcome of any scan is clinical, it is often argued that the assessment of scanner performance should also be based on users' judgement of the quality of the images they obtain. Although there is an obvious attraction to this approach, there are also several limitations:

1. There is likely to be a learning curve; i.e. the user's perception may change as he or she becomes more familiar with the scanner's controls.
2. Since the judgement is subjective, it may not be shared by other users.
3. Subtle changes in performance with time may be missed.
4. The quality of the clinical image depends on the patient. Since there cannot be a standard patient, there cannot be a standard image quality.
5. The quality of the image may be influenced by the skill of the operator.
6. Faults may be masked by modern image formation and processing techniques, e.g. compounding; adaptive image processing.

In response to these issues, techniques have been devised to allow QA to be carried out on the scanner in the absence of a patient. Some techniques are simple, requiring no test equipment, and others utilise tissue-mimicking test objects (TMTOs) designed to measure specific aspects of scanner performance. Key factors to consider in selecting QA methods are that they should be reproducible over a time scale of years and sensitive enough to detect a change before it becomes clinically significant.

There is evidence predating current guidelines that conventional visual assessment and measurement of TMTO features, e.g. resolution and contrast, are ineffective due to large intra- and inter-observer errors (Dudley et al. 2001). There is limited evidence for computerised measurement of TMTO features, showing that this may be useful for investigating suspected faults rather than as a routine test (Dudley and Gibson 2014).

Much of the recent evidence for ultrasound QA suggests that simple methods will detect the majority of faults. Hangiandreou et al. (2011) found that 91% of faults were detected by a visual inspection of the equipment and an assessment of image uniformity. Using similar methods Sipila et al. (2011) and Vitikainen et al. (2017) demonstrated 94% and 82% of faults, respectively. Vitikainen et al. (2017) also found that 18% of probes had sensitivity problems demonstrated by electronic testing and Dudley and Gibson (2017) found a reduction in sensitivity, measured using a TMTO, in 23% of probes in their QA program.

Standards and guidance relevant to the testing of ultrasound scanners in a hospital setting are produced by national professional bodies. In the United Kingdom these are Institute of Physics and Engineering in Medicine (IPEM) and British Medical Ultrasound Society (BMUS), in Europe it is European Federation of Societies for Ultrasound in Medicine and Biology (EFSUMB) and in the United States these are American Association of Physicists in Medicine (AAPM) and American Institute of Ultrasound in Medicine (AIUM). There is also guidance on periodic testing in an International Electrotechnical Commission (IEC) Technical Specification (IEC 2016).

The approach here is based on guidance produced within the United Kingdom (IPEM 2010; Dudley et al. 2014), which suggests simple tests including visual inspection, uniformity and sensitivity assessment, in keeping with the evidence previously outlined. This chapter details routine QA that is well within the capabilities of experienced sonographers and also describes methods suitable for acceptance testing and to inform faults management; the latter are often performed by physics or engineering staff. Finally some novel methods of potential value are described.

ROUTINE QUALITY ASSURANCE

Routine QA should be performed by clinical staff. This promotes and maintains an awareness of possible fault or damage conditions, increasing the chances of faults being reported and remedied outside the testing schedule. Tests should take no more than a few minutes, so should not have a significant adverse effect on clinical work. The scanner and all its probes should be tested. Table 15.1 lists the routine QA activities and suggested frequencies.

Table 15.1 Routine quality assurance activities and frequencies

Activity	Frequency
Cleaning	Probes, console and cables after each patient
	Scanner body and display daily
	Filters weekly
Functional checks	During clinical use
	During uniformity, sensitivity and noise tests
Visual inspection	During clinical use
	Formal inspection weekly
Uniformity	Daily for each probe in use
Sensitivity and noise	Monthly

Scanner set-up

Whatever test is being performed, it is important to have a clear and easily reproducible protocol for setting up the scanner. The following general advice is applicable to most tests and further detail is given for individual tests as appropriate.

When making measurements and observations of the image display it is important to ensure the correct ambient lighting conditions. Tests should be performed in a darkened room, with no sources of light that may reflect from the display or otherwise reduce visibility of low-level signals.

It is usual to begin with a preset which is compatible with the probe under test and its usual clinical application. Additional functions such as harmonics, compounding and advanced processing should be turned off for routine tests, as they may mask faults. Automatic gain and image optimization should be disabled as these functions may compensate for faults. The display depth is adjusted so that the useful range of the test image fills the screen. A single focus close to the probe should be used to minimise the transmit aperture, making element faults easier to see. Time-gain compensation (TGC) controls should be left at their default positions unless there are good reasons for adjustment; for slider controls this is the central position, where the control will often click into place. Once the baseline settings have been established at commissioning of a new scanner or probe it is good practice to save these as additional presets, so that they may be easily recalled during routine testing.

Cleanliness

It is good practice to clean ultrasound gel and body fluids from scanner console, transducers and cables after every patient, using manufacturer recommended cleaning materials and methods (Westerway and Basseal 2017). Professional bodies provide guidance on appropriate cleaning methods, e.g. Abramowicz et al. (2017). It is important to wipe, not rub, the probe lens as this may be easily damaged. Additionally the scanner body and display should be cleaned daily and filters should be cleared of dust weekly.

Functional checks

The most appropriate time to report functional issues is during clinical use of the equipment. This is when important controls will be regularly adjusted and any malfunctions should become evident. It is useful to incorporate a fault report form into the QA process in order to facilitate recording of faults and remedial actions. Additionally, any functional faults found during scheduled QA should be reported and remedied.

Functional checks should also include display adjustment, if necessary. In general the brightness and contrast of the display should be kept at settings established at commissioning, but if the greyscale bar is not fully displayed then adjustment and recording of the new settings are appropriate.

Visual inspection

A thorough visual inspection of the scanner, probes, cables and connections should be carried out. It is recommended to carry out and document a formal inspection weekly, but it is good practice to check the condition of probes and ensure proper stowage of probes and cables during daily use.

The visual inspection should ensure that there are no loose or damaged parts, wheels and brakes function correctly and there are no sharp edges that may cause injury. Electrical cables should be examined, ensuring that they are securely connected and undamaged. If there are any visible mechanical or electrical hazards the equipment should not be used clinically until they have been rectified. Inspection of probes should ensure that there is no physical damage to probes, cables or connectors, no visible wear or damage to lenses and that probe cases are intact with no opportunity for ingress of fluid. If a probe case is opening on a seam or cracked then an electrical safety test is essential. Physical defects which do not affect mechanical or electrical safety but affect, for example, the ability to clean the equipment, should be risk assessed and mitigating actions taken if necessary. Equipment should not be used if a risk assessment indicates a significant mechanical, electrical or infection risk to patients or personnel.

The inspection should also identify whether any parts, e.g. probes, are missing.

Uniformity

The simplest test of uniformity is assessment of the in-air reverberation pattern (Figure 15.1) and this may easily be carried out daily before the first use of each probe. The pattern arises from the large acoustic impedance mismatch between the transducer and the surrounding air giving rise to large multiple and regular reflections as described in Chapter 5. The uniformity test should be carried out using the highest available fundamental frequency, default output and gain settings, the shortest scale setting that still allows the full width and depth of the reverberation pattern to be seen, a single superficial focus and all advanced processing, e.g. compounding, adaptive filtering, turned off. The probe should be clean and free from coupling gel. The in-air reverberation pattern is then inspected for lateral uniformity; it should be a series of bright lines parallel to the probe surface and each of uniform brightness. Small lateral variations in brightness may be seen as a result of minor irregularities in the lens; with experience it should be possible to differentiate between this minor non-uniformity and unusual patterns of non-uniformity. Comparison with previous images may be helpful in identifying changes; it should be noted that small changes in the reverberation pattern may be seen as a result of temperature variations. Switching through all

available fundamental and harmonic frequencies can be useful as some anomalies are more evident at one frequency.

There are three fault conditions identifiable from the in-air reverberation pattern. The first is element failure, or dropout, which manifests as an axial band extending throughout the depth of the reverberations. Element failure can be confirmed by drawing the edge of a paper clip along the array, taking care to hold the paper clip perpendicular to the long axis of the array and maintain full contact across the width of the probe surface (Goldstein et al. 1989). The image will show reverberations from within the cross section of the paper clip extending axially; as the paper clip reaches the area of dropout the brightness of the reverberations will decrease if there is a faulty element in the array (Figure 15.2). A new or repaired probe should be rejected if there is any element failure.

A modified version of the paper clip test has been developed for phased arrays, where beam steering masks any dropout, and element failure is difficult to demonstrate (Dudley and Woolley 2016). Beam steering is turned off by using M-mode and a deep-scale setting and focus (say 10 and 5 cm) are chosen to ensure that the full aperture is active. With the M-line central in the image, the paper clip is drawn along the probe face and the image frozen. Any axial banding in the image represents element failure (see Figure 15.3).

A further test, if risk assessing the use of a probe with dropout, is to image a TMTO and consider the impact on clinical imaging of any shadowing seen. If two or more contiguous elements are faulty this is likely to affect Doppler and colour flow (Weigang et al. 2003; Vachutka et al. 2014); if this is suspected then electronic probe testing or testing with a Doppler test object will be necessary.

The second fault condition is delamination (Figure 15.4). This is where layers in the acoustic stack become detached; most commonly the lens separates from the layer below. This results in apparent disruption or blurring of the reverberation pattern. When pressure is applied the layers achieve full contact and an image may be formed with no apparent defects. In more severe cases there may be visible bulging of the lens. Delaminated probes should be replaced.

Figure 15.1 In-air reverberation pattern for a linear array. This shows reverberation, predominantly within the lens, when the probe is not coupled to a transmission medium.

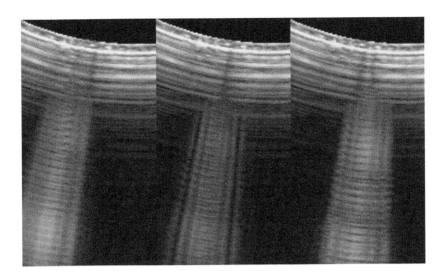

Figure 15.2 Dropout in the in-air reverberation pattern, confirmed using the paper clip test. Images show the paper clip being moved across the area of dropout. The intensity of the reverberation pattern from the paper clip is reduced at the location of the dropout, indicating a non-functioning element.

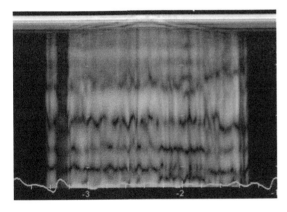

Figure 15.3 Dropout in a phased array using the M-mode paper clip test. A paper clip is drawn along the array and when in contact with a non-functioning element on the left a dark axial band is seen.

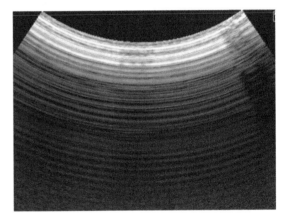

Figure 15.4 Delamination. The reverberation pattern on the right appears disrupted due to partial separation of layers within the probe.

The third fault condition is non-uniformity of lens thickness. This manifests in the in-air reverberation pattern as a departure from parallel lines. It may be inherent in the probe as a manufacturing defect or may be due to lens wear; the latter is more frequently seen at the ends of the array. The non-uniformity may be quantified by measuring the depth of a chosen reverberation line at points along the probe (Figure 15.5a). The effect of this deviation, noted on a new probe, is seen in Figure 15.5b and c. Figure 15.5b shows the sensitivity of individual elements using an electronic probe tester, where there is a clear gradient in sensitivity. Figure 15.5c shows the fractional bandwidth plot (the ratio of bandwidth to centre frequency) from the electronic probe tester; bandwidth is reduced at one end of the probe, presumably due to the change in lens thickness reducing its matching efficiency at some frequencies. In the authors' experience any

(a)

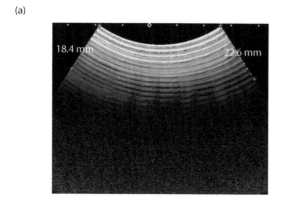

(b)

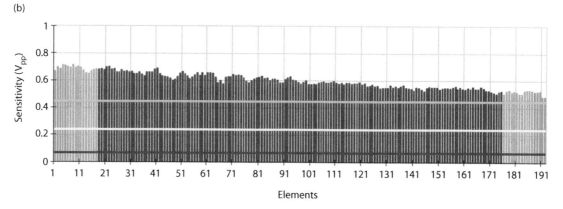

(c)

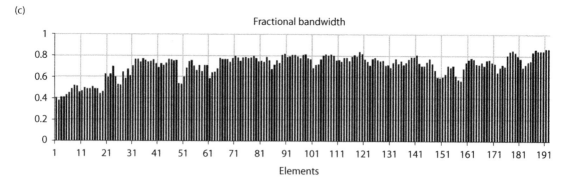

Figure 15.5 **(a)** In-air reverberation suggesting non-uniform lens thickness, demonstrated by measurements to the same reverberation plane on the left and right of the image; **(b)** sensitivity plot using an electronic probe tester, showing a reduction in sensitivity from left to right; **(c)** bandwidth plot using an electronic probe tester, showing reduced bandwidth in some areas, most notably on the left.

deviations of 10% or more are very clearly visible, have an impact on bandwidth and sensitivity, and such probes should be replaced.

The uniformity testing as described is effective for conventional array probes but may not be appropriate for detecting dropout in matrix arrays or any array with multiple rows of elements. However, failure of a single element in these probes is unlikely to have an effect on either the in-air reverberation pattern or clinical practice, as their individual contribution to the ultrasound beam is small. If a block of elements fail this is more likely to be seen as dropout or shadowing. Delamination and lens thickness variation affecting matrix arrays will have the appearances as previously described.

Sensitivity and noise

Tests for B-mode sensitivity and for B-mode, pulsed Doppler and colour flow noise should be carried out monthly (IPEM 2010).

BASELINE MEASUREMENTS

On commissioning of a new scanner or receipt of replacement or repaired probes, baseline measurements are required to establish tolerances for routine testing. The in-air reverberation image should be obtained as for uniformity testing. Overall gain should be increased to maximum. Scale should be adjusted to show the full depth of reverberation in 30%–50% of the image depth, together with some electronic noise at the bottom of the image, as shown in Figure 15.6a. If it is not easy to reproducibly determine the position of the deepest reverberation echo using the highest frequency, then toggling through the fundamental frequencies will help to identify the frequency where it is easiest and this frequency should be used.

If there is no noise visible in the distal image, the depth should be increased until noise is seen. If this results in the reverberation occupying less than 25% of the image, the depth is reduced again and the TGC set to maximum; noise should then appear. If TGC has been set to maximum, it may be necessary to toggle through the frequencies again

to find the one where it is easiest to reproducibly determine the position of the deepest reverberation echo. There may be occasions where the initial image is too bright and noisy and it is necessary to set TGC to minimum. It is important to note the position of the TGC controls and not to set them in positions that are not easily and exactly reproduced.

These settings should be recorded and saved as an additional preset, e.g. USER QA <PROBE id>. An image should be stored (preferably digitally) for future comparison.

Reverberation depth

The reverberation depth is measured vertically from the probe surface to the deepest visible reverberation line in the middle third of the image (ignore reverberations at the edge of the image), as shown in Figure 15.6a. On subsequent testing the loss or addition of a reverberation line will be reportable, so a tolerance of ± half the distance between the deepest reverberation and the line above is advised. This test is quantised, in that it requires a step change in performance of the system equal to or greater than the amplitude of the deepest reverberation in order for that reverberation to disappear. This limitation is addressed by the measurement of the reverberation threshold.

Reverberation threshold

The reverberation threshold is the overall gain setting at which the deepest reverberation line disappears across the whole width of the image, as shown in Figure 15.6b. The gain should be turned

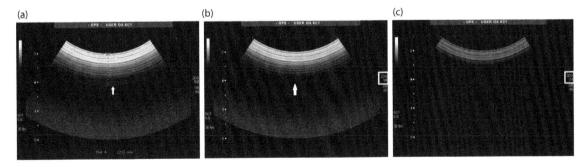

Figure 15.6 Measurement of (a) reverberation depth (arrow shows the deepest visible reverberation); (b) reverberation threshold (arrow shows the original position of the deepest reverberation, box highlights the recorded gain value at which the reverberation disappears); (c) noise threshold (box highlights the recorded gain value at which the noise disappears).

down until the reverberation line completely disappears and the gain noted. This may be checked by turning the gain up by one increment and the echo should reappear. The measurement should be repeated until the operator is confident that a reproducible result has been obtained.

B-mode noise threshold

The B-mode noise threshold is the overall gain setting at which the distal electronic noise in the image disappears, as shown in Figure 15.6c. The gain should be turned down until the noise completely disappears and the gain noted. This may be checked by turning the gain up by one increment and the noise should reappear. The measurement should be repeated until the operator is confident that a reproducible result has been obtained.

Pulsed Doppler and colour flow noise thresholds

The pulsed Doppler (PD) and colour flow (CF) noise thresholds are obtained in a similar fashion, operating these modes (separately), noting the transmission frequency, positioning the range gate or colour window centrally in the B-mode image, and reducing the gain until the PD or CF noise is eliminated.

Initially a tolerance for the reverberation and noise thresholds of ±4 gain increments is advised; with experience this may be adjusted upwards or downwards to encompass the random error within a series of measurements. Results should be recorded on a form designed for this purpose.

ROUTINE TESTING

The tests should be carried out on a monthly basis for all probes in use. In departments where probes are moved between scanners there are two options for testing: (1) return the probes to their original scanner for testing or (2) make baseline measurements for each probe on every scanner on which it is to be used.

For each probe, select the relevant preset, e.g. USER QA <PROBE id> and perform the tests as previously described, recording the results on a form (examples available in Dall et al. 2011). Measurements of reverberation depth, reverberation threshold, B-mode noise threshold, PD noise threshold and CF noise threshold should be made and compared with baseline tolerances. Any

measurement outside tolerance should be repeated, having first checked the scanner settings against those recorded at baseline (a common problem is that the TGC has not been set to the default position).

Any change in uniformity or measurements out of tolerance should be reported and appropriate action taken as outlined in the following sections. Results, risk assessments and mitigating actions should be documented.

Fault management

It is probably useful to define a fault and a failure. In a QA program, the aim is to identify any changes from baseline condition or performance. We call these changes 'faults'. Some faults may not require any immediate action if they are minor and are unlikely to have a significant clinical impact. Some changes may require immediate action, such as a repair or the replacement of a probe. We call these 'failures'.

Having identified a fault, management often requires experience, judgement and context. The latter is important as a failure in one context, e.g. radiology imaging, may be a minor fault in another context, e.g. line placement in theatres. There is very little published evidence to support fault management.

Actions depend on local circumstances and general guidance only is given here. It may be useful to use a traffic light system for grading faults, where green represents no fault, amber a fault that does not require immediate remedy but may require some action and red is a failure. Table 15.2 gives an example of possible categorisation of probe faults found by visual inspection or uniformity testing. Where quantitative user test results are out of tolerance further testing using a TMTO is required, as described in the following text; if this is not possible locally then the advice of the service agent should be sought.

Audit

Any QA system requires periodic audit; annually would be appropriate. Auditing routine QA should include answering the following questions: Have the tests been performed regularly? Are all results in tolerance? Have faults been reported, risk

Table 15.2 Example categorisation of probe faults for either visual inspection or uniformity assessment

Classification	Visual inspection	Reverberation	Action
Green	No fault	Uniform	None
Amber	External physical damage, no functional consequence	Single element failure Minor non-uniformity or lens wear	Risk assess and repair or monitor as appropriate
Red	Wear or damage to probe face	Cable fault (dropout) Large or multiple dropout Delamination	Replace

assessed and remedied? Is there evidence of servicing and electrical safety testing according to manufacturers' schedules? A negative answer to any of these questions may indicate the need for further training and a review of processes. If the audit is carried out independently, e.g. by a clinical scientist or clinical technologist, they should consider performing the tests themselves to check that their results agree with the routine results.

FURTHER TESTING

Where routine QA results suggest a change in sensitivity or an increase in noise, then further testing is required. While these tests should be within the capabilities of experienced sonographers, for a variety of reasons they are often performed by physics or engineering staff. In order for these further tests to be effective, baseline measurements and images are required at commissioning of a scanner and any new or repaired probe. The aim of further tests is to confirm a change in B-mode sensitivity and/or a reduction in B-mode signal-to-noise ratio. No reliable tests are available to assess the impact of increased PD or CF noise and so these cases will require discussion with the supplier or service agent.

Table 15.3 gives an example of possible categorisation of probe faults if quantitative user tests are out of tolerance and further testing is performed.

Tissue-mimicking test object sensitivity

Sensitivity may be measured as the depth of penetration (DOP), or low contrast penetration (LCP)

Table 15.3 Example categorisation of faults following further testing

Classification	Routine tests	Tissue-mimicking test object (TMTO) tests	Action
Green	Any results out of tolerance	Repeated user tests all in tolerance and images unchanged compared to reference images	None (false-positive user test)
Amber	B-mode results out of tolerance PD or CF noise results out of tolerance	TMTO sensitivity in tolerance and images unchanged compared to reference images None	Monitor user tests for trends Discuss with supplier or service agent
Red	Any results out of tolerance	TMTO sensitivity out of tolerance and images visibly changed compared to reference images	Consult service agent for scanner repair or replacement probe as appropriate

Note: There are many potential combinations of results, so limited examples are given here.

in a TMTO and as the mean grey level over depth in a uniform area of a TMTO; both are useful but it may not always be possible to measure DOP (Dudley and Gibson 2017). This test is performed using a preset consistent with the usual clinical application of each probe, with TGC and focus in the default positions, speed of sound correction, automatic gain and automatic image optimization disabled. Acoustic output should be set to maximum. An area of uniform tissue mimic within the TMTO should be imaged. Image scale and overall gain should be adjusted to show the depth in the TMTO where the unvarying speckle pattern disappears and the image becomes dominated by randomly varying noise (or in a system with very low noise, where the image becomes dark). The DOP is the distance from the TMTO surface to the depth at which speckle disappears (Figure 15.7).

Automated methods are available for this measurement (Gibson et al. 2001; Gorny et al. 2005) and these are preferred to subjective visual assessment. If visual assessment is the only method available it

is important to have a rigid protocol, clearly defining the method and location of DOP measurement.

Where the DOP is beyond the bottom of the TMTO, it may be possible to obtain a measurement by using a higher frequency, if available, or switching to an alternative mode, e.g. from harmonic to fundamental. It is good practice to make measurements in both fundamental and harmonic modes, as faults affecting bandwidth may affect sensitivity in one mode and not the other. At baseline testing the settings should either be saved as an additional preset or recorded in case the preset is later adjusted for clinical reasons.

Images should be stored (preferably digitally). If digital images are available then the mean grey level should be measured in a region of interest in or near the central axis of the TMM image extending from below the dead zone to just above the visible limit of speckle, avoiding any targets.

Tolerances for DOP of ±5% or ±5 mm, whichever is the greater (IPEM 2010), and for mean grey level of ±10% (Dudley and Gibson 2017) have been suggested.

Reference images

Reference images should be stored at commissioning (preferably digitally) of each available target type (e.g. filaments, greyscale, anechoic) in a TMTO, using a preset consistent with the usual clinical application of each probe (Figure 15.8a–c). TGC controls should be left at their default positions. The only adjustments necessary are to disable any speed of sound correction, as this will lead to image distortion, and adjust overall gain if necessary to achieve an appropriate level of brightness. The settings should either be saved as an additional preset or recorded in case the preset is later adjusted for clinical reasons.

QUALITY ASSURANCE PROGRAMME

The baseline, routine and further testing as previously described, together with fault reporting, remedial actions and audit almost complete a QA programme. The missing element is acceptance testing, which is an essential component. The outcome of acceptance testing should be a pass or

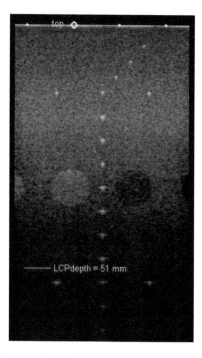

Figure 15.7 DOP/LCP measurement. In this automated method the DOP is marked at the point where speckle amplitude has reduced to twice the background noise level.

(a) (b) (c)

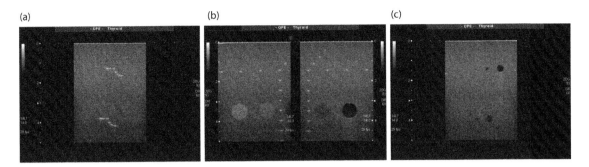

Figure 15.8 Reference images using a clinical setting: **(a)** resolution targets; **(b)** greyscale targets; **(c)** anechoic targets.

fail and should focus on safety, physical integrity, functionality and accuracy of the equipment and should be carried out on delivery of equipment. A restricted range of acceptance tests is also appropriate following a repair or upgrade, focusing on aspects of the scanner and its performance that may have been affected. Testing should be satisfactorily completed before a scanner is put into clinical use. Acceptance testing of absolute imaging performance is a controversial topic and is not addressed here.

Acceptance testing

The visual inspection previously detailed is sufficient to assure physical integrity and mechanical safety. Further tests are needed to assess electrical and acoustic safety, functionality and accuracy.

SAFETY

Electrical safety testing is required to assess compliance with international standards (IEC 2005). These standards apply to a wide range of medical equipment and in a hospital environment testing is likely to be carried out by specialist medical engineers following local protocols. For this reason detailed advice is beyond the scope of this text.

Acoustic safety is covered in detail in Chapter 16. Measurement of acoustic output requires expensive equipment and specialist skills not usually available in the hospital or clinic. Acceptance tests should include a visual assessment of displayed thermal and mechanical indices (TIs and MIs), referring to Chapter 16 for further guidance. Trained clinical staff should have the necessary knowledge and

experience of the relationship between scanner controls and the safety indices to do this. User manuals will show the conditions for maximum TI and MI values and the visual assessment should include attempting to reproduce these, as well as checking that TI and MI vary as expected when altering settings. Scanners where the safety indices do not exceed defined limits are not required to display them; this may be determined by inspection of the scanner user manuals. Unexpected results should be discussed with the supplier.

FUNCTIONALITY

Functionality testing should include the uniformity assessment described previously and the operation of the scanner should be tested while imaging a TMTO with a speed of sound of 1540 m s^{-1}. The operation of all scanner controls, including different scanning modes, should be assessed to ensure that they are functional and that they have the expected effect, e.g. moving the focal point should generate a visible change in the imaged width of nylon filament targets; reducing the imaging frequency should increase the depth of speckle visualisation.

Modern image enhancement techniques, e.g. compounding, adaptive image processing, should make a subjective improvement in image quality. This may be difficult to assess in a TMTO, which is unlike a patient in that there are no layers of fat and muscle to degrade the image of deeper areas of interest. It may be that any fault will manifest as a degradation of the image, e.g. switching on compounding generates blurring, rather than enhancement of, anechoic targets, as shown in

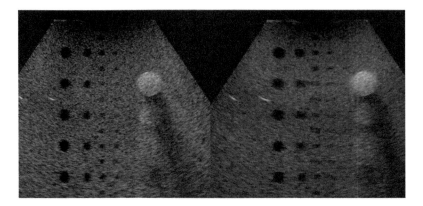

Figure 15.9 Images of a TMTO without (left) and with (right) compound imaging. The calibration speed of sound of the scanner and the speed of sound in the TMM are not well matched. The anechoic targets are blurred as the different imaging angles in the compound image are misregistered along the beam axis.

Figure 15.9 (it should be noted that in this example the blurring is due to the low speed of sound in the urethane test object rather than to a scanner fault). An example, from the authors' experience, of an unexpected degradation of the image due to a scanner fault is shown in Figure 15.10, where applying a speed of sound correction improves the lateral resolution of nylon filaments, where one would expect the best image to be obtained without correction as the speed of sound in the test object is 1540 m s^{-1}.

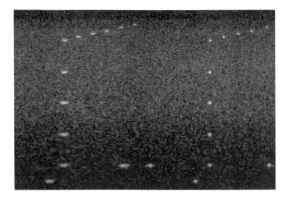

Figure 15.10 Images of a TMTO obtained with no speed of sound correction (1540 m s^{-1}; left of image) and with speed of sound correction set to 1500 m s^{-1} (right of image). It is expected that resolution in a TMTO will be better at 1540 m s^{-1} if the delay line algorithms correct appropriately for the lens/matching layer and the conventionally assumed speed of sound in human tissue.

Probes with dropout, delamination or non-uniform lens thickness should be rejected. Faulty controls or unexpected imaging results should be discussed with the supplier.

ACCURACY

Any measurements used to inform patient management must be sufficiently accurate, as discussed in Chapter 6. B-mode measurement accuracy, for most purposes, may be checked using a conventional TMTO containing filament targets arranged in rows and columns. Most commercial TMTOs have filaments with spacing appropriate for both larger-scale (>10 mm) and smaller-scale (0.25–10 mm) measurements. The majority of ultrasound measurements are performed in obstetrics, for fetal size and growth. Linear measurements should be checked at clinically relevant distances in clinically relevant directions, e.g. 50 mm laterally to represent fetal femur length; 2.5–5 mm axially to represent fetal nuchal translucency. Tolerances should be chosen to be smaller than the required clinical accuracy, in order to allow for other sources of error. In the examples given here tolerances of ±1 mm and ±0.1 mm, respectively, may be appropriate. In obstetrics it is widely recommended to derive circumferences from two orthogonal diameters, either by multiplying the mean diameter by pi or by fitting an ellipse. In either case accuracy may be checked using an orthogonal column and row of filaments in a TMTO (Figure 15.11). Measurements should be compared with the result

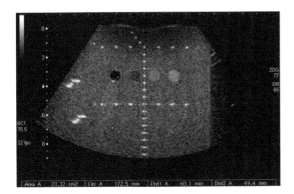

Figure 15.11 Checking the accuracy of circumference measurement from orthogonal diameters in a TMTO. (Circumference = π × [60.1 + 49.4] ÷ 2 = 172 mm; an acceptable result.)

calculated from the two diameters using the same method employed by the scanner, with a tolerance of ±1 mm.

A caveat in the test method is that if a convex probe is tested, there is the possibility of refraction errors caused by oblique incidence at the test object surface leading to an over-measurement (assuming that water is the coupling medium). This may be overcome either by using a matched coupling medium or by making measurement over a short distance at the centre of the field of view. A TMTO with a concave scanning surface has recently become available (Sono410 series, Gammex Inc., Middleton, Wisconsin) and overcomes this problem for many probes.

In other fields, where complex non-linear shapes are measured, a bespoke open topped test object may be appropriate, with nylon filaments arranged according to the shape of the anatomical structure to be measured. The test object should be filled with a medium with speed of sound 1540 m s^{-1} such as 9.5% ethanol by volume in water at 20°C (Martin and Spinks 2001). Tolerances should be chosen based on the clinical requirement and the estimated size of errors tabulated in Chapter 6.

In some applications of Doppler ultrasound, blood velocity estimates are used directly in patient management, e.g. peak systolic velocity to estimate percentage stenosis in the internal carotid artery. In many other applications indices and ratios are used. Where velocities are used absolute accuracy is important. Where indices or ratios are used linearity is important. Both should be assessed, ideally at clinically relevant velocities. Since clinically relevant velocities extend beyond 2 m s^{-1}, a moving string phantom may be the most appropriate test tool, as recommended by IPEM (2010). Flow phantoms are available offering mean velocities of up to about 0.7 m s^{-1}. Since both types of test object are expensive and some expertise is required in order to use them effectively, a pragmatic approach may be to ask the scanner supplier to provide results to verify accuracy and linearity over the required range.

Where a suitable test tool is available, accuracy may be assessed by comparing the stated constant velocity of the string or the mean velocity in an area of constant laminar flow with the mean velocity reported by the scanner. The spectral Doppler sample volume is positioned either at the level of the moving string, or covering the full width of the vessel in a flow phantom, using the B-mode image to guide placement. Spectral Doppler data are acquired. Most scanners have the facility to automatically trace maximum and mean velocities; otherwise the measured mean string velocity (but not necessarily the flow velocity) may be taken as the mean of the maximum and minimum velocities from manual calliper measurements. Figure 15.12 shows examples of the results of this test. When using a string phantom, care must be taken to use low output and gain to minimise saturation of the large signal obtained. Most guidance on checking velocity accuracy recommends comparing maximum velocity with string velocity, as this is measured clinically and modern commercial systems overestimate maximum velocity, as explained in Chapter 9. However, manufacturers calibrate their systems for mean velocity. For a string, the maximum velocity seen on the scanner is considerably greater than the true velocity due to spectral broadening. Spectral broadening may be regarded as a systematic error, as it can be predicted from the ultrasound beam geometry and the angle of insonation. Spectral broadening and its impact on clinical measurement is a complex issue requiring further investigation and understanding by the clinical community, and is not discussed further here. We suggest a tolerance for Doppler mean velocity accuracy of ±5%, as this is the smallest tolerance quoted by a manufacturer.

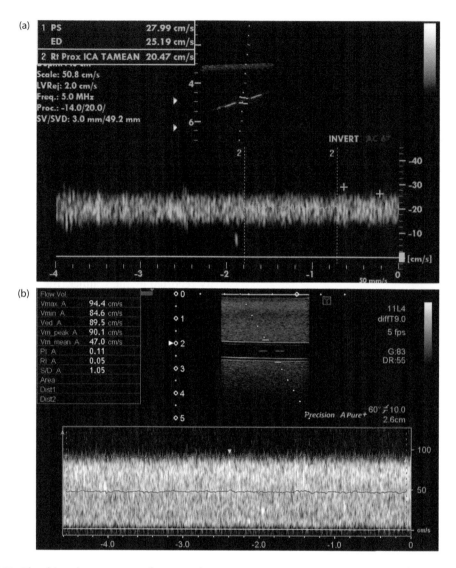

Figure 15.12 Checking the accuracy of mean velocity measurement using **(a)** a string phantom, where the automatic trace indicates a mean velocity of 20.47 cm s^{-1}, compared to the set velocity of 20 cm s^{-1}; **(b)** a flow phantom, where the automatic trace indicates a mean velocity of 47.0 cm s^{-1}, compared to the set mean flow velocity of 50 cm s^{-1}.

In B-mode and Doppler ultrasound, any parameters calculated by the scanner, e.g. estimated fetal weight, blood velocity indices, may be checked simply by recording the original measurements made by the scanner and calculating the result using the appropriate method. If the scanner is using the same method then the results will be identical.

Any results outside tolerance must be discussed with the supplier. There may be some difficulty

with this as manufacturers' tolerances are often wider than those required clinically, so it is wise to include clinical tolerances in pre-purchase specifications.

TEST OBJECTS

With the exception of complex non-linear measurements, all of the tests described in this chapter may be performed using commercially available test objects. A wide range is available, including some that are suitable for both B-mode and Doppler testing. B-mode and Doppler flow test objects are typically boxes filled with a tissue-mimicking material (TMM). A variety of materials is available, including aqueous gel, condensed milk and urethane rubber. Most TMMs tend to have a speed of sound close to 1540 m s^{-1} but they can dry out over time and require rejuvenation by the manufacturer. Urethane or other solid elastic materials are more stable over time, but their attenuation increases more rapidly with increasing frequency (Browne et al. 2003). Urethane has a speed of sound of approximately 1450 m s^{-1}, with a defocusing effect on the ultrasound beam. Both speed of sound and attenuation in urethane have a greater dependence on temperature than other TMMs, affecting the reproducibility of QA measurements (Dudley and Gibson 2010).

During use the test object should be placed on a firm, flat surface. If the test object has been stored in an unusually cool or warm environment, it may be wise to leave it for 30–60 minutes to temperature-equilibrate before use. If the test object has a well, this is filled with tap water. Scanning gel can be used if this is not practical, and may be preferred for linear and phased arrays. It should be noted that probes need to be in good acoustic contact over their whole length for some tests and this can be difficult for convex arrays if the well is shallow or insufficient gel is used.

Some practical points are worth mentioning here. If water is used in a test object well, it is common for a type of reverberation artefact to appear associated with echoes from the water surface. These are more irritating than misleading, but can be suppressed by using some damp paper or other absorber near the ends of the probe. A more important caveat is that the speed of sound in water is lower than that in most test objects, potentially leading to refraction as mentioned when discussing accuracy assessment. For the tests described here, other than accuracy, this will not cause a significant problem. Finally, unlike the clinical situation in which pressing harder may help by causing local distortion of the skin surface, it is important not to press on the top surface of the test object. These are typically made of low-density polythene and can

be distorted and damaged easily. Moreover, any targets inside the object can be pushed out of position if excess surface force is applied.

B-mode test objects

In order to create a speckle pattern similar to that of soft tissues, a suitable small-diameter scattering material is added, e.g. graphite powder. This is all that is required to facilitate sensitivity measurements. However, most TMTOs contain a variety of targets including small-diameter nylon filaments, cylinders containing material with differing backscatter properties than the background, spheres or other shapes. The filaments are designed to test calliper accuracy and the resolution of the system. Since resolution differs in three orthogonal planes (axial, lateral and slice thickness), and also alters with depth, the filament positioning needs to allow for this.

One drawback of using cylindrical targets is that the clarity of their display depends on the axial and lateral resolution of the scanner but is largely independent of the slice thickness. In clinical practice the orientation of a lesion is unlikely to be exactly transverse to the beam and hence test object targets of this type may flatter the scanner's capability. One way of addressing this is to use targets which are spherical rather than cylindrical since this achieves symmetry in all planes.

Doppler test objects

The design of test objects suitable for Doppler is more difficult than for B-mode imaging as the Doppler test object must include a moving target to simulate the moving blood. There are two types of moving-target test object which are available commercially: the string phantom and the flow phantom. A review of the design and application of Doppler test objects is given in Hoskins (2008).

STRING PHANTOM

In this device, the moving string simulates moving blood. The components of the string phantom are illustrated in Figure 15.13. Choice of the string is important as the scattering characteristics of blood need to be matched. Filaments, such as cotton and silk, are spiral wound, with a repeat pattern

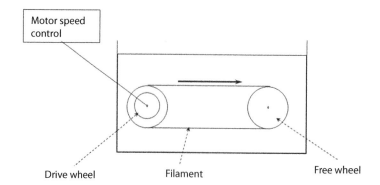

Figure 15.13 Components of a string phantom. The string is driven in a circuit by a drive wheel. The speed of the drive wheel may be controlled using an external computer to produce waveforms with a physiological appearance.

at distances that are comparable with the wavelength of ultrasound. This repeat pattern gives rise to high-amplitude scattering along certain directions, which distorts the Doppler spectrum, making this type of filament unsuitable (Cathignol et al. 1994). The use of commercial systems based on the use of spiral-wound filaments is not advised. A suitable filament is O-ring rubber, as this scatters ultrasound in all directions in a similar manner to blood (Hoskins 1994). The most important feature of the string phantom is that the velocity can be accurately measured. The true string velocity may be calculated from the speed of rotation of the drive wheel. The device is especially suited to the checking of velocity estimates made using Doppler.

The string phantom is a relatively straightforward system to set up and use, and its size enables it to be used as a portable test object. Guidance (IPEM 2010) on basic performance testing of Doppler systems is based on the use of a string phantom.

FLOW PHANTOM

This device simulates flow of blood within a vessel. The components of the flow phantom are illustrated in Figure 15.14. The main design criterion is that the acoustic properties of the tissue mimic, the blood mimic and the vessel must be matched to those of human tissue. The tissue mimic is usually a gel-based material, as previously described. The blood mimic must have the correct viscosity, as well as correct acoustic properties. A suitable

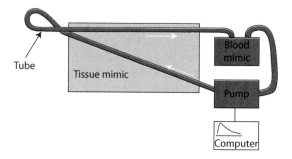

Figure 15.14 Components of a flow phantom. These are the tissue mimic, tube and blood mimics. The pump may be controlled using a computer to obtain physiological flow waveforms.

blood mimic has been described by Ramnarine et al. (1998), which is based on the use of nylon particles suspended in a solution of glycerol and dextran. Matching the acoustic properties of the artery is much more difficult. Rubber-based materials, such as latex (speed of sound $\sim$1600 m s^{-1}), have the correct acoustic velocity, but the attenuation is high and nonlinear. This mismatch of acoustic properties results in distortions in the shape of the Doppler spectrum obtained from flow within the tube. This is important when flow phantoms are used for calibration of mean velocity as this is overestimated as a result of the spectral distortion. Some commercially available flow test objects are based on the use of tubes of silicone (speed of sound $\sim$1000 m s^{-1}), and the calibration of mean velocity using these devices is, therefore, unreliable. Test object manufacturers continue to develop their products, and a flow phantom with nylon tubing with density, speed of sound and attenuation well matched to human soft tissue is available (Doppler 403 and Mini-Doppler 1430 Flow Phantoms, Gammex Inc. Middleton, Wisconsin). In research laboratories, flow phantoms can be designed which are correctly matched to human tissue (Ramnarine et al. 2001; Hoskins 2008).

MODERN SCANNING MODES

There is very limited professional guidance and no published evidence available for QA of other modes of scanning, including three-dimensional (3D) scanning, contrast imaging and elastography. The AIUM (2004) provide guidance on testing 3D measurement capabilities and a phantom is available to carry out these tests (3D Wire Test Object, CIRS, Norfolk, VA, USA). The AIUM (2007) discuss contrast agents in their guidance for testing Doppler ultrasound devices but provide no guidance on QA. There is no professional guidance on elastography QA, but phantoms are available commercially (CIRS, Norfolk, Virginia).

CIRS offer two elasticity QA test objects, containing spheres or stepped targets of known stiffness in a TMM background. The phantoms are sold as suitable for QA of both shear-wave and compression elastography. However, CIRS state that elasticity measurements will vary depending on the measurement system used, confirmed by Mulabecirovic et al. (2016), so the phantom may not be suitable for the assessment of absolute accuracy. Mulabecirovic et al. (2016) also showed that intra-observer variability could be high (coefficient of variation ranging from 1% to 19%) making the interpretation of serial QA measurements difficult.

RECENT DEVELOPMENTS

There is very little evidence to support the use of TMTOs containing resolution and contrast targets (Dudley and Gibson 2014). In response to this, researchers have devised new methods which are aimed at providing a single figure of merit depending on resolution, contrast and noise, or targeted at specific types of fault, e.g. deterioration of beam profile; element dropout. Two interesting examples of each approach are given here.

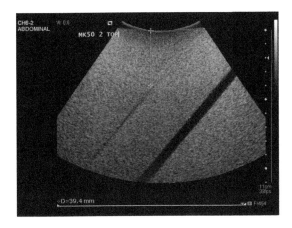

Figure 15.15 Image obtained from the Edinburgh Pipe Phantom.

Edinburgh pipe phantom

Pye and Ellis (2004) devised a measurement called the resolution integral (R) which is essentially the ratio of penetration to lateral resolution in an ultrasound image. They have implemented this measurement using a test object, the Edinburgh Pipe Phantom, which contains a number of anechoic pipes running diagonally in a conventional TMTO. The pipes are created with a range of diameters to cover the resolution of scanners over a wide frequency range spanning clinical and pre-clinical imaging. Imaging performance is characterised in terms of R and two related parameters, depth of field and characteristic resolution (MacGillivray et al. 2010). There is an increasing body of evidence that these measurements can demonstrate improvements in imaging technology over time (Pye and Ellis 2011), distinguish probes suited to particular clinical applications (Pye and Ellis 2011) and reliably detect faults (Moran et al. 2014). The Edinburgh Pipe Phantom is commercially available (CIRS, Norfolk, Virginia). A typical image is shown in Figure 15.15.

The random void phantom

Doblhoff et al. (2017) discussed the use of the random void phantom (Tissue Characterization Consulting, Timelkam, Austria; IEC 2011), consisting of an open pore foam with random pore sizes and filled with degassed saline solution. This is another attempt to combine the various spatial and greyscale parameters into a single value which may be better related to clinical performance than any single one of them alone. Software records a series of images as the ultrasound probe is moved over the surface of the phantom. A 3D data set is formed and the visibility and contrast of voids is calculated and graphically displayed.

Cross-filament phantom

Doblhoff et al. (2017) also introduced the cross-filament phantom (Tissue Characterization Consulting, Timelkam, Austria) in an attempt to address the important problems of side lobes in the elevation direction and side and grating lobes in the lateral direction. The phantom consists of two orthogonal columns of nylon filaments, allowing construction of beam profiles in the lateral and elevation directions once the probe has been scanned across the surface. Beam profiles may be affected by, for example, non-functioning elements or a damaged or poorly repaired lens.

Electronic probe testers

The FirstCall (Unisyn, Golden, Colorado) electronic probe tester was developed to test the relevant acoustic and electrical parameters of ultrasound probes. Testing is performed by attaching the probe connector to a dedicated adapter and mounting the probe at the surface of a water bath parallel to a steel reflecting plate. Three plates are available: a flat plate for linear and phased arrays, a plate with a large radius of curvature matched to typical convex arrays for abdominal use and a more tightly curved plate matched to typical endocavity probes. The entire array is pulsed, one element at a time, and a sensitivity plot produced. The system then measures the capacitance of each element circuit and displays a capacitance plot; there are a number of probes where the FirstCall cannot measure capacitance. The capacitance results allow the user to determine whether low sensitivity is due to a short circuit, open circuit or damaged element. Additionally the system provides plots of pulse width, centre frequency and fractional bandwidth for each element and pulse shapes and frequency spectra for three user selected elements.

A similar device, with some added features, is available (ProbeHunter: BBS Medical AB, Stockholm, Sweden).

Dudley and Woolley (2017) carried out a blinded study comparing FirstCall with visual evaluation of the in-air reverberation pattern. They showed that, with careful choice of scanner settings, inspection of the reverberation pattern is an excellent first-line test for detecting uniformity faults. An electronic probe tester is required only if detailed evaluation of faults is required, e.g. in a probe repair laboratory.

QUESTIONS

Multiple Choice Questions

Q1. There is good evidence to support routine testing to include:
   a. Visual inspection of equipment
   b. Electronic probe testing
   c. Resolution measurement
   d. Uniformity assessment
   e. Sensitivity measurement

Q2. For routine uniformity assessment (in-air reverberations) it is important to use:
   a. Harmonic frequencies
   b. Fundamental frequencies
   c. Superficial focal depth
   d. Compounding
   e. Advanced processing

Q3. Faults requiring immediate action include:
   a. Multiple dropout
   b. Minor lens wear
   c. Cable damage
   d. Dented console panels
   e. Needle damage to lens

Q4. Acceptance testing should include:
   a. Physical inspection
   b. Image quality measurement
   c. Functional checks
   d. Safety testing
   e. Accuracy measurements

Short-Answer Questions

Q1. Describe a simple in-air test for sensitivity which may be performed using a pre-defined preset.

Q2. Describe the expected appearance of the in-air reverberation pattern and name three possible fault conditions.

Q3. What is penetration depth and how would this be measured?

Q4. Outline the acceptance testing for acoustic safety.

Q5. Describe the main components of a string phantom.

Q6. Describe the measurement of Doppler velocity accuracy using a string phantom.

REFERENCES

Abramowicz JS, Evans DH, Fowlkes JB, Maršal K, ter Haar G on behalf of the WFUMB Safety Committee. 2017. Guidelines for cleaning transvaginal ultrasound transducers between patients. *Ultrasound in Medicine and Biology*, 43, 1076–1079.

American Institute of Ultrasound in Medicine (AIUM). 2004. *Standard Methods for Calibration of 2-Dimensional and 3-Dimensional Spatial Measurement Capabilities of Pulse Echo Ultrasound Imaging Systems*. Laurel, MD: AIUM.

American Institute of Ultrasound in Medicine (AIUM). 2007. *Performance Criteria and Measurements for Doppler Ultrasound Devices: Technical Discussion* – 2nd Edition. Laurel, MD: AIUM.

Browne JE, Ramnarine KV, Watson AJ, Hoskins PR. 2003. Assessment of the acoustic properties of common tissue-mimicking test phantoms. *Ultrasound in Medicine and Biology*, 29, 1053–1060.

Cathignol D, Dickerson K, Newhouse VL, Faure P, Chaperon JY. 1994. On the spectral properties of Doppler thread phantoms. *Ultrasound in Medicine and Biology*, 20, 601–610.

Dall B, Dudley N, Hanson M, Moore S, Seddon D, Thompson W, Verma P. 2011. *Guidance Notes for the Acquisition and Testing of Ultrasound Scanners for Use in the NHS Breast Screening Programme. NHS Breast Screening Programme Publication No. 70*. Sheffield: NHS Cancer Screening Programmes.

Doblhoff G, Satrapa J, Coulthard P. 2017. Recognising small image quality differences for ultrasound probes and the potential of misdiagnosis due to undetected side lobes. *Ultrasound*, 25, 35–44.

Dudley NJ, Gibson NM. 2010. Practical aspects of the use of urethane test objects for ultrasound quality control. *Ultrasound*, 18, 68–72.

Dudley NJ, Gibson NM. 2014. Early experience with automated B-mode quality assurance tests. *Ultrasound*, 22, 15–20.

Dudley NJ, Gibson NM. 2017. Is grey level a suitable alternative to low contrast penetration as a serial measure of sensitivity in computerized ultrasound quality assurance. *Ultrasound in Medicine and Biology*, 43, 541–545.

Dudley NJ, Griffith K, Houldsworth G, Holloway M, Dunn MA. 2001. A review of two alternative ultrasound quality assurance programmes. *European Journal of Ultrasound*, 12, 233–245.

Dudley NJ, Harries D, Wardle J. 2017. The feasibility of implementation of ultrasound equipment standards set by United Kingdom professional bodies. *Ultrasound*, 25, 25–34.

Dudley N, Russell S, Ward B, Hoskins P. 2014. The BMUS guidelines for regular quality assurance testing of ultrasound scanners by sonographers. *Ultrasound*, 22, 8–14.

Dudley NJ, Woolley DJ. 2016. A simple uniformity test for ultrasound phased arrays. *Physica Medica*, 32, 1162–1166.

Dudley NJ, Woolley DJ. 2017. A blinded comparison between an in-air reverberation method and an electronic probe tester in the detection of ultrasound probe faults. *Ultrasound in Medicine and Biology*, 43, 2954–2958.

Gibson NM, Dudley NJ, Griffith K. 2001. A computerised quality control testing system for B-mode ultrasound. *Ultrasound in Medicine and Biology*, 27, 1697–1711.

Goldstein A, Ranney D, McLeary RD. 1989. Linear array test tool. *Journal of Ultrasound in Medicine*, 8, 385–397.

Gorny KR, Tradup DJ, Hangiandreou NJ. 2005. Implementation and validation of three automated methods for measuring ultrasound maximum depth of penetration: Application to ultrasound quality control. *Medical Physics*, 32, 2615–2628.

Hangiandreou NJ, Stekel SF, Tradup DJ, Gorny KR, King DM. 2011. Four-year experience with a clinical ultrasound Quality Control program. *Ultrasound in Medicine and Biology*, 37, 1350–1357.

Hoskins PR. 1994. Choice of moving target for a string phantom. I. Backscattered power characteristics. *Ultrasound in Medicine and Biology*, 20, 773–780.

Hoskins PR. 2008. Simulation and validation of arterial ultrasound imaging and blood flow. *Ultrasound in Medicine and Biology*, 34, 693–717.

International Electrotechnical Commission (IEC). 2005. *60601-1. Medical Electrical Equipment. General Requirements for Basic Safety and Essential Performance.* Geneva: IEC.

International Electrotechnical Commission (IEC). 2011. *TS 62558. Ultrasonics – Real-Time Pulse-Echo Scanners – Phantom with Cylindrical, Artificial Cysts in Tissue-Mimicking Material and Method for Evaluation and Periodic Testing of 3D-Distributions of Void-Detectability Ratio.* Geneva: IEC.

International Electrotechnical Commission (IEC). 2016. *TS 62736. Ultrasonics – Pulse-Echo Scanners – Simple Methods for Periodic Testing to Verify Stability of an Imaging System's Elementary Performance.* Geneva: IEC.

Institute of Physics and Engineering in Medicine (IPEM). 2010. *Report 102. Quality Assurance of Ultrasound Imaging Systems.* York: IPEM.

Martin K, Spinks D. 2001. Measurement of the speed of sound in ethanol/water mixtures. *Ultrasound in Medicine and Biology*, 27, 289–291.

MacGillivray TJ, Ellis W, Pye SD. 2010. The resolution integral: Visual and computational approaches to characterising ultrasound images. *Physics in Medicine and Biology*, 55, 5067–5088.

Moran CM, Inglis S, Pye SD. 2014. The resolution integral – A tool for characterising the performance of diagnostic ultrasound scanners. *Ultrasound* 22, 37–43.

Mulabecirovic A, Vesterhus M, Gilja OH, Havre RF. 2016. In vitro comparison of five different elastography systems for clinical applications, using strain and shear wave technology. *Ultrasound in Medicine and Biology*, 42, 2572–2588.

Pye SD, Ellis W. 2004. Assessing the quality of images produced by an ultrasound scanner. UK Patent Application BG2396213.

Pye SD, Ellis W. 2011. The resolution integral as a metric of performance for diagnostic grey-scale imaging. *Journal of Physics Conference Series*, 279, 012009.

Ramnarine KV, Anderson T, Hoskins PR. 2001. Construction and geometric stability of physiological flow rate wall-less stenosis phantoms. *Ultrasound in Medicine and Biology*, 32, 245–250.

Ramnarine KV, Nassiri DK, Hoskins PR, Lubbers J. 1998. Validation of a new blood mimicking fluid for use in Doppler flow test objects. *Ultrasound in Medicine and Biology*, 24, 451–459.

Sipila O, Mannila V, Vartiainen E. 2011. Quality assurance in diagnostic ultrasound. *European Journal of Radiology*, 80, 519–525.

Vachutka J, Dolezal L, Kollmann C, Klein J. 2014. The effect of dead elements on the accuracy of Doppler ultrasound measurements. *Ultrasonic Imaging*, 36, 18–34.

Vitikainen A, Peltonen JI, Vartiainen E. 2017. Routine ultrasound quality assurance in a multi-unit radiology department: A retrospective evaluation of transducer failures. *Ultrasound in Medicine and Biology*, 43, 1930–1937.

Weigang B, Moore GW, Gessert J, Phillips WH, Schafer M. 2003. The methods and effects of transducer degradation on image quality and the clinical efficacy of diagnostic sonography. *Journal of Diagnostic Medical Sonography*, 19, 3–13.

Westerway SC, Basseal JM. 2017. The ultrasound unit and infection control – Are we on the right track? *Ultrasound*, 25, 53–57.

Safety of diagnostic ultrasound

FRANCIS DUCK

INTRODUCTION: RISK AND HAZARD

The words 'risk' and 'hazard' are emotive terms when used commonly. They are sometimes used to imply that an action should be avoided so as to ensure that there are no risks involved. Strictly, hazard describes the nature of the threat (e.g. burning, electrocution) while the associated risk takes into account the potential consequences of the hazard (e.g. death, scarring) and the probability of occurrence. It also takes account of what may be at risk (e.g. embryo, adult brain) and whether the effect is immediately obvious, or only appears later. Ultrasound scanning is potentially hazardous, but the real questions are: 'Is there any risk for the patient?' And if so, 'what is the correct way to manage this risk?' The purpose of this chapter is to explain the scientific basis informing the responses to these safety questions, and to describe the ways by which the enviable safety record of diagnostic ultrasound may be maintained.

Whenever an ultrasound scan is carried out, some part of the patient is exposed to the energy in the ultrasound beam. As the ultrasound pulse travels through the body, some of its energy is transferred to the tissue, in ways giving either a transient or lasting biological effect, depending on whether the exposure is sustained and of sufficient strength. For instance, it is well known that elevated temperature affects normal cell function and that the risk associated with this particular hazard is dependent on the extent of the temperature increase, the duration for which the elevation is maintained and the nature of the exposed tissue. During every scan with every transducer, some of the ultrasound energy is converted to heat and causes temperature elevation; the ultrasound is therefore a source of thermal hazard. The degree of elevation (i.e. the severity of the hazard) will vary throughout the region of the scan and will depend on many properties of the ultrasound field and the exposed tissues. If the maximum temperature increase within the exposed region lies within the range normally occurring in tissue, the hazard may be considered to be negligible, and so may the risk to the patient. If the maximum temperature increase is outside the normal physiological range, factors such as the duration of the elevation and the sensitivity of the tissue to damage must be borne in mind when assessing risk.

Temperature increase is the dominant hazard during diagnostic ultrasound procedures. However, a second hazard arises from the presence of gas within soft tissues. Gas may occur naturally, for instance air in the alveoli of lungs or gas in the intestines. Alternatively gas bubbles may be introduced deliberately in the form of gas contrast agents. Finally, gas nuclei may exist in crevices on the surface of solid bodies such as renal calculi. When such 'gas bodies' are exposed to ultrasound they can induce a variety of local mechanical effects that may cause damage to cells or tissue structures. The oscillation of the gas surface that causes the mechanical effects is termed 'acoustic cavitation' for a free bubble, or 'gas body activation' for the more general case. In these cases the hazard, a bubble caused to oscillate by the ultrasonic wave, gives a risk of tissue damage that will vary depending on the size of the oscillation, where the gas is, and what cellular changes result.

The prudent sonographer, therefore, should consider the safety aspects of each examination, and only undertake it if the benefits to the patient from the expected diagnostic information outweigh the risks. This chapter aims to provide the sonographer with the information needed to assess the risks associated with ultrasound examinations. It first reviews how ultrasound output and exposure are related to hazard and risk. Thermal and mechanical processes and effects are then reviewed, leading to a discussion of the safety indices now presented on scanners to assist users in making risk/benefit judgements. A brief review of epidemiological evidence is given. The partnership between manufacturers and users in managing safety is reviewed, including an overview of standards and regulations. Finally, situations giving rise to specific safety issues are discussed.

Those wishing to gain further background knowledge, or explore the topics in greater depth are recommended to refer to other texts on ultrasound safety (Barnett and Kossoff 1998; McKinlay 2007; AIUM 2008; Duck 2008; ter Haar 2012). Tutorials on a number of safety topics have also been prepared by the European Committee of Medical Ultrasound Safety (ECMUS) on behalf of the European Federation of Societies for Ultrasound in Medicine and Biology (EFSUMB), and are freely available on the EFSUMB Web site www.efsumb.org/blog/archives/869

WHAT DOES ULTRASOUND EXPOSURE MEAN?

During an ultrasound examination, tissues are exposed to ultrasound beams and pulses (the ultrasound field) used to acquire the image or Doppler waveform. To measure the ultrasound exposure in the tissue, it is necessary to characterise the ultrasound field in terms of a number of standard quantities. We have seen in Chapter 2 that ultrasound is a longitudinal pressure wave and so, as the ultrasound wave passes a particular point, the pressure at that point will increase and decrease in a cyclic manner. Figure 16.1 shows a measurement of the acoustic pressure variation with time for a typical pulsed Doppler ultrasound pulse near to the focus in water. The very rapid change from negative to positive pressure and the larger positive than negative pressure are normal for these pulses. There are many different properties that can be measured. The main quantities used to describe the beam and how they are measured are described in Appendix B.

From a safety perspective, we want to know what is happening inside the exposed tissue. Obviously this is not an easy thing to do, so, instead of using tissue, measurements are made in a substitute medium which is well characterised and reproducible: the medium that is currently chosen is water.

Seventy per cent of most soft tissues consists of water and, in some respects, ultrasound travels through water in a similar way to how it travels through soft tissue. It travels at about the same speed, vibrates internally in the same way (i.e. it has a similar acoustic impedance) and so reflects and refracts in a similar way. The main difference is that water generally absorbs very little ultrasound at these frequencies compared to tissue. As a result, measured values of acoustic quantities (see Appendix B) in water are higher than would be expected in tissue. Measurements made in water are called 'acoustic output measurements' (or, more correctly, 'free-field acoustic output measurements'). Using this term allows the concept of

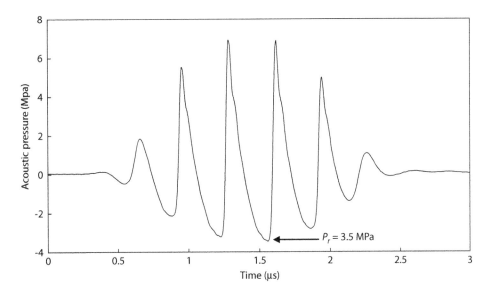

Figure 16.1 Variation of acoustic pressure with time in a pulsed Doppler ultrasound pulse near the focus in water.

'exposure measurements' and 'exposure parameters' to be reserved for what happens in tissue rather than what happens in a tank of water. As explained later, estimates of exposures in tissue made from these, usually referred to as 'estimated in-situ values', are approximations derived from simple models of tissue properties and structure.

The relationship between risk, hazard, exposure and acoustic output is outlined in Figure 16.2. It can be seen that as we move from actual measurements

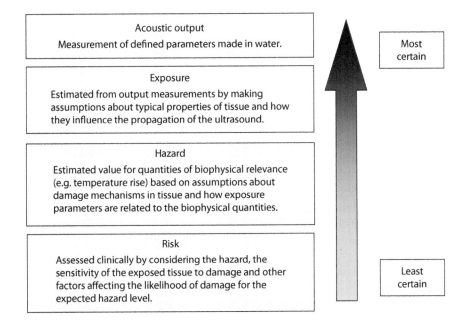

Figure 16.2 Relationship between acoustic output, exposure, hazard and risk.

of acoustic output parameters in water to estimates of exposure, hazard and risk, our assessments become less and less certain. A semi-formal evaluation of hazard and risk in diagnostic ultrasound has been published elsewhere (Duck 2008).

WHAT HAPPENS TO TISSUE EXPOSED TO ULTRASOUND: AND DOES IT MATTER?

Thermal effects

As noted above, tissue heating is the most important and general hazard for diagnostic ultrasound. When an ultrasound pulse travels through tissue, some of the energy in the pulse is absorbed by the tissue and is converted to heat, which in turn produces a temperature rise.

The rate at which energy is absorbed per unit mass q_m depends on the amplitude absorption coefficient α_0 and density ρ of the tissue and on the intensity I:

$$q_m = 2\alpha_0 I / \rho \qquad (16.1)$$

The factor 2 arises because the intensity absorption coefficient is twice the amplitude absorption coefficient. The absorption coefficient of soft tissue depends on the ultrasound frequency, being higher at higher frequencies than at lower frequencies. This means that the energy deposited in the tissue for a 10-MHz scan is about two times greater than that at 5 MHz, if the intensities are the same.

The quantity q_m has been termed the 'acoustic dose rate', and is measured in watts per kilogramme (Duck 2009). It is a similar unit of dosimetry to the specific absorption rate (SAR) for electro-magnetic radiation, used for example on all magnetic resonance imaging scanners.

The intensity varies throughout the field, being high at the beam focus and lower elsewhere. In addition, the absorption coefficient and density vary, particularly between bone and soft tissue. As a result, the rate at which energy is absorbed, and hence the temperature rise, also varies throughout the exposed region.

To discuss the effects on tissue temperature in more detail, it is easiest to consider a fixed beam mode, such as a pulsed Doppler beam interacting with soft tissue, in which the temperature is initially uniform. When the field is first applied, energy will be absorbed at a rate proportional to the local intensity, which means that the temperature will increase fastest at the focus. As time goes on, the temperature will continue to increase as more energy is absorbed, but the regions where the increase has been greatest (e.g. at the focus) will start to lose some of their heat by conduction to neighbouring cooler regions, and so the rate of increase will begin to slow. In soft tissues, the rate of heat loss by conduction becomes about the same as the rate of energy absorption within about a minute. In practice, the scan head must be held stationary for this time to reach the greatest temperature; any movement allows the tissue to start to return to its normal temperature again. In wider regions of the beam, such as near to the transducer, very little heat will be lost by conduction initially and so the temperature will continue to increase for a longer time. But the rate of energy deposition is lower because the intensity is lower.

In summary, at the focus, heating is fast but heat loss is high; near the transducer, heating is slow but the heat loss is less. As a result, the greatest temperature is typically reached in soft tissue about midway between the skin and the focus.

That is true when a beam is held stationary, such as in a pulsed Doppler examination. But during any scanning (scanned imaging, Doppler imaging or plane-wave imaging), the rate of energy deposition is greatest at the surface, so this is where the greatest increase in temperature from the ultrasound beam occurs.

In practice, a second source of heat is also important; as it generates ultrasound, the transducer itself begins to heat up. This heat conducts into the tissue, enhancing the temperature rise near the transducer caused by the absorption of ultrasound. Commonly, even for pulsed Doppler, the temperature near the transducer exceeds the focal temperature. Nevertheless, remembering this is the hazard rather than the risk, the focal temperature rise may be more important, since this is where sensitive tissue (such as the embryo) is most likely to be found.

The presence of bone in the field increases the temperature rise. Bone strongly absorbs ultrasound at all frequencies and so the ultrasound energy is almost completely absorbed in a very small volume at the bone surface. So the temperature increases more rapidly than for soft tissue. Because the absorption takes place in a small region, the temperature gradients and also the conductive heat losses will become large. This means that the temperature will quickly approach its final value, typically within a few seconds.

Some heat is removed by perfusion; that is, blood flowing through the tissue acts to flush the heat away. This is an important cooling mechanism in highly vascularised organs such as the heart, and perhaps the liver. But it is relatively unimportant for most soft-tissue scanning. Additionally, hyperthermic reactive perfusion, whereby perfusion may increase in response to local temperature rise, occurs only at much higher sustained temperature rises. For the assessment of risk it is best to assume that there is no significant increase in perfusion that results from heating caused by diagnostic ultrasound.

TEMPERATURE PREDICTIONS

In principle it is possible to predict the temperature distribution using theoretical models, if enough is known about the *in situ* intensity distribution and the properties of the tissue. In practice, accurate predictions are difficult since the *in situ* intensity can only be poorly predicted from acoustic output measurements. In addition, measured tissue properties have demonstrated their wide variation, not only varying with age, tissue type and species, but also showing a range of values for each 'normal adult' tissue. This means that all estimates of temperature rise in tissue can only be uncertain predictions of particular situations. Consequently, alternative approaches have been developed. These include the use of tissue-mimicking phantoms to make measurements of temperature rise, and the development of simplified, approximate methods of estimating the maximum temperature increase, such as thermal indices.

TEMPERATURE MEASUREMENT

Ideally, of course, it would be possible to measure the temperature rise in the patient during the ultrasound examination. Unfortunately, with current technology, this is not yet possible. However, it is possible to measure the temperature rise in a tissue-mimicking phantom, also called a thermal test object (TTO) (Shaw et al. 1999). Figure 16.3 shows a schematic diagram of one of these TTOs. This is a much more direct way of evaluating the heating potential of ultrasound scanners which automatically caters for most of the complexities which are ignored by other methods. While TTOs have been available commercially and have provided important results (e.g. Shaw et al. 1998), it is more common for investigators to construct their own test objects.

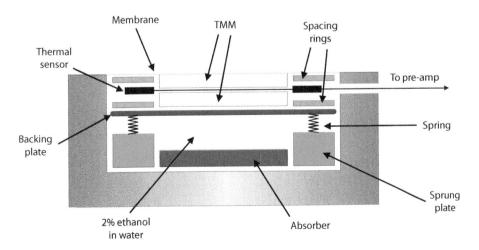

Figure 16.3 A thermal test object. The temperature rise in the middle of the tissue-mimicking material (TMM) is measured with a small thermocouple of less than 0.5 mm in diameter.

The most important application of TTOs is in determining the contact temperature between the transducer and the patient. IEC60601–2–37 (2007) places a limit on the temperature of the transducer face; this limit is 50°C when the transducer is radiating into air (to replicate the case where the scanner is active even though the patient is not being scanned) and 43°C when the transducer is in contact with a suitable TTO. Companies who claim compliance with this standard when seeking CE marking or other regulatory approval must undertake type-testing to ensure that their scanners meet this limit.

THERMAL INDICES

Since it is impossible for the user to know how much temperature increase is occurring in the body, the thermal index (TI) is a useful alternative to provide some guidance. A TI is a rough estimate of the greatest increase in temperature that could occur in the region of the ultrasound scan. Thus, a TI of 2.0 means that you might expect a temperature rise of about 2°C. However, the TI must not be considered as a remote sensing thermometer. It is much more helpful to note whether the TI value is increasing or decreasing as the scan progresses, so noting whether the overall thermal hazard is increasing or decreasing. TI values can provide extra information to help weigh up the risks and benefits to an 'average' patient as the examination progresses. They are not supposed to apply to any particular patient and other factors such as the physical state of the patient must be considered when making a risk assessment. Originally developed by the American Institute of Ultrasound in Medicine, these indices are now incorporated into IEC Standards IEC60601–2–37 (2007) and IEC62359 (2010).

The TI itself has a very simple definition. It is the ratio between two powers. The first is the ultrasound power exposing the tissue W. The second is the power W_{deg} required to cause a maximum temperature increase of 1°C anywhere in the beam, with identical scanner operating conditions. This temperature is reached with the beam held stationary under 'reasonable worst-case conditions' allowing thermal equilibrium to be achieved, and is intended to be a temperature which could be approached but never exceeded *in vivo*:

$$TI = \frac{W}{W_{deg}}$$

The difficulty with calculating TI lies mostly in the estimation of W_{deg} the power giving a worst-case 1°C rise. In order to simplify the problem, three tissue models have been chosen to distinguish three applications. They give rise to three thermal indices: the soft-tissue thermal index (TIS), the bone-at-focus thermal index (TIB) and the cranial (or bone-at-surface) thermal index (TIC). Figures 16.4 through 16.6 show these three conditions. The tissue models are based on simple assumptions about the acoustic and thermal properties of tissue, including the assumption that the tissue attenuation coefficient is uniform at 0.3 dB cm^{-1} MHz^{-1}. A small but finite perfusion is assumed. For bone, a fixed proportion of the incident power is assumed to be absorbed in the bone layer. All three TI values

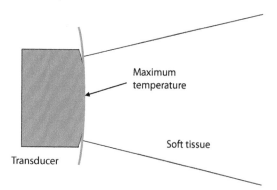

Figure 16.4 Conditions for the soft-tissue thermal index (TIS) for scanning. These conditions are also assumed to apply for the calculation of bone-at-focus thermal index (TIB) for scanning.

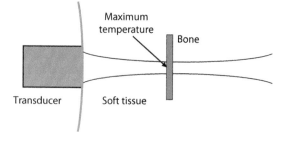

Figure 16.5 Conditions for the bone-at-focus thermal index (TIB).

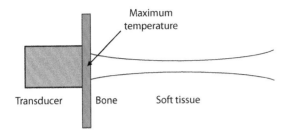

Figure 16.6 Conditions for the cranial (bone-at-surface) thermal index (TIC).

depend linearly on the acoustic power emitted by the transducer. If the machine controls are altered in such a way as to double the power output, the TI displayed will increase by a factor of 2.

In addition to the dependence on power, one of the thermal indices, soft tissue (TIS), also depends on the frequency of operation of the transducer. This is because the absorption coefficient of soft tissue depends on frequency and hence, so does the heating.

There is a subtle but important caveat: the extra temperature rise caused by the heated transducer is not included in the TI calculation and must be assessed in a different way, for example by using a thermal test object. This deficiency is less important than it seems. Theory shows that the temperature rise within the body during scanning, for example within fetal tissue, should never exceed the temperature at the surface, ignoring transducer heating. This means that heating at depth can only exceed that indicated by the TI in highly unusual conditions, such as those indicated later in the text.

These TI values are more helpful than any other information you are likely to see and you should take note of the values when you can. On current-generation equipment, TI values can usually be found around the edge of the scanner screen – often in the top right corner – indicated by the letters TIS, TIB or TIC followed by a number that changes when the scanner controls are altered. If they are not visible, it may be necessary to turn on the display of these values. Check the manual or ask your supplier how to do this. Smaller, portable or low-output equipment may not display TI because it is always very low. Again, check the manual or ask your supplier if you are not sure.

DOES TEMPERATURE RISE MATTER?

What is the result of an increased tissue temperature? According to Miller and Ziskin (1989) the range of 'normal' core temperatures is approximately 36°C–38°C in man, and a temperature of 42°C is 'largely incompatible with life'. To maintain life requires the careful balance of many chemical processes. Changing cellular temperature not only changes the rate at which these reactions take place, it also affects the equilibrium position between competing reactions.

Clearly, it is undesirable to elevate the whole body above its natural temperature, but during an ultrasound examination only a small volume of tissue is exposed. Does it matter if this is affected? Generally, the human body is quite capable of recovering from such an event. There are, however, several regions that do not tolerate temperature rise so well. These include the reproductive cells, the unborn fetus and the central nervous system (brain and spinal cord). In view of the large number of obstetric examinations carried out, risk to the embryo and fetus has been subjected to the greatest emphasis in safety discussions.

Clinical ultrasound scanners operating towards the top end of their power range are capable of producing temperature rises above the recommendations of the World Federation for Ultrasound in Medicine and Biology (WFUMB). These recommendations, now two decades old, and constituting the most enduring and well-based scientific advice underpinning ultrasound safety discussions, were based on a thorough review of thermal teratology, the investigation of genetic abnormalities from heat.

The two most important recommendations are:

- A diagnostic exposure that produces a maximum temperature rise of no more than 1.5°C above normal physiological levels (37°C) may be used clinically without reservation on thermal grounds.
- A diagnostic exposure that elevates embryonic and fetal *in situ* temperature above 41°C (4°C above normal temperature) for 5 minutes or more should be considered potentially hazardous.

When applying these recommendations, it should be remembered that they were developed to assess risk for the most critical target, the embryo, exposed during the first trimester, the period of organogenesis. Any use of these limits for other organs should bear this in mind, recognising that the risks from these temperature rises in a different organ, or even in the fetus at a different stage of development, would be different, and probably lower.

Nevertheless, it is true that temperatures reaching and exceeding those given in the WFUMB recommendations have been demonstrated in clinical beams. In one study (Shaw et al. 1998), the temperature rise was measured in tissue-mimicking phantoms caused by a range of clinical pulsed Doppler ultrasound systems, at conditions of maximum output.

Under most situations, it appeared unlikely that a temperature rise of more than 1.5°C would occur in soft tissue in the absence of bone or other strongly absorbing material. However, conditions were identified for which caution was advised. Exposure of the first-trimester fetus through the full bladder could cause a possible temperature rise of up to 3°C. The use of intra-cavity probes was also noted, where prolonged exposure may result in higher temperature rises within 1 cm of the transducer face (Calvert et al. 2007). When bone is in the ultrasound beam, even if it is not the subject of the examination, the temperature rise may be much higher. Seventy-five per cent of the systems studied produced a temperature rise in a bone mimic of between 1.5°C and 4°C, with even higher temperatures considered possible if these same fields were used for examinations through the full bladder, for instance for the third-trimester fetus. Although it is important to remember that the exact study conditions are unlikely to be replicated in most examinations, 50% of the systems gave temperature rises in excess of 4°C and 15% in excess of 8°C.

Non-thermal mechanisms and effects

There are two mechanisms other than heating which must also be considered to gain a complete view of safety. The first includes those arising when there is gas adjacent to tissue exposed in the beam. These are often referred to loosely as cavitation effects, although as is shown, they include a

rather wider range of conditions. This class is most important at the lower end of the ultrasonic frequency spectrum. The second class includes radiation force and becomes increasingly important at the highest ultrasonic frequencies.

CAVITATION AND OTHER GAS-BODY MECHANISMS

Acoustic cavitation refers to the response of gas bubbles in a liquid under the influence of an acoustic (ultrasonic) wave. It is a phenomenon of considerable complexity (Leighton 1994, 1998), and for this reason it is common to simplify its description into two categories. Stable (or non-inertial) cavitation refers to a pulsating or 'breathing' motion of a bubble whose diameter follows the pressure variation in the ultrasonic wave. During the compression in the wave, the bubble contracts in size, expanding again as the rarefaction phase in the wave passes. Over time the bubble may grow by a process of 'rectified diffusion', or it may contract and dissolve back into the liquid. Cavitation only occurs if suitable cavitation nuclei exist to seed the cavitation process. These nuclei are typically microbubbles or solid particles in suspension, or bubbles trapped in crevices at a solid surface. For safety discussions, the most important sources of cavitation nuclei are those introduced *in vivo* in the form of contrast materials, and these are discussed in the text that follows.

Free, solitary bubbles have a resonant frequency f_r depending on their radius R_0, which for a spherical air bubble in water is given by:

$$f_r R_0 \approx 3\,\mathrm{Hz\,m}$$

This expression is a reasonable approximation for bubble radii greater than 10 μm, but also suggests that typical diagnostic frequencies cause resonance in bubbles with radii of the order of 1 μm.

Of course, the existence of an oscillating bubble (i.e. a hazard) does not automatically result in a risk. It is necessary to relate its behaviour with something that could be potentially damaging. When a suspension of cells is exposed to ultrasound and stable cavitation occurs, shear stresses may be sufficient to rupture the cell membranes. Shear is a tearing force, and many biological structures are much more easily damaged by tearing than by compression or tension. Destruction of blood cells in suspension by ultrasound may occur in this way provided that the acoustic pressure is high enough, and this has been shown *in vitro* for erythrocytes, leucocytes and platelets. In addition, the existence of a second effect, not causing cell destruction, may occur. Gaps can open transiently in the cell membrane during ultrasound exposure, allowing the passage of larger biomolecules such as DNA, an effect known as 'sonoporation'.

The second class of cavitation activity has been termed 'inertial cavitation'. It is more violent and potentially more destructive than stable cavitation, occurs at higher peak acoustic pressures, and can be generated by short ultrasound pulses. The bubble undergoes very large size variations, and may collapse violently under the inertia of the surrounding liquid (hence the term 'inertial cavitation'). As with stable cavitation, local shear stresses are generated which may cause lysis of adjacent cells. Acoustic shocks are generated which can propagate stress waves outwards. In addition, enormously high, localised pressures and temperatures are predicted to occur, and the energy densities are sufficient to cause free radical (H^+ and OH^-) generation local to the collapsing bubble.

While bubbles will undergo stable (and probably non-harmful) oscillations in ultrasound beams at very low pressures, the onset of inertial cavitation only occurs above a threshold for acoustic pressure. An analysis by Apfel and Holland (1991) quantified this threshold, assuming that a full range of bubble sizes were present to provide nucleation centres. This resulted in the formulation of a 'mechanical index' (MI), which was intended to quantify the likelihood of onset of inertial cavitation:

$$MI = \frac{p_r}{\sqrt{f}}$$

where p_r is the peak rarefaction pressure *in situ*, and f is the ultrasound frequency. The MI is displayed on the screen of most scanners in the same way that TI is displayed (see previous discussion). For MI < 0.7 the physical conditions probably cannot exist to support bubble growth and collapse. Exceeding this threshold does not, however, mean there will automatically be a biological effect caused by cavitation.

The destructive outcome of cavitation on cells in culture is well documented. *In vivo*, there is evidence for acoustic cavitation when using extracorporeal lithotripsy. However, there is no evidence that diagnostic pulses cause cavitation within soft tissues under the majority of conditions. There are only three specific conditions when the presence of gas *in vivo* can alter this. These are with the use of contrast materials, for the exposure of soft tissues which naturally include gas, such as in the lung and in the intestine, and when stable microbubbles are trapped in fissures on the surface of solid concretions such as renal stones.

CONTRAST MATERIALS

Much of the previous discussion is of direct relevance to the use of contrast materials. When injected into the bloodstream, ultrasound contrast agents introduce a plentiful supply of cavitation nuclei, which in principle could result in the generation of free radicals, cell lysis or sonoporation as previously described. Extrapolation from studies carried out *in vitro* to what may happen *in vivo* is, however, notoriously difficult. For example, free-radical scavengers in blood strongly limit the lifetime of radicals generated *in vivo*, and hence the potential for damage. Most contrast agents break up when they are exposed to ultrasound. Nevertheless, there is good evidence that bubble fragments continue to circulate for a considerable time after the initial disintegration. For this reason it is considered unwise to use contrast materials during lithotripsy or at any time during the preceding day in order to minimise the chance of the contrast fragments acting as cavitation nuclei in the lithotripsy field. More seriously, microvascular damage may often result from exposure of gas-filled contrast agents *in vivo* to pulsed ultrasound at *in situ* rarefaction pressures of less than 1 MPa, well within the diagnostic range (Miller and Quddus 2000). International bodies have published reviews and guidelines for the safe use of contrast agents (Barnett et al. 2007; AIUM 2008; EFSUMB 2008).

LUNG CAPILLARY DAMAGE

Exposure of the pleural surface of lungs to diagnostic levels of ultrasound has been consistently shown to cause alveolar capillary bleeding. This has been observed in experiments on a variety of small mammals. Pressure thresholds for damage in small animals are about 1 MPa, which is well within the diagnostic range. Physical damage occurs when the local stress exceeds the ultimate tensile strength of the alveolar membrane. Similar small haemorrhages (petechiae) have also been observed in the exposed intestines of small animals and related to the presence of intestinal gas. Whatever the exact cause, diagnostic pulse amplitudes can cause damage when fragile tissue structures are exposed to pulsed ultrasound while being adjacent to gas bodies.

Lung and intestinal bleeding from ultrasound exposure has never been observed in humans. It has been suggested that the lungs of smaller animals are more susceptible to this form of damage than those of larger mammals, where the pleural layers and perhaps the alveolar membranes are thicker and stronger. If so, the relevance of these animal studies to clinical scanning may be restricted to studies in neonates, particularly cardiac examinations. Even so, such slight alveolar bleeding would generally be almost without clinical importance, provided it was limited in extent.

CAVITATION AT SOLID SURFACES

A twinkling artefact may be observed during Doppler examinations of solid inclusions such as calcifications and renal calculi. The cause of this artefact has been much discussed, and is now convincingly attributed to the activation of cavitation nuclei stabilised in crevices in the surface of the solid. Stable cavitation is induced, in which a microbubble is caused to breathe, and possibly grow, under the influence of the ultrasound beam. The biological significance of this phenomena remains to be investigated.

RADIATION FORCE

Radiation force is the second non-thermal mechanism which results in energy transfer to tissue, and which could theoretically result in a biological outcome. Radiation force is exerted on all tissues in an ultrasound beam, acting in a direction away from the transducer along the ultrasound beam.

It is a very small force. It depends upon the acoustic dose rate q_m (Equation 16.1) and is, therefore, more strongly related to heating than to cavitation. Unlike cavitation, and like heating, it is always present and accompanies temperature rise at all times.

The most obvious and visible outcome of radiation force is acoustic streaming. In liquids such as amniotic fluid, there is a bulk movement of the liquid in the direction of the beam, pushed by the radiation force. This movement is most obviously generated *in vivo* in pulsed Doppler studies, when it may be demonstrated using Doppler imaging. When exerted on soft tissues, they move in the direction of the beam. This mechanism forms the basis for shear-wave elastography (see Chapter 14).

No evidence has been found that fluids can be moved within adult soft tissue by radiation force at diagnostic intensities and frequencies. Diffusion of intra- and extra-cellular fluids is too constrained by other forces. While soft tissues will move, the forces do not exceed those needed to cause damage. There remains a caution, though, that these forces may be capable of disturbing cellular structures in the developing embryo, although no evidence has ever been presented to support this possibility.

Epidemiological evidence for hazard

Epidemiology is concerned with the patterns of occurrence of disease in human populations and of the factors influencing these patterns. Epidemiological methods have also been used to search for risks associated with any external agents, such as diagnostic ultrasound, and to look for associations between exposure and unwanted outcome. There have been a number of studies to investigate the possible association between exposure to ultrasound *in utero* and childhood mal-development.

This section summarises briefly the outcome of the more important epidemiological studies into ultrasound exposure *in utero*. Fuller reviews may be found elsewhere (Salvesen 2007), in which the details of the literature may be found.

BIRTH WEIGHT

There have been several investigations into a possible association between exposure *in utero* and subsequent birth weight. No consistent results have emerged from these studies, which showed both increased and decreased birth weight depending on the study design. In view of the conflicting evidence presented by these studies, it is currently concluded that there is no evidence to suggest an association between exposure to ultrasound and birth weight.

CHILDHOOD MALIGNANCIES

There have been three well-managed case-control studies into ultrasound and childhood malignancies, all of which were of sufficient size to have statistical validity. No association with childhood malignancy was found in any study.

NEUROLOGICAL DEVELOPMENT

A range of neurological functions have been examined and no association between ultrasound exposure *in utero* and subsequent hearing, speech development, visual acuity, cognitive function, dyslexia or behaviour has been confirmed. Several studies have reported on the left- or right-handedness of children and its association with exposure to ultrasound *in utero*. These studies have been subjected to meta-analysis, concluding that there was a small but significant trend away from right-handedness associated with ultrasound exposure, especially in boys. No convincing bioeffects mechanism has been postulated for these outcomes.

In summary, there is no independently verified evidence to suggest that ultrasound exposure *in utero* has caused an alteration in the development and growth of the fetus. All studies have either proved to be negative, or when positive findings have appeared they have not been verified, or have been shown to result from poorly designed studies. New studies will be difficult to structure, because of the difficulty of finding an unexposed control group, because of the widespread use of ultrasound during pregnancy throughout the world. It is necessary to sound a note of caution, however. Details of ultrasound exposure are missing in many of the studies, which commonly record neither exposure intensities nor dwell time. There are no studies into the outcomes following exposure to pulsed Doppler or Doppler imaging, where intensities and powers are known to be higher than in pulse-echo imaging. While the results of epidemiological studies so far are comforting, they cannot be used to support an argument that it is safe

to extend exposure *in utero* to include the higher levels associated with pulsed Doppler studies. Further epidemiological studies focused specifically on Doppler exposure would be needed before such confidence can be claimed.

HOW IS SAFETY MANAGED?

The successful management of safety for medical ultrasound involves everyone: manufacturers, users and other experts. Manufacturers must comply with standards intended to make the equipment safe. Users must make sure that they use the equipment in an appropriate and safe manner. Both groups, together with other experts such as embryologists, biochemists and physicists, must contribute to the maintenance of national and international standards to keep ultrasound safe without unduly restricting its use and benefits.

Manufacturers' responsibility

The Medical Device Directive (MDD) in Europe and the Food and Drug Administration (FDA) in the United States both make demands of manufacturers regarding safety of their scanners and provision of information to purchasers and users. The standards supporting these directives are generated by the International Electrotechnical Commission (IEC). The IEC has also published a number of measurement and performance standards.

US FOOD AND DRUG ADMINISTRATION

Any equipment sold in the United States must meet the US FDA regulations requiring that manufacturers supply information on acoustic output and ensure that certain derated acoustic parameters do not exceed allowable levels. Most commonly, manufacturers follow the exposure limits that are set out in Table 16.1. This is done in conjunction with on-screen indication of thermal and mechanical indices according to the methods set out by the IEC. The limits set by the FDA strictly apply only to the sale of equipment in the United States. Nevertheless, even without their formal adoption by the international community, there is no evidence that manufacturers commonly make scanners for sale outside the United States that would not conform to the limits set by the FDA.

Table 16.1 The upper limits of exposure specified by the US Food and Drug Administration

	Derated I_{spta} (mW cm^{-2})	Derated I_{sppa} (W cm^{-2})	Mechanical index (MI)	Thermal index (TI)
All applications except ophthalmology	720	190	1.9	(6.0)
Ophthalmology	50	Not specified	0.23	1.0

Note: The upper limit of 6.0 for thermal index is advisory. At least one of the quantities MI and I_{sppa} must be less than the specified limit.

EUROPEAN MEDICAL DEVICE DIRECTIVE

The European Community (EC) Medical Device Directive (MDD) (2017) requires that all medical devices (except custom-made and devices intended for clinical investigation) meet essential requirements for safety and performance and carry a CE mark before they are placed on the market in the EC. The CE mark is the manufacturer's declaration that the device conforms to these requirements. The MDD includes requirements for the visual display or warning of emission of potentially hazardous radiation, and for the indication of accuracy of equipment with a measuring function. The normal method for a manufacturer to demonstrate that they have complied with these requirements is for them to follow procedures laid down in international standards. Standards recognised as demonstrating compliance with particular safety requirements are listed in the *Official Journal of the European Community*. At the time of writing, there is no clarity about the status of European regulations in the United Kingdom once the Brexit negotiations have been completed. Nevertheless, it may be assumed that a comparable regulatory system will be implemented.

Since ultrasound scanners are pieces of electro-medical equipment, all relevant parts of the IEC 60601 series (Safety and Essential Performance) of standards apply. The international standard that is specific to diagnostic ultrasound is IEC60601-2-37, 'Medical Electrical Equipment, Part 2–37: Particular requirements for the basic safety and essential performance of ultrasonic medical diagnostic and monitoring equipment'. This standard requires manufacturers to provide information about the acoustic output of their scanners in water under conditions that produce the maximum indices (TI, MI) for each operational mode (B-mode, M-mode, colour flow, etc.).

Equipment which cannot generate indices of 1.0 or above, or which meets certain other low-output criteria, need not provide such detailed output information (usually, this applies to fetal Doppler and peripheral vascular Doppler devices only). In addition, 60601-2-37 sets an upper limit of 43°C for the temperature of the transducer surface in contact with the patient; the user should be aware that this limit is only intended to be safe for contact with adult skin. Transducers that are left running in free air are allowed to reach 50°C according to the standard.

Users' responsibility

The role of the sonographer in managing safety is obviously vital. The users must ensure that they are properly trained and that they keep their knowledge up to date with changing technology and practices. They must choose and use scanners that are appropriate for the type of examination and the condition of the patient. They must follow good practice for reducing the risk to the patient. They must take part of the responsibility for ensuring that the equipment they use is properly maintained and meets necessary standards.

PROPER TRAINING

It is obvious, but worth repeating, that all users need to be trained in and familiar with both the relevant anatomy and physiology for the examinations they make and the strengths and weaknesses of the equipment they use. As new equipment and techniques are developed, further training will be needed to ensure that, for instance, different imaging artefacts are understood and an accurate diagnosis is made. The greatest risk to the patient still comes from misdiagnosis and that should never be overlooked.

To complement this, users need also to appreciate the current thinking with respect to hazard posed by the ultrasound exposure itself and should seek training either through their own department, their professional body, the British Medical Ultrasound Society (BMUS) or published literature. Various committees and professional bodies have offered guidance to the sonographer when it comes to assessing the possible risks (Barnett 1998; NCRP 2002; AIUM 2008). A summary of the various positions taken by national and international bodies has been given by Barnett et al. (2000). The BMUS has published guidelines for safety, prepared by the BMUS Safety Group (www.bmus.org/policies-statements-guidelines/safety-statements).

The BMUS guidelines include recommended action levels associated with values of MI and TI under some specific conditions. Nelson et al. (2009) have put forward further suggested guidelines to be used for setting and monitoring acoustic output during scanning. For prenatal scanning, TI = 0.5 is recommended as a general upper limit for extended use, with graded time limits for higher values, reaching a limit of 1 minute for TI > 2.5. For post-natal examinations, the equivalent limits are TI < 2 for extended scanning, time-limited to 1 minute for TI > 6. Under all conditions, MI should be limited to 0.4 if gas bodies may be present, and be increased as needed in their absence.

APPROPRIATE EQUIPMENT

If a probe is intended for a specific purpose, it should only be used for that purpose because choice of working frequency, acoustic output levels and field geometry are tailored to the intended use. A probe that is perfectly safe for adult cardiac use may not be safe for an obstetric or neonatal cephalic examination. If possible, use equipment that provides safety information to the user – this will generally be in the form of TI and MI displays – and take note of this information and your knowledge of the patient when planning and carrying out the scan.

In general, you should be less concerned about examining 'normal', non-pregnant adults. You should take more care when examining:

• The fetus, or the maternal abdomen near the fetus

• Neonates
• Expectant mothers with fever or elevated core temperature
• Patients under anaesthesia or during surgery
• During the use of injected ultrasound contrast agents

In some types of examination, TI and MI may underestimate the heating or cavitation. In general, you should take more care for:

• Fetal examination through the full bladder
• The use of trans-vaginal, trans-rectal or other internal probes
• Transcranial examination

GOOD SAFETY PRACTICE

Keeping up to date with current thinking on ultrasound safety and risk minimisation will allow the sonographer to make the best decisions on how to maximise the benefit to the patient while reducing any risks. The previous section on use of appropriate equipment, in conjunction with the following list, gives some simple guidelines:

• Only carry out an examination if there are clinical grounds to do so.
• Make sure you get the information you need to make a good diagnosis. There is more chance of causing harm by misdiagnosis than through exposure to ultrasound.
• Reduce the amount of time that the probe is in contact with the patient. You can do this by stopping the examination as soon as you have the information you want and by removing the probe during the examination if you need to talk to the patient or to a colleague and are not able to concentrate on the image on the scanner.
• Generally, Doppler modes (PW, colour or power Doppler) are more likely to cause heating. So, use greyscale imaging to find the clinical site and only use Doppler modes if they are necessary and when you have found the site.
• The most cautious approach is to display TIB most of the time. Only display TIC for transcranial examinations.

PROPER MAINTENANCE

The final point is that users are not stuck with the equipment they are given. They can and must play an active role in making sure that the equipment is properly maintained and repaired, and they can actually help develop better and safer equipment by making safety issues an important issue with manufacturers. They can do this both by questioning manufacturers directly (and, of course, a trained and knowledgeable sonographer will be able to ask more challenging questions) and by taking active part in the debates within BMUS, the professional bodies, BSI and IEC. It is these debates that, in the long run, produce better international standards with which the equipment bought by individual hospitals and departments will comply.

Is exposure altering?

An important practical question in discussions of ultrasound safety is whether current clinical practice causes higher ultrasound exposure to the patient population than was so in the past. There are three overlapping but distinct questions:

1. Do the acoustic pressures, intensities or powers used today differ significantly from those used in the past?
2. Has the introduction of any new scanning modes (such as plane wave imaging or shear-wave elastography) or transducers (such as those for trans-vaginal or trans-oesophageal scanning) been accompanied by altered exposure?
3. Has the clinical use of ultrasound altered over the years, by the numbers of scans being carried out, and/or by the exposure dwell time used?

INCREASING OUTPUT FROM DIAGNOSTIC SCANNERS

During the 1990s, evidence was published for a continuing trend towards higher ultrasound output. The most probable cause at that time was the relaxation of regulatory limits applied by the FDA in the United States (see previous discussion) in particular related to the allowed intensity for non-cardiovascular scanning, including obstetric scanning. As a result, manufacturers increasingly used the higher intensities to exploit new methods of scanning, with the result that output generally tended to move towards the limits set by the FDA. No evidence has been published more recently that would suggest any continuing increase similar to that observed two decades ago.

Unfortunately, there remains an unresolved measurement problem, called 'acoustic saturation', which may result in increases in MI values remaining unreported. This is because there is a limit to the acoustic pressure in water no matter how strongly the transducer is driven (Duck 1999). As a result *in situ* exposure is underestimated, especially at higher frequencies. Indeed, there are some conditions when the FDA limit for MI of 1.9 can never be exceeded, however hard the transducer is driven. This counter-intuitive outcome arises because of energy lost to water by non-linear processes during measurement. An IEC method is available to resolve this problem, (IEC 61949, 2007) but it is unclear how widely it is being adopted. As a result, any further upward trends in pulse amplitude, if they are occurring, may remain hidden.

OTHER DIAGNOSTIC MODES OF OPERATION

Overall, the greatest intensities are associated with pulsed Doppler operation, although some settings for Doppler imaging can give rise to intensities commonly associated only with pulsed Doppler. These occur when Doppler imaging is used with a narrow colour box, high line density and high frame rate. All forms of Doppler imaging, whether colour-flow imaging or 'power Doppler', use the same range of acoustic output.

While Doppler modes are associated with the highest intensities, the highest pulse amplitudes (given by either rarefaction pressure p_r or MI) for Doppler applications are broadly similar to those used for imaging. On average, maximum rarefaction pressures are about 2.5 MPa. Harmonic imaging would be expected to use amplitudes confined to the upper range, since it is only at such pressures that harmonics are generated significantly in tissue.

Plane-wave and zonal imaging may be expected to reduce the highest intensities and acoustic pressures, because beam focusing on transmission is not used. Other developments may result in an increase in energy deposition, heating and radiation force. This is particularly true with the development of higher-frequency probes, and those for intraluminal use, such as trans-rectal, trans-oesophageal and trans-vaginal transducers. The output from such transducers is ultimately limited by FDA regulations in the United States; most manufacturers also work to these same limits in other parts of the world, although this is not generally a legal requirement. However, the FDA regulations do not set limits based on predicted temperature rise in tissue, apart for ophthalmic use (see previous discussion). Since interstitial transducers generally operate at higher ultrasound frequencies, they are capable of causing greater tissue heating, while still operating within the regulations, than is true for transducers operating at a lower frequency.

INCREASE IN USE OF DIAGNOSTIC ULTRASOUND

In England, there were almost 10 million ultrasound scans reported to the Department of Health in the year April 2013–March 2014, the last year for which such statistics were gathered at the time of writing this chapter. This may be compared with about 6 million scans a decade earlier. Much of this increase arose from a considerable increase in non-obstetric scans. It is clear from these and other data that exposure to ultrasound, quantified in terms of proportion of the population exposed, is growing, and may be expected to continue to grow. Within this group the majority of scan times will be short, not least because of time pressures within ultrasound clinics. Nevertheless, there are reasons why some patients may be exposed for more extended periods. This may arise because the procedure requires it (fetal breathing studies, flow studies or guided interventional procedures, for example) or because there is an increased usage of ultrasound by novice practitioners who are yet to develop mature skills and confidence in ultrasound scanning. In general it may be concluded that overall exposure of the population to ultrasound for medical diagnostic reasons is currently growing strongly, and may be expected to continue to do so for the foreseeable future.

SAFETY FOR SPECIFIC USES OF DIAGNOSTIC ULTRASOUND

Reviews of the safety of ultrasound often explore the exposure and the mechanisms for biological effects, but only occasionally review the safety aspects of the specific tissues being exposed (see, e.g. Barnett et al. 1997). Often it is assumed that embryological and fetal exposure is the only concern. Some general recommendations are based upon a rationale specific to exposure *in utero*. This is true of the World Federation recommendations on heating (Barnett 1998), which were developed from investigations only into thermal teratology. It is helpful, therefore, to consider separately the safety concerns relating to a few particular uses of diagnostic ultrasound.

Diagnostic ultrasound during the first trimester

Probably the most critical question concerns the exposure of the embryo during the early stages of pregnancy. This is a period of rapid development and complex biochemical change, which includes organ creation and cell migration. There is widespread evidence that during this period the developing embryo is particularly sensitive to external agents, whose effect on subsequent development may range from fatal developmental malformation to minor and subtle biochemical disturbance. There are gaps in our knowledge and understanding of the way in which ultrasound may interact with embryonic tissue, and whether any adverse effect may result in developmental problems because of the particular sensitivity of the tissue at this time. Moreover, this sensitivity may be cyclic, with some tissues being sensitive only during particular time-bands of rapid cell development and differentiation. Heat is a teratogen, and any temperature increase from the absorption of ultrasound can disturb subsequent development, if of sufficient magnitude and maintained for sufficiently long. Fortunately the tissue with the greatest tendency to heat, bone, only starts to condense at the end of the first trimester. In the absence of bone, present evidence suggests that temperature rises of more than 1.5°C will not occur within embryonic tissue at present diagnostic exposures. While this suggests that significant developmental changes may not occur, the kinetics of biochemical processes are known to be temperature-sensitive, and little research has investigated the influence of small temperature changes induced locally on membranes and signal-transduction pathways.

There is no evidence that cavitation occurs either, since there are no gas bubbles to act as nucleation sites within the uterus. Radiation pressure is exerted on embryonic tissue during exposure, however, and certainly amniotic fluid is caused to stream around the embryo during exposure. Acoustic streaming itself seems to present no risk. Nevertheless the force that causes streaming is also exerted on embryological tissues. A report by Ang et al. (2006), showing evidence of neuronal migration in mouse embryos from exposure to diagnostic ultrasound, attributed to radiation force, has never been independently confirmed. Thus although our present understanding suggests that current practice is safe, there is sufficient uncertainty about the detailed interaction processes to advise caution.

Diagnostic ultrasound during the second and third trimesters

Bone ossification is the main developmental change during the second and third trimesters of pregnancy of significance to ultrasound safety. As bone condenses it forms local regions of high absorption. Ultrasound energy is absorbed more by the fetal skeleton than by fetal soft tissues, preferentially causing heating. This is important in part because soft tissues alongside this bone will also be warmed by thermal conduction, reaching a higher temperature than expected from ultrasound absorption alone. Neurological tissues are known to be particularly sensitive to temperature rise, and the development of brain tissue, and of the spinal cord, could be affected if adjacent skull or vertebral bone were heated too much. Within the fetal haematopoietic system, the bone marrow is the main site of blood formation in the third trimester of pregnancy. Abnormal cell nuclei in neutrophils in guinea pigs have been reported after 6 minutes exposure at 2.5°C temperature elevation. This temperature increase in bone is within the capability of modern pulsed Doppler systems.

Cavitation is unlikely to occur during these later stages in pregnancy in diagnostic fields, as is the case during the first trimester, because of the absence of available nucleation sites. Radiation pressure effects can occur, and streaming will be caused within fluid spaces *in utero*. It seems unlikely that streaming, or other radiation pressure effects,

have safety significance at this stage, because the development of collagenous structures and an extracellular matrix gives increased strength to fetal tissues.

Obstetric scanning on patients with fever

It is noted in the WFUMB recommendations that 'care should be taken to avoid unnecessary additional embryonic and fetal risk from (heating due to) ultrasound examinations of febrile patients'. If a mother has a temperature, her unborn child is already at risk of mal-development as a result of the elevated temperature. This being so, it is sensible not to increase this risk unnecessarily. This does not mean withholding obstetric scanning from patients if they have a temperature. The methods of limiting exposure, including minimising TI, limiting the duration of the scan and avoiding casual use of Doppler techniques, should be employed with particular vigilance in these cases.

Neonatal scanning

Two particular concerns regarding neonatal scanning have been raised: These are neonatal head scanning and cardiac scanning. The reasons for caution are different for each, so they will be dealt with separately. Often, scanning is carried out when neonates are seriously ill, and the paramount needs for diagnosis should be judged against any potential for hazard and consequent risk.

NEONATAL HEAD SCANNING

The exposure of the neonatal head to diagnostic ultrasound carries with it the same safety issues as for second- and third-trimester scanning. Neuronal development is known to be particularly sensitive to temperature. Temperature increases of a few degrees can occur in bone when exposed to diagnostic ultrasound. This means that there is the potential for brain tissue close to bone to experience temperature elevations by thermal conduction. The temperature increases may occur either close to the transducer or within the skull. At the surface there are two potential sources of heat. First, the ultrasound beam is absorbed by the skull bone. The TIC provides the user with an estimate of this temperature rise. Second, transducer self-heating can further increase the surface temperature, whether or not the transducer is applied over bone or the anterior fontanelle. Transducer heating increases the surface temperature by up to a few degrees when transducers are operated at the highest powers.

NEONATAL CARDIAC SCANNING

Neonatal cardiac scanning, including Doppler blood-flow imaging, may expose surrounding pleural tissue. There is good experimental evidence for alveolar capillary damage both in adult small animals and in juvenile larger animals, although there is no evidence for lung damage in humans – even in neonates. The experimental evidence suggests that as pleural tissue matures it thickens and becomes stronger, so preventing capillary rupture. Immature (and hence structurally weak) neonatal pleura may be vulnerable to stress caused by diagnostic ultrasound pulses. The relevant safety index is the mechanical index (MI), and this should be kept to a minimum appropriate to effective diagnosis.

Ophthalmic scanning

The eye is the only tissue identified separately for regulation by the US FDA, and the only case for which TI is used for regulation. The limit for derated time-averaged intensity is 50 mW cm^{-2} (in comparison with 720 mW cm^{-2} for all other applications), and 0.23 for MI (compared with 1.9) (see Table 16.1). The TI limit is 1.0. The reason for the regulatory caution is the particular sensitivity of parts of the eye to potential damage. The cornea, lens and vitreous body of the eye are all unperfused tissues. This means that they dissipate heat only by means of thermal conduction. Furthermore, the lack of blood perfusion limits the ability to repair any damage arising from excess exposure. The acoustic attenuation coefficient of the lens is about 8 dB cm^{-1} at 10 MHz, and since a lens has a greatest thickness of about 4 mm, about one-half of the incident acoustic power may be deposited there. A further specific concern is that the lens lies very close indeed to the transducer under normal scanning conditions. Transducer self-heating becomes an important secondary source of heat under these circumstances. For all these reasons, particular care is exercised in the design of systems specifically for ophthalmic use. If general-purpose scanners are used for eye studies, great care needs to be exercised to reduce to an absolute minimum any chance of lens heating.

SUMMARY AND CONCLUSIONS

Ultrasound has an enviable record for safety. All the epidemiological evidence points to the conclusion that past and current practice has presented no detectable risk to the patient and can be considered safe. Nevertheless there is ample evidence that modern scanners, designed in accordance with national and international standards and regulations, can warm tissues by several degrees under some circumstances. If gas bubbles or other pockets of gas lie in the ultrasound field, the tissues may be damaged from stresses from cavitation-like oscillations. Radiation pressure, sufficient to cause acoustic streaming *in vivo*, is exerted on all exposed tissues. These are important facts underpinning the responsibilities of the manufacturers to produce safe equipment to use, and for the users of ultrasound in managing their scanning practice.

Ultrasound equipment now can display safety indices that give users feedback so that more informed safety judgments can be made. Greatest intensities and powers, and the greatest potential heating, are generally associated with pulsed Doppler modes. There is considerable overlap between modes, however. Particular safety concerns pertain during all obstetric scanning, during neonatal scanning, during ophthalmic scanning and when using contrast materials.

QUESTIONS

Multiple Choice Questions

Q1. Which is the correct formula for mechanical index (MI)?

a. $\dfrac{p_r}{\sqrt{f}}$

b. $\dfrac{W}{W_{\text{deg}}}$

c. $\dfrac{p_r}{f}$

d. $\dfrac{P_r{}^2}{f}$

e. $\dfrac{p}{\sqrt{f\alpha}}$

Q2. What is the limit set by the Food and Drug Administration in the United States for the maximum derated time-averaged intensity?
   a. 190 W/cm²
   b. 1.9 W/cm²
   c. 190 mW/cm²
   d. 720 W/cm²
   e. 720 mW/cm²

Q3. What is the maximum fetal temperature elevation, recommended by WFUMB, which may be sustained without reservation on thermal grounds?
   a. 4°C
   b. 0.5°C
   c. 1.5°C
   d. 0.1°C
   e. 10°C

Q4. Which of the following hazards is indicated by the mechanical index (MI)?
   a. Stable non-inertial cavitation
   b. Lung capillary bleeding
   c. Radiation force
   d. Inertial cavitation
   e. Temperature rise

Q5. Which of the following national and international bodies sets upper limits on the allowed ultrasound output of diagnostic ultrasound equipment?
   a. World Federation for Ultrasound in Medicine and Biology (WFUMB)
   b. International Electrotechnical Commission (IEC)
   c. US Food and Drug Administration (FDA)
   d. European Medical Devices Agency (MDA)
   e. UK British Standards Institute (BSI)

Q6. Select the conditions that are likely to give a high temperature rise *in vivo*:
   a. Lung scanning
   b. Trans-cranial Doppler
   c. First trimester scanning
   d. Vascular scanning
   e. Ophthalmic scanning

Q7. For which of the following conditions should safety be of particular concern?
   a. During Doppler procedures
   b. During first-trimester scanning
   c. During adult cardiac scanning
   d. During contrast studies
   e. During neonatal cranial scanning

Q8. By which of the following means can safe practice be ensured, consistent with obtaining the required clinical outcome?
   a. Minimising the thermal index
   b. Performing regular equipment maintenance
   c. Displaying the safety indices on the screen
   d. Removing the transducer from the skin when not in use
   e. Training in safety awareness

Q9. Which of the following have been shown consistently to be associated with obstetric ultrasound scanning?
   a. Reduced birth weight
   b. Childhood malignances
   c. Dyslexia
   d. A trend towards left-handedness
   e. Retarded speech development

Q10. Why is the FDA limit for ophthalmic scanning lower than for all other tissues?
   a. Heat cannot be dissipated easily by perfusion in the eye.
   b. Cavitation is easily caused in the eye by ultrasound.
   c. There is a risk of retinal detachment.
   d. Heat from the transducer easily conducts to the eye.
   e. The eye is close to brain tissue.

Short-Answer Questions

Q1. What are the meanings of the words 'hazard' and 'risk', and what is the connection between them?

Q2. What is the meaning of estimated *in situ* exposure, and how is it usually calculated? (It is also called 'derated exposure'.)

Q3. What are the three forms of the thermal index, and what conditions do they relate to?

Q4. Apart from ultrasound absorption, describe the other main source of heat during ultrasound scanning, and the conditions where it may be of particular significance.

Q5. Explain why MI should be used for control of scanner output for safety purposes during neonatal cardiac investigations.

ACKNOWLEDGEMENT

This chapter is based upon similar chapters in earlier editions of this book. The considerable contributions to these chapters on safety in earlier editions by Adam Shaw are gratefully acknowledged.

STANDARDS AND REGULATIONS

European Community 2017. Regulation (EU) 2017/745 of 5 April 2017 on medical devices. *Official Journal of the European Community*, L117/1 5/5/2017.

IEC 60601–2–37 Ed 2 2007. Amendment AMD1 (2015). Medical Electrical Equipment: Particular requirements for the safety of ultrasound diagnostic and monitoring equipment. Geneva: International Electrotechnical Commission.

IEC 62359 2010. Amendment AMD1 Ed2. (2017) Ultrasonics – Field characterisation – Test methods for the determination of thermal and mechanical indices related to medical diagnostic ultrasonic fields. Geneva: International Electrotechnical Commission.

IEC 61949 2007. Ultrasonics – Field characterisation – in-situ exposure estimation in finite-amplitude ultrasonic beams. Geneva: International Electrotechnical Commission.

REFERENCES

American Institute of Ultrasound in Medicine (AIUM). 2008. Bioeffects consensus report. *Journal of Ultrasound in Medicine*, 27, 503–632.

Ang E, Gluncic V, Duque A, Schafer M, Rakic P. 2006. Prenatal exposure to ultrasound waves impacts neuronal migration in mice. *Proceedings of the National Academy of Sciences of the USA*, 103, 12903–12910.

Apfel RE, Holland CK. 1991. Gauging the likelihood of cavitation from short-pulse, low-duty cycle diagnostic ultrasound. *Ultrasound in Medicine and Biology*, 17, 179–185.

Barnett SB, ed. 1998. WFUMB symposium on safety of ultrasound in medicine. *Ultrasound in Medicine and Biology*, 24 (Suppl 1).

Barnett SB, Kossoff G, eds. 1998. *Safety of Diagnostic Ultrasound*. Progress in Obstetric and Gynecological Sonography Series. New York: Parthenon.

Barnett SB, Rott H-D, ter Haar GR, Ziskin MC, Maeda K. 1997. The sensitivity of biological tissue to ultrasound. *Ultrasound in Medicine and Biology*, 23, 805–812.

Barnett SB, ter Haar GR, Ziskin MC et al. 2000. International recommendations and guidelines for the safe use of diagnostic ultrasound in medicine. *Ultrasound in Medicine and Biology*, 26, 355–366.

Barnett SB, Duck F, Ziskin M. 2007. WFUMB symposium on safety of ultrasound in medicine: Conclusions and recommendations on biological effects and safety of ultrasound contrast agents. *Ultrasound in Medicine and Biology*, 33, 233–234.

Calvert J, Duck F, Clift S, Azaime H. 2007. Surface heating by transvaginal transducers. *Ultrasound in Obstetrics and Gynecology*, 29, 427–432.

Duck FA. 1999. Acoustic saturation and output regulation. *Ultrasound in Medicine and Biology*, 25, 1009–1018.

Duck FA. 2008. Hazards, risks and safety of diagnostic ultrasound. *Medical Engineering and Physics*, 30, 1338–1348.

Duck FA. 2009. Acoustic dose and acoustic dose rate. *Ultrasound in Medicine and Biology*, 35, 1679–1685.

European Federation of Societies for Ultrasound in Medicine and Biology (EFSUMB) Study Group. 2008. Guidelines and good clinical practice recommendations for contrast enhanced ultrasound (CEUS) – Update 2008. *Ultrasound in Medicine and Biology*, 29, 28–44.

Leighton TG. 1994. *The Acoustic Bubble*. London: Academic Press.

Leighton TG. 1998. An introduction to acoustic cavitation. In Duck FA, Baker AC, Starritt HC (Eds.), *Ultrasound in Medicine*. Bristol: Institute of Physics Publishing, pp. 199–223.

McKinlay A. (guest editor) 2007. Effects of ultrasound and infrasound relevant to human health. *Progress in Biophysics and Molecular Biology*, 93, 1–420.

Miller DL, Quddus J. 2000. Diagnostic ultrasound activation of contrast agent gas bodies induces capillary rupture in mice. *Proceedings of the National Academy of Sciences of the USA*, 97, 10179–10184.

Miller MW, Ziskin MC. 1989. Biological consequences of hyperthermia. *Ultrasound in Medicine and Biology*, 15, 707–722.

National Council for Radiation Protection and Measurements (NCRP) 2002. *Report 140. Exposure Criteria for Medical Diagnostic Ultrasound: II. Criteria Based on All Known Mechanisms*. Bethesda, MD: NCRP.

Nelson TR, Fowlkes JB, Abramowicz JS, Church CC. 2009. Ultrasound biosafety considerations for the practicing sonographer and sonologist. *Journal of Ultrasound in Medicine*, 28, 139–150.

Salvesen KA. 2007. Epidemiological prenatal ultrasound studies. *Progress in Biophysics and Molecular Biology*, 93, 295–300.

Shaw A, Bond AD, Pay NM, Preston RC. 1999. A proposed standard thermal test object for medical ultrasound. *Ultrasound in Medicine and Biology*, 25, 121–132.

Shaw A, Pay NM, Preston RC. 1998. *Assessment of the likely thermal index values for pulsed Doppler ultrasonic equipment – Stages II and III: experimental assessment of scanner/transducer combinations. NPL Report CMAM 12.* Teddington, UK: National Physical Laboratory.

ter Haar G. ed. 2012. *The Safe Use of Ultrasound in Medical Diagnosis*. 3rd ed., London: British Institute of Radiology.

Appendix A: The decibel

The decibel (dB) is the unit which is normally used to compare ultrasound intensities at different depths in tissue, e.g. due to attenuation. It is also used as a measure of the dynamic range of echo amplitudes that are displayed in the image. In practice, the absolute amplitude of an echo signal is rarely of interest. It is more useful to know how echoes compare with one another. The ratio of two intensities or amplitudes can be expressed in decibels. As the decibel is used only for ratios, there are no other units involved (e.g. mW, volts).

The decibel is a logarithmic scale, so in simple terms, if the ratio of two values is 1,000:1 or 1,000,000:1, rather than writing out lots of zeros, the decibel scale effectively counts the zeros, rather like expressing these numbers as 10^3 or 10^6. On this logarithmic scale, a ratio of 10:1 can be expressed as 1 Bel, a ratio of 1,000:1 as 3 Bel, and a ratio of 1,000,000:1 as 6 Bel. In practice, the unit of 1 Bel (ratio of 10) is often too large and it is more useful to use the decibel, which is one-tenth of a Bel. The logarithmic nature of the decibel also means that the attenuation of the ultrasound pulse in each centimetre travelled into tissue can be added rather than multiplied.

As an ultrasound pulse propagates through tissue and is attenuated, the ratio of the intensities within the pulse at two different depths can be expressed in decibels. If the intensity associated with an ultrasound pulse at depth 1 in tissue is I_1 and that at a greater depth 2 is I_2, the ratio of the intensities I_2/I_1 in dB is given by:

$$\frac{I_2}{I_1} (dB) = 10 \log_{10} \left(\frac{I_2}{I_1} \right) \qquad (A.1)$$

As described in Chapter 2, the intensity of a pulse is proportional to the square of the acoustic pressure, i.e. $I \propto p^2$. So the ratio of two intensities:

$$\frac{I_2}{I_1} = \frac{p_2^2}{p_1^2} = \left(\frac{p_2}{p_1} \right)^2$$

That is, the intensity ratio is equal to the pressure ratio squared. The intensity ratio in decibels can be expressed in terms of the pressure ratio:

$$\frac{I_2}{I_1} (dB) = 10 \log_{10} \left(\frac{p_2}{p_1} \right)^2$$

This equation can be rewritten as:

$$\frac{I_2}{I_1} (dB) = 20 \log_{10} \left(\frac{p_2}{p_1} \right) \qquad (A.2)$$

This equation is used whenever two amplitudes (pressure, voltage) are compared, whereas Equation A1 is used to compare power or intensity levels.

Table A1 shows a range of intensity and amplitude ratios with equivalent values expressed in decibels. There are several things to note:

a. When the ratio is greater than 1, that is I_2 in Equation A1 is greater than I_1, the decibel value is positive. When the ratio is less than 1, that is I_2 is less than I_1, then the decibel value is negative. This implies that the intensity or amplitude is getting smaller.

Table A1 Intensity and amplitude ratios on the decibel scale

dB	Intensity ratio	Amplitude ratio
60	1,000,000	1,000
40	10,000	100
30	1,000	31.6
20	100	10
10	10	3.16
6	4	2
3	2	1.41
0	1	1
-3	0.5	0.71
-6	0.25	0.5
-10	0.1	0.316
-20	0.01	0.1
-60	0.000001	0.001

b. It can be seen that it is more convenient to express large ratios with lots of zeroes in decibels. An intensity ratio of 1,000,000:1 can be written as 60 dB.

c. Large ratios can be made by multiplying smaller ratios together. On the decibel scale, the values are simply added. For example, an intensity ratio of 2 is equivalent to 3 dB. A ratio of 4 (2×2) is equivalent to 6 dB ($3 + 3$). Other values can be created from the table. An intensity ratio of 400:1 ($2 \times 2 \times 100$) is 26 dB ($3 + 3 + 20$).

d. For a given value in decibels, the intensity ratio is equal to the amplitude ratio squared.

Appendix B: Acoustic output parameters and their measurement

ACOUSTIC OUTPUT PARAMETERS

Many different parameters have been defined in order to characterise medical ultrasound fields. Due to the complexity of these fields, it is not possible to describe the parameters fully and unambiguously here. Definitions of some terms are given; the more interested reader can find further formal definitions and description in IEC (2007a), IEC (2004) and IEC (2010). There are five acoustic parameters which seem to be most important to safety: peak negative pressure, pulse average intensity, temporal average intensity, total power and acoustic frequency (Figure B1).

Peak negative acoustic pressure

As described in Chapter 2, when an ultrasound wave passes a point in a medium, the particles of the medium are alternately compressed together and pulled apart, leading to oscillations in the local pressure. The local pressure variation due to the passage of a typical ultrasound pulse is illustrated in Figure B1. Before and after the passage of the pulse, the pressure in the medium is the local static pressure. The maximum value of pressure during the passage of the pulse is the peak positive acoustic pressure (p_+) or peak compression (p_c). The peak negative acoustic pressure (p_-), also called the peak rarefactional pressure and given the symbol p_r, is the most negative pressure that occurs during the pulse. This is an important parameter because it relates to the occurrence of cavitation. Negative acoustic pressure means that the acoustic wave is trying to pull the water molecules apart; the water molecules resist this separation but if the pressure is sufficiently negative and lasts for long enough, it is possible to produce a small void, a cavitation bubble. If there are pre-existing gas bubbles or dust particles in the water, cavitation occurs more easily. A similar effect can occur in tissue. Figure B2 shows the pressure waveform measured near the focus of a pulsed Doppler beam. The pulse has become distorted due to non-linear propagation and the compressional half cycles are taller and narrower than the rarefaction half cycles. The peak rarefaction pressure is about 3.5 MPa.

Acoustic frequency

As the wave passes a point, the water molecules are squeezed together during the high-pressure periods and stretched apart during the low-pressure

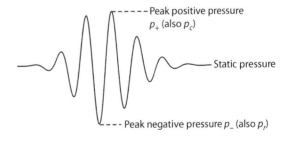

Figure B1 The peak positive and peak negative pressures are the maximum and minimum values of pressure in the medium during the passage of an ultrasound pulse.

periods. The acoustic frequency is essentially the rate at which this squeezing and stretching takes place. In Figure B2, the time between consecutive high-pressure periods is about 0.3 microseconds and so the acoustic frequency is approximately $1/(0.3 \, \mu s) = 3.3 \, \text{MHz}$. Frequency is important to safety because the absorption coefficient of most soft tissues increases with frequency, leading to energy being absorbed in a smaller volume and producing higher temperature rises. Additionally, cavitation is more likely to occur at lower frequencies because the periods of negative pressure last for longer and cavitation nuclei have more time to grow.

Pulse average intensity

As well as the acoustic pressure, we can consider the intensity of the ultrasound field at a point. Intensity is a measure of the rate at which energy is flowing through a small area around the measurement point. Intensity varies with time as the ultrasound pulse passes the point and for most practical purposes can be derived from the pressure variations using the following equation:

$$I = \frac{p^2}{\rho c}$$

Here p is the pressure, c is the speed of sound in the medium and ρ is its density. Figure B3a shows the intensity waveform corresponding to the pressure waveform in Figure B1. Note that because the values of pressure at each point in the waveform are squared, all values of intensity are positive. Each

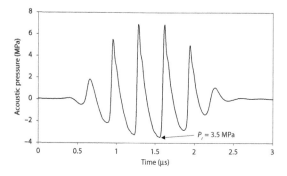

Figure B2 Variation of acoustic pressure with time in a pulsed Doppler ultrasound pulse near the focus.

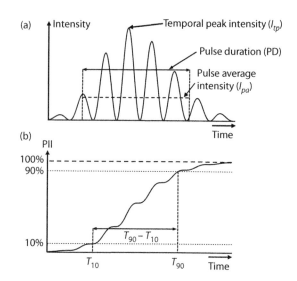

Figure B3 **(a)** The intensity is related to the pressure squared and is always positive. The temporal peak intensity is the maximum value during the pulse. The pulse average intensity is the average value over the duration of the pulse (PD). **(b)** The pulse duration (PD) is defined in terms of the pulse intensity integral (PII). The PII takes the form of a rising staircase from zero to the final value of the integral (PII) as the pulse passes. The times at which the waveform reaches 10% and 90% of its final value are labelled as T_{10} and T_{90}. The pulse duration (PD) is defined as 1.25 times the time interval $T_{90} - T_{10}$. The pulse average intensity is defined as PII/PD.

positive and negative peak in the pressure waveform results in a positive peak in the intensity waveform. The maximum value of intensity during the passage of the pulse is the temporal peak intensity (I_{tp}).

The time taken for the pulse to pass a point in the medium is the pulse duration (PD). As the start and finish of the intensity waveform can be difficult to define, the accepted method for calculating the pulse duration involves integrating the pulse waveform to give the pulse intensity integral (PII) (IEC 2004). This takes the form of a rising staircase from zero to the final value of the PII as the pulse passes (Figure B3b). The time interval between the points at which the staircase reaches 10% and 90% (T_{10} and T_{90}) of the final value is measured. The pulse duration is then given by:

$$PD = 1.25(T_{90} - T_{10})$$

The pulse average intensity I_{pa} is the average value of intensity during the pulse. This is calculated from:

$$I_{pa} = \frac{PII}{PD}$$

Spatial distribution

Figure B4 shows the spatial variation in intensity in a transverse plane through a circular beam, with the beam axis in the centre. The maximum value of intensity in the plane is on the beam axis and is referred to as the spatial peak (I_{sp}). The intensity may also be averaged over the cross-sectional area of the beam to give the spatial average intensity (I_{sa}). The spatial peak values of acoustic parameters, including PII, are often found at or near the beam focus and this is where many measurements are made. The maximum value of I_{pa} in the field is the spatial peak pulse average intensity, I_{sppa}, which is a parameter limited by the US Food and Drug Administration (FDA) regulations (FDA 2008).

Temporal average intensity

The time-averaged value of the intensity at a particular point in the field is called the 'temporal average intensity' and given the symbol I_{ta}. Imaging scanners generally produce a large number of short pulses (approximately 1 μs long) separated by relatively long gaps (perhaps 200 μs long). In a stationary beam, the intensity waveform is repeated at the point in the medium with each pulse-echo cycle. The rate at which the waveform is repeated is the pulse repetition frequency (PRF). The intensity waveform may be averaged over the duration of one complete pulse-echo cycle to give the temporal average intensity I_{ta} (Figure B5). I_{ta} is much smaller than I_{pa} because the average includes the relatively long time between pulses.

In a real-time imaging system, the imaging beam is swept through the tissue cross section, so the pressure waveform at a particular point in the tissue is repeated once in every sweep. Figure B6 shows the pressure waveform at a point in the field of a real-time scanning system. There is normally some overlap between consecutive beam positions so several pulses are seen as the beam sweeps past, with the highest pressure when the beam axis aligns with the point of interest. To obtain I_{ta} in this case, the intensity waveform must be averaged over a complete scan repetition period or over a time which includes many scan repetition periods.

I_{ta} is relevant because its spatial distribution is one of the main factors governing temperature rise in tissue. The maximum value of I_{ta} in the field is called the 'spatial peak temporal average intensity' I_{spta}. This is one of the parameters on which the acoustic output is regulated in the United States by the Food and Drug Administration (FDA 2008).

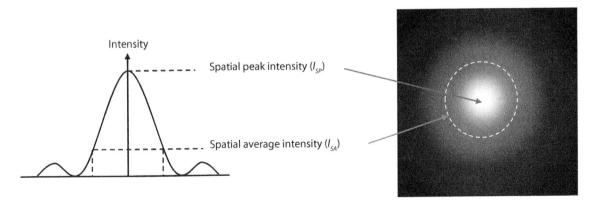

Figure B4 The values of the various intensity parameters change with position in the beam also. The highest value in the beam is the spatial peak intensity. The average value over the area of the beam is the spatial average intensity.

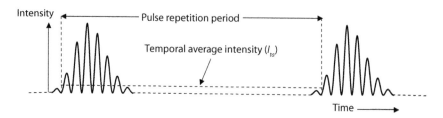

Figure B5 The intensity waveform is repeated with every pulse-echo cycle. The temporal average intensity is the average value over a complete pulse-echo cycle and is much lower that the pulse average intensity.

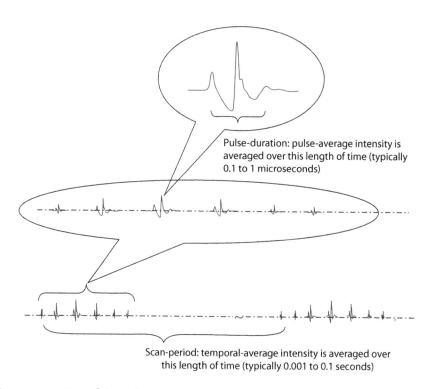

Figure B6 A representation of the pulse sequence in an imaging scan frame at one point in the field. Temporal average intensity is averaged over the repeating period of the sequence, including 'dead time' between pulses.

Total power and output beam intensity

Total power is the amount of ultrasonic energy radiated by the transducer every second. This is an important parameter for safety because it is relevant to heating and is used to calculate thermal indices. It is worth noting that, for a typical imaging transducer, the total radiated power is substantially less than the total electrical power supplied to the transducer and most of the electrical energy is converted to heat in the transducer. This 'self-heating' is an important factor in determining the hazard to tissues close to the transducer. Dividing the total power by the output area of the transducer gives the output beam intensity, I_{ob}. This is the mean temporal average intensity at the transducer face.

CALCULATION OF *IN SITU* EXPOSURE PARAMETERS

As ultrasound travels through tissue it is attenuated. That is, some of the energy is lost from the beam by absorption or scatter. This means that the *in situ* intensity (i.e. the intensity at some point in tissue) is less than the intensity measured in water at the same point in the field.

The propagation of ultrasound of the type used in medical imaging is an extremely complex phenomenon and there is no accepted way of deriving truly reliable estimates of the *in situ* exposure levels from the acoustic output measurements made in water. The most widely used method is that of 'derating', used in the US FDA regulations (see later), and the International Electrotechnical Commission (IEC)/National Electrical Manufacturers' Association (NEMA) Output Display Standard. Derating involves multiplying the value of an acoustic property measured in water by a logarithmic attenuation factor. This factor is always frequency dependent and is often distance dependent. The most widely used theoretical attenuation model comes from the IEC/NEMA Output Display Standard, and IEC 60601-2-37 (IEC 2007b), and uses a derating factor of 0.3 dB cm^{-1} MHz^{-1} to estimate exposure in tissue. The assumption is that soft tissue fills the path between the transducer and the point where the exposure level is required. The value of 0.3 dB cm^{-1} MHz^{-1} is lower than the attenuation of most soft tissues (which is typically closer to 0.44 dB cm^{-1} MHz^{-1}) to allow for the possibility of some fluid in the path (as would be the case for examination of a fetus through the bladder). The value of 0.3 dB cm^{-1} MHz^{-1} means that at 3 cm from a 3.3-MHz transducer, the derated temporal average intensity, $I_{ta.3}$, is 3 dB less than (e.g. half of) the value measured in water; the derated peak negative pressure, $p_{r.3}$, is 70% of the value in water. This derating factor is used in calculating thermal and mechanical indices and, if you see the term 'derated' being used without the derating factor being specified, it is generally this IEC/NEMA value which has been used. Other attenuation models have been proposed using different attenuation factors that may give more realistic estimates of *in situ* acoustic pressure (Nightingale et al. 2015).

Safety indices

The acoustic output parameters described are useful in characterising the acoustic field that the patient is exposed to and some are used in the regulation of acoustic outputs. On their own, however, they may not be good indicators of the risks of adverse biological effects due to ultrasound. Safety indices, as described in Chapter 16, have been developed to give a more immediate indication to the user of the potential for such effects. Some of the acoustic parameters described are used in the calculation of the mechanical index (MI) and the thermal index (TI).

The MI is defined by the equation:

$$\mathrm{MI} = \frac{p_{r0.3}}{\sqrt{f}}$$

Here, $p_{r0.3}$ is the derated peak rarefactional pressure measured at the point in the acoustic field where it is greatest. The frequency f is the centre frequency of the pulse.

Thermal index (TI) is defined by the equation:

$$\mathrm{TI} = \frac{W}{W_{\mathrm{deg}}}$$

Here, W is the current acoustic output power from the transducer and W_{deg} is the power required to raise the temperature of the tissue by 1°C. Hence TI is taken to give an estimate of the likely maximum temperature rise in degrees Celsius (°C) under the current operating conditions.

A number of thermal models are used to calculate values for W_{deg} from measured acoustic output parameters. These differ according to the particular thermal index, that is, soft tissue thermal index (TIS), bone-at-focus thermal index (TIB) and cranial (or bone-at-surface) thermal index (TIC) (see Chapter 16) and whether the beam is stationary or scanned. Standard measurement and calculation methods are described in international standards (IEC 2004, 2010). The simplest model used is that for calculation of the TIS in a scanned beam. In this mode, the current acoustic power is defined as that passing through a 1-cm wide aperture at the face of the transducer (W_{01}) and the value of W_{deg} is given simply by 210/f. Hence the value of TIS is given by:

$$\mathrm{TIS} = \frac{fW_{01}}{210}$$

Calculation methods for alternative TI models are more complex and may involve measurement of various intensity parameters in the beam (IEC 2004, 2010).

HOW DO YOU MEASURE ULTRASOUND?

Two types of devices are used to measure the acoustic output of imaging and other medical ultrasonic equipment. The first is the hydrophone; the second is the radiation-force balance (sometimes called a power balance, a power meter or a radiation-pressure balance).

Hydrophones

Hydrophones are underwater microphones. Usually they are made of a piezoelectric material that converts the rapid pressure changes in the ultrasound pulse to an electrical signal that can be measured with an oscilloscope. If the hydrophone is calibrated, the pressure waveform can be calculated from the voltage waveform measured by the oscilloscope. In order to make measurements of real ultrasound fields, the hydrophone element must be small (<1 mm) and it must be mounted in a positioning

system that allows it to be moved to different positions in the field. This allows the acoustic quantities to be measured throughout the field and the spatial peak values determined. In principle, any of the acoustic properties described can be measured with a hydrophone but this type of measurement is really only possible in a laboratory or medical physics department with substantial experience and equipment for characterising ultrasound fields. There are two types of hydrophone in general use: the membrane hydrophone and the probe hydrophone (see Figure B7). The membrane hydrophone (Figure B7a) consists usually of a circular membrane of piezoelectric polymer material stretched across a frame. It has a small region in the centre which is activated, allowing spatial variations in acoustic pressure to be measured. Probe hydrophones (Figure B7b) have the active element mounted at the tip of the probe and are available with a range of element sizes. Probe hydrophones are generally cheaper to buy and are available with smaller active elements. In general, membrane hydrophones are to be preferred because they have a smoother frequency response than probes, especially at frequencies below 4 MHz, and are more stable over time.

Radiation-force balances

Radiation-force balances (RFBs) are much easier to use than hydrophones and are used to measure

(a)

(b)

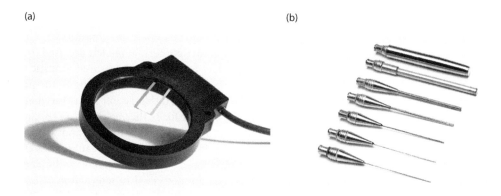

Figure B7 **(a)** A membrane hydrophone. The membrane is 80 mm in diameter but has an active region which is just 0.4 mm across in the centre. This allows spatial variations in acoustic pressure to be measured. **(b)** In probe hydrophones, the active element is mounted at the tip of the probe. Probe hydrophones are available with a range of element sizes, for example from 4 to 0.04 mm. (Both images courtesy of Precision Acoustics Ltd.)

only one property of the ultrasound field – the total ultrasonic power radiated by the transducer. RFBs work by measuring the force exerted on a target when it absorbs or reflects an ultrasound beam. The relationship between the measured force and the incident power depends on the design of the balance and so, although this relationship can be calculated approximately, the balance should be calibrated. For most designs of balance in common use, a power of 1 mW produces a force equivalent to approximately 69 μg. Measuring power from diagnostic equipment is complicated by the fact that output powers are relatively low and the fields are focused and scanned. The resulting low forces require considerable sensitivity for the balance design. The focused and scanned nature of the fields means that the relationship between the measured force and the total power can only be determined approximately (in other words, two transducers which generate the same ultrasonic power may produce different readings on the RFB).

REFERENCES

International Electrotechnical Commission (IEC). 2004. NEMA UD 3-2004 (R2009). *Real-Time Display of Thermal and Mechanical Acoustic Output Indices on Diagnostic Ultrasound Equipment*. Arlington, Virginia: National Electrical Manufacturers Association. NEMA, 2010.

International Electrotechnical Commission (IEC). 2007a. *62127–1, Ultrasonics – Hydrophones – Part 1: Measurement and Characterization of Medical Ultrasonic Fields up to 40 MHz*. Geneva, Switzerland: IEC.

International Electrotechnical Commission (IEC). 2007b. *IEC 60601-2-37. Medical Electrical Equipment, Part 2-37: Particular Requirements for the Basic Safety and Essential Performance of Ultrasonic Medical Diagnostic and Monitoring Equipment* (2nd ed.). Geneva, Switzerland: IEC.

International Electrotechnical Commission (IEC). 2010. *IEC 62359. Ultrasonics: Field Characterization – Test Methods for the Determination of Thermal and Mechanical Indices Related to Medical Diagnostic Ultrasonic Fields* (2nd ed.). Geneva, Switzerland: IEC.

Nightingale KR, Church CC, Harris G et al. 2015. Conditionally increased acoustic pressures in non-fetal diagnostic ultrasound examinations without contrast agents: A preliminary assessment. *Journal of Ultrasound in Medicine* 34, doi: 10.7863/ultra.34.7.15.13.0001

US Food and Drug Administration (FDA). 2008. *Guidance for Industry and FDA Staff Information for Manufacturers Seeking Marketing Clearance of Diagnostic Ultrasound Systems and Transducers*. Rockville, MD: FDA.

Glossary

1D 1-dimension; referring to a line in space.

2D 2-dimensions; referring to an area in space.

3D 3-dimensions; referring to a volume in space.

4D 4-dimensions; referring to the movement of a volume through time.

A to D conversion *See* 'analogue-to-digital conversion'.

A to D converter *See* 'analogue-to-digital converter'.

A-line Single ultrasound scan line, usually in a 2D ultrasound image consisting of many A-lines.

A-mode Display of echo amplitude with depth for a single ultrasound line.

Absorption Mechanism by which energy is transferred from the ultrasound wave to the tissue resulting in reduction in wave amplitude and heating of the tissues. *See* 'absorption coefficient', 'attenuation'.

Absorption coefficient The property of a medium which describes the rate at which an ultrasound energy is converted to heat. Usually the absorption coefficient is expressed in dB cm^{-1} at a particular frequency, or in dB cm^{-1} MHz^{-1}. *See* 'attenuation coefficient'.

Acoustic Relating to sound or ultrasound. *See* 'sound', 'ultrasound'.

Acoustic cavitation The mechanical response of gas cavities to a sound field. Acoustic cavitation is divided into two classes, inertial cavitation (when the cavity eventually collapses violently) and non-inertial cavitation (when the cavity grows and contracts cyclically). *See* 'inertial cavitation', 'non-inertial cavitation'.

Acoustic impedance Fundamental property of a medium in the context of sound wave propagation.

Acoustic intensity *See* 'intensity', 'I_{sppa}', 'I_{spta}'.

Acoustic lens Curved region of the face of a transducer used to focus the beam in the tissue; usually used in single-element transducers, or in array transducers to provide focusing in the elevation plane.

Acoustic pressure The amount by which the pressure exceeds the ambient pressure at any instant. In the compression part of the wave, the acoustic pressure is positive, and in the rarefactional phase, the acoustic pressure is negative. The unit of measurement is megapascal, MPa. *See* 'pressure'.

Acoustic radiation force imaging (ARFI) Ultrasound elastography method in which tissue displacement is produced at the focal spot by the radiation force generated from a high-output beam (the 'pushing beam'), with measurement of tissue displacement performed by several imaging lines using cross-correlation techniques. *See* 'elastography'.

Acoustic shock An abrupt alteration in acoustic pressure in a propagating acoustic wave. Typically a high positive peak pressure follows immediately behind a decompression.

Activation The process of reconstituting an ultrasound contrast agent prior to clinical use. *See* 'contrast agent'.

Active group Group of elements of the transducer which are used to produce the current ultrasound beam. *See* 'aperture'.

Adaptive image processing Image processing where the degree of processing, e.g. edge enhancement or noise reduction, is adjusted at each location in the image dependent on the local image content. *See* 'edge enhancement', 'image processing', 'noise reduction'.

ADC *See* 'analogue-to-digital converter'.

AIUM (American Institute of Ultrasound in Medicine) American ultrasound professional body that, among other things, produces standards documents and codes of practice for ultrasound.

Aliasing Incorrect estimation of Doppler frequency shift occurring in pulsed Doppler systems when the pulse repetition frequency is too low or the target velocity is too high. *See* 'pulse repetition frequency', 'pulsed-wave Doppler'.

Amplifier Device used to increase the size of the received ultrasound signal; the amplification for ultrasound is depth-dependent in order to boost the smaller echoes received from greater depths or area of low scatter/backscatter. *See* 'time-gain control'.

Amplitude Maximum excursion of a wave in either the positive or negative direction.

Amplitude demodulation Method used in B-mode imaging to extract the envelope of the received ultrasound signal, involving rectification and smoothing of the signal. *See* 'demodulation'.

Amplitude modulation An imaging technique for detecting ultrasound scattering from non-linear scatterers, e.g. contrast agents. Generally, three consecutive pulses are emitted in the incident pulse sequence – a half-amplitude pulse, followed by a full-amplitude pulse followed by a second half-amplitude pulse. When received, the scattered responses of the half-amplitude pulses are subtracted from the scattered response of the full-amplitude pulse – responses from linear scatterers will cancel, those from non-linear scatterers will not. *See* 'contrast-specific imaging'.

Analogue Relating to devices and electrical signals in which the voltage may vary continuously. *See* 'digital'.

Analogue-to-digital conversion Conversion of an analogue signal into a digital signal. *See* 'analogue', 'analogue-to-digital converter', 'digital'.

Analogue-to-digital converter Device for converting analogue (continuously varying voltage) to digital (voltage levels with only two allowed values corresponding to '0' and '1') signals. *See* 'analogue', 'digital'.

Angle General: a measure of the orientation or amount of rotation between two lines. In Doppler ultrasound: abbreviation for the angle between the Doppler ultrasound beam and the direction of motion of the moving blood or tissue. *See* 'angle cursor'.

Angle cursor Relating to spectral Doppler ultrasound: marker, such as a line, on the ultrasound B-mode image, present when pulsed-wave spectral Doppler is activated, which can be rotated in order to align it with the vessel wall. This is performed for the purpose of providing the ultrasound machine with information on the beam-vessel angle, to enable the machine to convert Doppler frequency shift to velocity using the Doppler equation. *See* 'beam-vessel angle', 'Doppler equation'.

Angle dependence Relating to Doppler ultrasound or colour flow; variation of the Doppler ultrasound frequency shift with the angle between the beam and direction of motion of the target, leading to a reduction in Doppler frequency shift as the angle approaches 90°. *See* 'cosine function', 'Doppler equation'.

Annular array Transducer consisting of several ring elements arranged concentrically.

Annular array scanner Mechanical scanner consisting of an annular array transducer. *See* 'annular array', 'mechanical scanner'.

Aperture Portion of the piezoelectric plate used to generate the current ultrasound beam; note the position of the aperture on the plate will change as different beams are formed; e.g. in a linear array the aperture is made up of a group of elements, with the elements forming the aperture changing in order to sweep the beam through the tissues. *See* 'piezoelectric plate'.

Apodization Technique used to improve beam formation and spatial resolution; by excitation of elements non-uniformly in transmission, and non-uniform amplification of received echo-signals in reception. *See* 'beam-former'.

Area Quantity representing the 2D size of an object or image; often measured using the calliper system for tissues imaged in cross section using an ultrasound system. *See* 'calliper system'.

ARFI *See* 'acoustic radiation force imaging'.

1D array Array consisting of a single line of elements (e.g. 128), allowing electronic focusing in the scan plane, with focusing in the elevation plane achieved through the use of an acoustic lens. *See* 'array'.

1.5D array Array consisting of several lines of elements arranged in a rectangular grid (e.g. 128 by 5), allowing electronic focusing in the scan plane, with some limited electronic focusing in the elevation plane in order to improve elevation resolution. *See* 'array'.

2D array Array consisting of a square of elements (e.g. 64 by 64), allowing electronic focusing and beam-steering in 3D. *See* 'array'.

Array Piezoelectric plate consisting of many elements; typically 128 for a 1D array, 192–640 for a 1.5D array and several thousand for a 2D array. *See* '1D array', '1.5D array', '2D array'.

Artefact Error in the displayed image or measurement, the source of the error arising from the patient, the ultrasound machine or the operator. In general artefacts arise when the assumptions used by the machine to create images or calculate quantities are not valid; e.g. 'mirror image artefact'. *See* 'error', 'measurements'.

Attenuation The loss of energy from an ultrasound wave as it propagates: lost energy may be absorbed and generate heat or may be scattered and generate new ultrasound waves. Attenuation is commonly given in a logarithmic scale in decibels (dB). It is the opposite of amplification. *See* 'absorption', 'attenuation coefficient'.

Attenuation artefact Increased or decreased brightness on the ultrasound image occurring when the attenuation in one part of the image does not match the assumed value set by the TGC control. *See* 'enhancement', 'TGC'.

Attenuation coefficient Quantity describing the attenuation of plane waves through tissue; unit is dB cm^{-1} in most soft tissues, attenuation coefficient increases linearly with frequency.

2D autocorrelation Relating to colour flow; improved autocorrelation method for estimation of mean Doppler frequency in which some account is taken of the variation in mean RF frequency of received echoes with position in the field of view. *See* 'autocorrelation'.

Autocorrelation Relating to colour flow; a method for estimation of mean Doppler frequency based on a small number (typically 3–20) of received echoes. *See* '2D autocorrelation', 'autocorrelator', 'colour flow'.

2D autocorrelator Relating to colour flow; improved autocorrelator for use in colour flow imaging, widely used in commercial scanners. *See* 'autocorrelator'.

Autocorrelator A device for use in colour flow systems which estimates the quantities displayed in the colour flow image; mean Doppler frequency, power and variance. This device and modifications of the device are used as the basis for virtually all commercial colour flow systems. *See* '2D autocorrelator', 'colour flow', 'mean frequency (2)', 'power Doppler', 'variance'.

AVI (audio video interleave) File format used for digital video files.

Axial (1) Relating to the ultrasound beam; the direction along the beam axis.

Axial (2) Relating to blood flow; the direction along the vessel axis.

Axial resolution *See* 'spatial resolution'.

Azimuthal resolution *See* 'spatial resolution'.

B-flow B-mode-based method for visualization of blood flow, where the amplitude of blood flow signals is boosted by the use of coded pulse techniques. *See* 'coded pulse techniques', 'colour B-flow'.

B-mode Ultrasound imaging mode in which the received echo amplitude is displayed in 2D, usually in greyscale; short for 'brightness mode'.

Backing layer A component of most ultrasound transducers, designed to damp motion of the element in order to prevent ringing, and to produce a short ultrasound pulse. *See* 'element', 'ringing', 'transducer'.

Backscatter The portion of a scattered or reflected ultrasound beam which travels in the direction back to the transducer; i.e. where the angle between incident waves and scattered/reflected waves is 180°.

Bandwidth The range of frequencies present within a signal, or that a transducer can respond to e.g. the bandwidth of a 6-MHz transducer is 4–8 MHz.

Baseline (1) Relating to Doppler spectral display or colour Doppler display where there are both positive and negative Doppler frequency shifts displayed; the line or marker indicating zero Doppler frequency shift. *See* 'colour Doppler', 'spectral Doppler'.

Baseline (2) Machine control enabling the operator to reallocate the displayed Doppler frequency shift range so that more or less is given to positive or negative Doppler shifts. *See* 'colour Doppler', 'spectral Doppler'.

Beam The region of space within the tissue which contributes to the formation of an ultrasound line; the beam is a composite of the 'transmit beam' and the 'receive beam', and may have a different shape for different modalities (B-mode, colour flow, etc.).

Beam axis Line from the transducer face passing through the middle of the beam at each depth.

Beam-former Part of the ultrasound system that deals with formation of the ultrasound beam. *See* 'receive beam-former', 'transmit beam-former'.

Beam forming Formation of the ultrasound beam using the beam-former. *See* 'beam-former'.

Beam steering Capability of an ultrasound system where the direction of the beam is able to be altered; required in e.g. phased-array and mechanical transducers for collection of sector data, and linear arrays for collection of Doppler data from steered angles and for compound scanning. *See* 'compound scanning', 'Doppler ultrasound', 'mechanical scanner', 'phased array'.

Beam-steering angle Machine control enabling the operator to adjust the direction of the beam, in, e.g. Doppler ultrasound.

Beam-steering array Transducer consisting of a phased array, in which the beam is steered through space by adjustment of delays to the elements. *See* 'phased array'.

Beam-stepping array Transducer consisting of a linear or curvilinear array, in which the beam is stepped along the face of the array by activation of different groups of elements. *See* 'curvilinear array', 'linear array'.

Beam-vessel angle Relating to spectral Doppler ultrasound; angle between the Doppler ultrasound beam and the vessel wall; measured using the angle cursor. *See* 'angle cursor'.

Beam width Dimensions of the beam in the lateral and elevation directions. *See* 'elevation plane', 'spatial resolution'.

Bifurcation A section of the arterial tree in which an artery has split into two sub-branches; e.g. common carotid divides into the internal carotid and external carotid.

Binary Generally: phenomena with two outcomes, '0' or '1', or 'true' or 'false'. In signal processing: relating to digital signals. *See* 'digital'.

Blood Fluid suspension of mainly red cells contained in the vascular system.

Blood flow Flow of blood, through the heart, arteries, veins or micro-circulation.

Blood mimic Suspension of particles in a fluid used in a flow phantom; the blood mimic may be designed to match the acoustic and viscous properties of human blood so that the spectral Doppler and colour flow signals produced are more realistic. *See* 'flow phantom'.

Blood-mimicking fluid (BMF) *See* 'blood mimic'.

Blood-tissue discriminator Component of the colour flow processor; which decides whether a pixel of the colour flow image is coded for tissue using the B-mode data or for blood flow using the colour flow data; in other words the unit attempts to put colour only in regions of true blood flow. *See* 'colour gain', 'colour-write priority'.

BMUS (British Medical Ultrasound Society) UK professional body which, among other things, issues guidance on the safe use of ultrasound. *See* 'BMUS guidelines'.

BMUS guidelines Set of guidelines issued by BMUS on the safe use of ultrasound.

Bolus Injection of a contrast agent in a short time. *See* 'infusion'.

Bone-at-focus thermal index (TIB) The thermal index calculated using a model which includes bone in the focal zone of the transducer. *See* 'thermal index'.

Bubble *See* 'microbubble'.

Bubble destruction The disruption of the microbubble shell when subject to an ultrasound beam.

Bubble population The group of microbubbles within a contrast agent solution.

Bulk modulus A measure of the ability of a material to withstand a change in pressure; equal to the change in pressure divided by the fractional change in volume; unit is pascals (Pa); e.g. the bulk modulus of most soft tissues is in the range 2–5 GPa.

C-scan Image plane parallel to the face of the transducer; may be obtained using 3D ultrasound. *See* '3D ultrasound'.

Calibration The act of checking or adjusting the accuracy of a machine which makes a measurement, usually performed with the use of phantoms. *See* 'phantom', 'quality control'.

Calliper system System of on-screen cursors for the measurement of distance, area and velocity; controlled by the operator allowing positioning of cursors or lines at specific points on the ultrasound image, with on-screen readout of the measured value.

Cathode ray tube display Image display monitor, not used in modern ultrasound systems. *See* 'flat-screen display'.

Cavitation A general term for the behaviour of gas cavities within a liquid when subjected to mechanical stress. The stress may be due to an acoustic wave (e.g. ultrasound), the rapid movement of a surface (e.g. the propeller of a boat) or a decompression.

Ceramic Inorganic solid that may be manufactured with piezoelectric properties, for use in the manufacture of the piezoelectric plate of an ultrasound transducer. *See* 'piezoelectric plate'.

CFI Colour flow imaging. *See* 'colour flow'.

Charts Tables or graphs showing the variation of a quantity with time; e.g. abdominal circumference with gestational age.

Cine loop Series of ultrasound frames stored in memory, able to be replayed immediately after acquisition, and in most cases able to be stored in longer-term memory for archival purposes.

Circumference Distance around the edge of a 2D object or image; often measured using the calliper system for an object imaged in cross section using an ultrasound system.

Classic beam forming Beam forming based on building up an image line by line, with each line produced by transmission along a single beam followed by reception along the same beam.

Clutter Relating to Doppler ultrasound detection of blood flow; ultrasound signals from tissue, with an amplitude typically 30–40 dB greater than that from blood. *See* 'clutter filter', 'wall filter'.

Clutter breakthrough Tissue motion giving rise to Doppler frequencies above the wall filter or clutter filter, resulting in their display as an artefact on spectral Doppler or colour flow, usually with complete loss of the true blood flow Doppler signal. *See* 'clutter', 'clutter filter', 'wall filter'.

Clutter filter Relating to Doppler ultrasound detection of blood flow; signal-processing step which attempts to remove the clutter signal, leaving behind the Doppler ultrasound signal from the moving blood. This term is usually reserved for colour flow. *See* 'colour flow', 'wall filter'.

Coded excitation Method for improving the penetration depth, but without loss of spatial resolution, by the use of ultrasound pulses which are longer, and have embedded within them a short code of zeroes and ones. The received echo(es) are compared against a decoding filter. This may involve a single ultrasound pulse or pairs of pulses. *See* 'Golay pair', 'matched filter'.

Coded pulse techniques *See* 'coded excitation'.

Colour B-flow Adaptation of the B-flow technique in which two pairs of coded pulses are used to obtain a rough indication of velocity of blood which is then used to colour code the B-flow image. *See* 'B-flow'.

Colour blooming artefact In this artefact, the colour Doppler signal (colour) extends beyond the boundaries of the imaged vessel. This artefact is commonly seen in vessels as contrast enters the vessels. It can generally be corrected by reducing the colour gain. *See* 'colour flow', 'contrast-specific imaging'.

Colour box 2D region superimposed on the B-mode image within which a colour-coded flow image is visualized. *See* 'colour flow'.

Colour Doppler 2D imaging of blood flow where the mean Doppler frequency shift is displayed (i.e. imaging of blood velocity). *See* 'power Doppler'.

Colour flow General term describing a range of techniques which are used for 2D imaging of blood flow using methods based on the Doppler equation. *See* 'colour Doppler', 'directional power Doppler', 'power Doppler'.

Colour flow signal processor Part of an ultrasound system which processes the ultrasound signals when the machine is used in colour flow mode. *See* 'colour flow'.

Colour gain Relating to colour flow; machine control enabling the operator to adjust the Doppler amplitude threshold above which colour is displayed. *See* 'blood-tissue discriminator'.

Colour speckle Variations in the received Doppler signal, arising as a result of changes in the number and relative orientation of red cells within the sample volume, which give rise to noise on the colour flow image. *See* 'speckle'.

Colour-write priority Relating to colour flow; machine control enabling the operator to adjust the B-mode echo amplitude threshold, above which colour is not displayed and below which colour data are displayed. *See* 'blood-tissue discriminator'.

Component of velocity *See* 'velocity component'.

Compound error Relating to the calculation of a quantity from two or more other quantities; where the final error is a combination of errors from the other quantities. *See* 'error (2)'.

Compound scanning Scanning in which images are acquired with ultrasound beams from several (e.g. five to nine) different directions, reducing speckle and improving image appearance through reduction in the directional nature of the image.

Compression (1) Regions in which the local volume of tissues decreases due to locally increased pressure; in the context of a wave, the high-pressure part of the wave.

Compression (2) Reduction in the dynamic range of received signals through the use of a non-linear amplifier (in which smaller signals are boosted in size). *See* 'dynamic range'.

Compressional wave A name for a sound or ultrasound wave characterized by local changes in density arising through local changes in pressure. *See* 'sound wave', 'ultrasound'.

Constructive interference The combination of two waves which are in phase at a specific time and location, with the crest of one wave coinciding with the crest of the other wave, resulting in doubling of the net amplitude. *See* 'destructive interference', 'in phase'.

Continuous-wave (CW) Doppler Referring to a Doppler ultrasound system in which ultrasound is transmitted continuously; therefore requiring separate element(s) to receive the return echoes.

Continuous-wave (CW) Doppler signal processor Part of the machine which processes the ultrasound signals when the machine is used in CW Doppler ultrasound mode.

Continuous-wave duplex system Component of an ultrasound system, consisting of a combination of a B-mode imaging system and a continuous-wave spectral Doppler system; mainly used in cardiology for the measurement of high-velocity intra-cardiac blood-flow jets. *See* 'duplex system'.

Continuous-wave (CW) ultrasound An ultrasound system in which ultrasound is generated (and received) continuously by the transducer. *See* 'continuous-wave (CW) Doppler'.

Contrast (1) Fractional difference in image values for two image regions, e.g. organ tissue and tumour; in general detection of lesions is easier when contrast is higher, and very difficult when there is no contrast.

Contrast (2) Relating to contrast agents. *See* 'contrast agent'.

Contrast agent A substance which is administered, typically intravenously, to enhance differentiation between tissues and to visualize blood flow for diagnostic purposes. *See* 'contrast-specific imaging', 'microbubble'.

Contrast-specific imaging Imaging techniques developed to improve the detection of ultrasound contrast agents for the purposes of improved clinical diagnosis. *See* 'amplitude modulation', 'flash imaging', 'harmonic imaging', 'intermittent imaging', 'non-linear imaging', 'pulse-inversion imaging', 'second-harmonic imaging'.

Contrast-to-tissue signal ratio Ratio of the backscattered signal from a contrast agent to the backscattered signal from tissue. *See* 'backscatter'.

Conventional beam forming *See* 'classic beam forming'.

Cosine function Mathematical function whose value is dependent on the angle, varying from 1 at an angle of 0°, to 0 at angle of 90°, to −1 at an angle of 180°. *See* 'angle dependence', 'Doppler equation', 'sine function'.

Cosine wave Same as sine wave, but shifted by a quarter of a wavelength (or by a phase of 90°). *See* 'sine wave'.

Cranial (or bone-at-surface) thermal index (TIC) The thermal index calculated using a model in which bone lies at the surface. *See* 'thermal index'.

Cross-correlation Method for estimation of displacement from ultrasound A-lines, used in strain elastography. *See* 'acoustic radiation force imaging', 'strain elastography'.

Curvilinear array Array transducer in which the array is curved in order to generate a diverging field of view. *See* 'array'.

Curvilinear format Arrangement of scan lines from a curvilinear array, in which the scan lines diverge with distance from the transducer producing a wider field of view with depth. *See* 'curvilinear array'.

Cut-off filter *See* 'wall filter'.

Cut-off frequency Value of the wall-filter in hertz. *See* 'wall filter'.

CW *See* 'continuous-wave ultrasound'.

Decibel The scale used to measure, for example, the change in ultrasound wave intensity; calculated as the logarithmic ratio of one intensity to another; e.g. the acoustic intensity decreases by 1 dB if an ultrasound beam passes through 2 cm of tissue that has an attenuation coefficient of 0.5 dB cm^{-1}.

Delay and sum Commonly used method for focusing in reception involving delaying received signals so that all signals post-delay are in phase.

Delay line Analogue method for delaying an electrical signal; used in analogue beam-formers, but mostly not used in modern ultrasound systems with digital beam forming. *See* 'digital beam-former'.

Demodulation Part of the signal processing which is performed on the received ultrasound signal, where the underlying RF signal is removed (i.e. extraction of the modulating signal). *See* 'amplitude demodulation', 'Doppler demodulation'.

Demodulator Part of the ultrasound system which performs demodulation. *See* 'demodulation'.

Density Mass of a material or tissue per unit volume, unit is kg m^{-3}; e.g. the density of most soft tissues is in the range 1040–1060 kg m^{-3}.

Depth of field Maximum depth within tissue for which ultrasound is displayed on the screen of the ultrasound system. *See* 'field of view'.

Derating A standard means for compensating for attenuation, used to estimate *in situ* exposure. A common derating model for soft tissues is a homogeneous medium having an attenuation coefficient of 0.3 dB cm^{-1} MHz^{-1}.

Destructive interference The combination of two waves which are out of phase at a specific time and location, with the crest of one wave coinciding with the trough of the other wave, resulting in zero net amplitude. *See* 'constructive interference', 'out of phase'.

Diameter Distance from one side to the other of a (roughly) circular structure, such as an artery in cross section.

DICOM (Digital Imaging and Communications in Medicine) Standard for storing and transmitting images, including specification of a file format. *See* 'picture archiving and communication system'.

Diffraction Change in the direction of a wave as it passes through an opening or around a barrier.

Diffuse reflection Similar to 'reflection', but occurs when the dimensions of the interface or irregularities on its surface are comparable to the wavelength, resulting in strong reflections over a narrow range of angles. *See* 'reflection'.

Digital Description of electrical devices or electrical signals, in which the signals have two discrete voltage levels corresponding to '0' and '1'. *See* 'analogue'.

Digital beam-former Beam-former which processes ultrasound signals which have been digitized. *See* 'beam-former'.

Digital zoom Relating to ultrasound images which are stored in digital form; where the image is magnified on the screen by using a more limited part of the stored data. *See* 'write zoom'.

Digitization Conversion of analogue (continuously varying voltage) signals to digital (two discrete voltage levels corresponding to '0' and '1') signals, using an analogue-to-digital converter. *See* 'analogue', 'analogue-to-digital converter', 'digital'.

Directional power Doppler 2D imaging of blood flow where the power of the Doppler signal is displayed, colour coded with the direction of blood flow. *See* 'colour Doppler', 'power Doppler'.

Display Monitor on which the ultrasound images are viewed by the operator. *See* 'liquid crystal display'.

Disturbed flow Relating to blood flow; where there are regions within a vessel or the heart in which there are periodic changes in the direction and magnitude of blood velocity, often appearing as local circulating flow called 'vortices', often seen immediately distal to a narrowing such as a stenosis. The vortices may be shed downstream where they die out after a few diameters. In clinical ultrasound there is generally no distinction made between disturbed flow and turbulence. It may lead to changes in the Doppler spectrum including spectral broadening and high-frequency spikes. *See* 'spectral broadening', 'turbulence', 'vortices'.

Doppler angle cursor *See* 'angle cursor'.

Doppler aperture Portion of the piezoelectric plate used to generate and receive the Doppler beam. *See* 'aperture'.

Doppler beam The region of space within the tissue from which the Doppler signal can arise. *See* 'sample volume'.

Doppler demodulation Method used in pulsed Doppler systems to remove the high-frequency RF component of the received signal, leaving behind the Doppler shift frequencies. *See* 'demodulation'.

Doppler equation Equation describing the relationship between the Doppler shift frequency, the target velocity and the direction of motion of the blood or tissue. *See* 'Doppler ultrasound (1) and (2)'.

Doppler gain Machine control enabling the operator to alter the amplification of the Doppler ultrasound signals, resulting in increase or decrease of brightness of the spectral Doppler display. *See* 'spectral display'.

Doppler shift Difference in ultrasound frequency between that of the transmitted ultrasound and the received ultrasound; e.g. the typical maximum Doppler shift from a healthy femoral artery is 2–4 kHz. *See* 'Doppler equation'.

Doppler signal processor Part of the ultrasound machine whose input is the RF signal from the Doppler beam, and whose output is the estimated Doppler signal; note that there are separate processors for spectral Doppler and for colour flow. *See* 'colour flow', 'spectral Doppler'.

Doppler speckle Variations in the received Doppler signal amplitude, arising as a result of changes in the number and relative orientation of red cells within the sample volume, which give rise to noise on the Doppler spectral data. *See* 'speckle'.

Doppler tissue imaging (DTI) Doppler-based ultrasound imaging method which provides 2D display of the velocities and other derived indices (e.g. strain) within moving tissue; mostly used in cardiac imaging.

Doppler tissue signal processor Part of the machine which processes the ultrasound signals when the machine is used in Doppler tissue imaging mode. *See* 'Doppler tissue imaging'.

Doppler ultrasound (1) Ultrasound which has been frequency-shifted as a result of scattering from a moving target. *See* 'Doppler equation'.

Doppler ultrasound (2) General term used to describe ultrasound systems whose design is based on use of the Doppler equation. *See* 'Doppler equation'.

Drop-out Relating to colour flow images; loss of colour due to the variable nature of the calculated mean frequency or power. Occurs when an incorrectly low value of mean Doppler frequency or power is estimated, which is below the threshold value used in the blood-tissue discriminator; is most marked at low velocities and in small vessels.

DTI *See* 'Doppler tissue imaging'.

Duplex system Ultrasound system, or component of an ultrasound system, consisting of a combination of a B-mode imaging system and a pulsed-wave spectral Doppler system. *See* 'continuous-wave duplex system'.

Dynamic elastography *See* 'shear-wave elastography'.

Dynamic focusing Progressive shift in the focal depth during receive mode, in order to produce best spatial resolution as a function of depth.

Dynamic range Ratio of largest to smallest signal that the ultrasound system is capable of processing; this will be less than the range of received echoes so these must be compressed. *See* 'compression (2)'.

Echo (1) Generally, following an initial sound, a delayed sound arriving back at the source, after reflection from an object.

Echo (2) In pulsed ultrasound; the ultrasound signal detected by the transducer consisting of reflections and scattering from tissues along the line of the beam.

Echo-ranging Method used to measure the depth of water beneath a boat, by timing a short burst of ultrasound from transmission to reception. *See* 'pulse-echo'.

ECMUS European Committee of Medical Ultrasound Safety.

Edge enhancement Image-processing method that accentuates the appearance of edges, usually at the expense of increase in the noise level. *See* 'image processing'.

Edge shadowing artefact Shadows commonly seen below the edges of cystic structures, occurring as a result of refraction.

EFSUMB European Federation of Societies for Ultrasound in Medicine and Biology.

Elastic Behaviour of a material which stretches or deforms in response to an applied force, returning immediately to its resting state when the force is removed. *See* 'elastic modulus', 'viscoelastic'.

Elastic modulus Property of an elastic tissue which describes its response to a deformation; a quantity describing the change in dimensions (strain) that occurs due to an applied force (stress); units are pascals (Pa). *See* 'bulk modulus', 'elastic', 'shear modulus', 'Young's modulus'.

Elastography Ultrasound techniques that provide information related to the stiffness of tissues. *See* 'shear-wave elastography', 'strain elastography'.

Electromagnetic tracking system *See* 'magnetic tracking system'.

Electronic beam forming Formation of the beam by a transducer consisting of elements in which there is control of the timing and amplitude of the excitation voltage applied to each element during transmission, and of the delays and amplification of the voltages applied to each element during reception. This is used as the basis for beam forming in modern ultrasound systems. *See* 'beam forming', 'digital beam-former'.

Electronic focusing Focusing of the beam using electronic beam forming. *See* 'electronic beam forming'.

Electronic noise Random noise produced from the ultrasound system, which is displayed when the gain is set too high, or when there is no ultrasound signal in deep tissues to which the ultrasound beam cannot penetrate. In B-mode and spectral Doppler the noise appears as a random greyscale pattern which changes rapidly with time. In colour Doppler the noise results in random estimates of mean frequency shift which appear as a mosaic of colours which change rapidly with time.

Element A part of the transducer which converts electrical energy into sound waves and vice versa; there may be one element (in a single-element transducer) or, more commonly in modern transducers, there are many elements (in phased and linear arrays). *See* 'piezoelectric plate', 'transducer'.

Elevation Relating to an ultrasound beam; the direction which is perpendicular to the 2D scan plane. *See* 'axial', 'lateral'.

Elevation plane The plane perpendicular to the scan plane for a 2D ultrasound system. *See* 'scan plane'.

Elevation resolution *See* 'spatial resolution'.

3D endoprobe Endoprobe which has the capability of collecting 3D data, e.g. by mechanical retraction of the transducer within the endoprobe housing.

Endoprobe Transducer designed for insertion into a body cavity or surgical wound.

Endoprobe 3D ultrasound system 3D ultrasound imaging system based on the use of a 3D endoprobe. *See* '3D endoprobe'.

Enhancement Increased brightness seen in B-mode imaging below a cystic structure. The TGC control default position assumes uniform attenuation with depth. The attenuation within the cyst is much lower than the machine expects, resulting in a brighter area displayed below the cyst.

Epidemiology The study of the distribution and occurrence of disease, typically based on statistical analyses of large population samples.

Error (1) Generally in life and in ultrasound: a mistake, a condition of non-functioning, a situation where the expected outcome is not obtained, an artefact. *See* 'artefact'.

Error (2) Relating to the measurement of quantities such as distance, area, volume, velocity; the difference between the true value of the quantity and that measured. *See* 'measurements', 'random error', 'systematic error'.

Exposure levels A description of the ultrasound field under standard conditions, typically in water or estimated *in situ*, such as rarefaction pressure, acoustic intensity and acoustic power.

Extracorporeal lithotripsy A therapeutic ultrasound method which uses high-amplitude acoustic shocks to cause destruction of renal calculi.

Far-field Term applied to that part of the acoustic field from an unfocused transducer beyond the region where destructive interference may occur. *See* 'near-field'.

Fast Fourier transform (FFT) Computationally efficient version of the Fourier transform, especially suited for implementation in hardware or software. *See* 'Fourier transform'.

FDA Food and Drug Administration; the agency in the United States with responsibility for licensing the manufacture and sale of ultrasound scanners.

Field of view Depth and width of the displayed ultrasound image. *See* 'depth of field'.

Filament Component of a string phantom, where the movement of the filament gives rise to the Doppler signals, simulating those produced from blood. *See* 'O-ring rubber', 'string phantom'.

Filter *See* 'wall filter'.

Filter value *See* 'wall filter value'.

Flash artefact Relating to colour flow systems; generation of colour within the tissue as a result of movement of the transducer or patient.

Flash filter Component of the colour flow signal processor which suppresses flash artefacts. *See* 'flash artefact'.

Flash imaging Contrast-imaging technique in which a low mechanical index (MI) pulse is used to image the contrast agent wash-in to a region of interest. By emitting a high-MI pulse, the microbubbles in the region collapse, releasing free gas which gives an enhanced backscattered signal. When imaging the enhanced signal from collapsing microbubbles is seen as a bright flash on the monitor, hence the term 'flash' imaging.

Flash pulse A high-MI pulse (or several pulses) which destroys the contrast microbubbles in the scan plane. *See* 'contrast-specific imaging'.

Flat-screen display Image display monitor, which is used as standard on modern ultrasound systems, which is very thin compared to older cathode ray tubes. *See* 'liquid crystal display'.

Flow Movement of a fluid such as blood.

Flow phantom Device used to test spectral Doppler and colour flow systems, in which the flow of blood in an artery is mimicked by the flow of a fluid in a vessel commonly embedded in tissue mimic. *See* 'blood mimic', 'tissue-mimicking material'.

Flow rate Volume of a fluid passing through a vessel or region per unit time; units are $L\ s^{-1}$, or $mL\ s^{-1}$; e.g. the average flow rate in the common carotid artery in adults is about $6\ mL\ s^{-1}$.

Flow rate (of blood in an artery) Quantity which may be measured using an ultrasound system; e.g. using a combination of mean velocity obtained from Doppler ultrasound data (usually the time-averaged maximum velocity) and the diameter obtained from the B-mode image. *See* 'diameter', 'time-average velocity (TAV)'.

Fluorocarbon (also known as perfluorocarbons) The gas which is generally found within commercially available contrast agents. Fluorocarbons have very low water solubility compared to air, thus delaying bubble dissolution. *See* 'gas', 'microbubble'.

Focal depth (1) Distance from the transducer where the spatial resolution of a beam has its minimum value. *See* 'spatial resolution'.

Focal depth (2) Machine control allowing the operator to adjust the position of the beam focus.

Focus Position in an ultrasound beam where the spatial resolution has its minimum value.

Forward flow Relating to blood flow in arteries; flow in the direction away from the heart, the dominant situation for flow in arteries. *See* 'reverse flow'.

Fourier transform Method for estimating the frequency components within a signal. *See* 'fast Fourier transform'.

Frame averaging A processing feature relevant to the displayed image, in which the displayed image is a weighted average of the current image and several previous images; used for the purposes of noise reduction, but also results in blurring of objects which are moving rapidly.

Frame rate Number of ultrasound image frames acquired per second.

Free gas The gas released from a contrast microbubble after the shell has been compromised. The gas which escapes is no longer surrounded by shell. *See* 'gas', 'microbubble'.

Free radical A chemical species that is highly reactive.

Freehand 3D ultrasound system System for acquiring 3D ultrasound data, based on acquisition of 2D images while the probe is moved by hand. *See* '3D endoprobe', 'magnetic tracking system', 'mechanically steered array', 'optical tracking system'.

Freehand ultrasound Movement of the ultrasound transducer by hand, as opposed to movement by a mechanical system; the basis for virtually all modern ultrasound scanning.

Freeze Machine control which results in suspension of real-time imaging with the last acquired frame displayed on the screen.

Frequency Property of a wave; the number of oscillations passing a given point per second; unit is hertz or Hz; e.g. ultrasound transducers typically have frequencies between 1 and 18 MHz.

Fundamental imaging Conventional B-mode imaging in which the ultrasound beam is transmitted and received over the same frequency bandwidth (frequency range) to form the B-mode image. *See* 'harmonic imaging'.

Gain Machine control enabling the operator to alter the amplification of the received ultrasound signals, resulting in increase or decrease of brightness on the display.

Gamma curve Relationship between the image value and displayed grey level; a smooth curve which is described by a specific function ($V_{out} = V_{in}{}^{\gamma}$) with a single variable γ. Adjustment of γ can be used to change the grey-level appearance. *See* 'grey-level curve'.

Gas Material encapsulated by the shell within microbubbles; the gas can be air or a heavier gas such as fluorocarbon. *See* 'fluorocarbon', 'free gas', 'microbubble'.

Gas body A general term used to represent any collection of gas or vapour. Gas bodies are designed and manufactured to be the active ingredient of ultrasound contrast agents. *See* 'gas'.

Gas body activation The interaction of acoustic waves with gas bodies. Applied, especially, to non-inertial cavitation involving gas bodies in biological systems.

Gate Range of depths from which Doppler signals are obtained using pulsed-wave Doppler. *See* 'sample volume'.

Gate position Relating to Doppler spectral display; machine control enabling the operator to adjust the depth of the region from which Doppler signals are acquired. *See* 'gate size'.

Gate size Relating to Doppler spectral display; machine control enabling the operator to adjust the length of the region from which Doppler signals are acquired. *See* 'gate position'.

Geometric spectral broadening Relating to spectral Doppler, where a target with a single velocity will give rise to a range of Doppler frequency shifts due to the finite size of the Doppler aperture. Ultrasound is received from the target with a range of angles; commonly resulting in overestimation of maximum blood velocity. *See* 'intrinsic spectral broadening'.

Gigapascal (GPa) The unit of measurement commonly used for elastic modulus in hard tissues such as bone and tooth enamel, equivalent to 10^9 pascal. *See* 'elastic modulus'.

Golay pair In coded excitation, a pair of ultrasound pulses whose codes are designed to eliminate range artefacts when the received echoes are detected, filtered and combined. *See* 'coded excitation', 'range artefact'.

Grating lobe Weaker beams, produced by linear arrays on either side of the main lobe, giving rise to reduction in image contrast and other artefacts.

Grey-level curve Relationship between the image pixel value and the displayed grey level. *See* 'gamma curve'.

Harmonic Additional frequencies, usually multiples of the base frequency F_o, e.g. $2F_o, 3F_o, 4F_o$; in ultrasound the received echo may contain harmonics arising from non-linear effects within the tissue, especially during imaging of contrast agents. *See* 'contrast agent', 'harmonic imaging'.

Harmonic imaging Mode of ultrasound imaging in which the receive beam-former is adjusted to receive harmonics of the transmit frequency, usually for the purpose of improved image contrast. *See* 'harmonic'.

Hazard Any actual or potential source of harm.

Hertz (Hz) Unit of frequency. *See* 'frequency'.

High-frame-rate colour flow Colour flow method involving the use of plane waves.

High-frame-rate imaging Ultrasound imaging in which beam forming produces higher than usual frame rates (up to several thousand Hz), usually with sacrifice of field of view and/or image quality. *See* 'plane-wave imaging', 'synthetic-aperture imaging'.

High-frame-rate spectral Doppler Spectral Doppler method involving the use of plane waves.

High-MI techniques Contrast-specific imaging techniques which rely upon the destruction of microbubbles within the scan plane. *See* 'contrast-specific imaging'.

High-pass filter *See* 'wall filter'.

Human error Relating to the measurement of a quantity (e.g. distance, volume); where part of the error arises from variations from one time to another by the operator.

IEC (International Electrotechnical Commission) An international body with responsibility to set technical and safety standards for the manufacture of goods, including medical ultrasound equipment. *See* 'standards'.

Image processing Changes made to an image in order to improve its appearance and to help improve the detection of abnormalities. *See* 'adaptive image processing', 'edge enhancement', 'noise reduction'.

In phase Relating to two waves of the same frequency, for which wave peaks occur at the same time or distance (phase difference $= 0°$). *See* 'out of phase'.

Incompressible Material or tissue whose density does not change when compressed. *See* 'bulk modulus', 'pressure wave', 'shear modulus', 'shear wave', 'Young's modulus'.

Inertial cavitation The class of acoustic cavitation for which a bubble undergoes very large cyclic changes in volume, dominated by the inertia of the surrounding liquid, and associated with very high transient internal temperatures and pressures. It commonly results in bubble instability and violent collapse.

Infusion Method for injecting contrast agents at a set rate, over time, usually by means of an infusion pump. *See* 'bolus'.

Insonation Application of ultrasound to a region of tissue.

Intensity The power (e.g. of an ultrasound wave) flowing through unit area; unit is $W\,cm^{-2}$; e.g. the intensity (spatial peak pulse average) for B- and M-mode imaging is in the range 14–933 $W\,cm^{-2}$. *See* 'I_{sppa}', 'I_{spta}'.

Interference The combination of two or more waves, with the amplitude at any one position and time being the combination of the amplitudes from each of the individual waves. *See* 'constructive interference', 'destructive interference'.

Intermittent imaging Contrast-specific imaging technique for imaging contrast agents over a predefined time sequence or triggered from an ECG. *See* 'contrast-specific imaging'.

Intravascular ultrasound (IVUS) Imaging of arteries or the heart using a transducer which is located within the artery or heart, where the transducer is attached to a catheter which is fed into the artery via an arterial puncture, usually of the femoral artery.

Intrinsic spectral broadening Relating to spectral Doppler, where a target with a single velocity will give rise to a range of Doppler frequency shifts; due to the characteristics of the scanner rather than the blood flow. The dominant cause is geometric spectral broadening. *See* 'geometric spectral broadening'.

Invert Machine control enabling the operator to turn the display upside down (for spectral Doppler), and to reallocate the colour scale (for colour Doppler), e.g. red to blue.

Inverted mirror-image artefact Relating to spectral Doppler, where the true Doppler waveform appears inverted in the opposite channel, occurring when the Doppler gain is set too high.

IPEM (Institute of Physics and Engineering in Medicine) UK professional body which produces (among other things) guidance on quality assurance of ultrasound systems.

I_{sppa} Spatial peak pulse average intensity. *See* 'intensity'.

I_{spta} Spatial peak temporal average intensity. *See* 'intensity'.

JPEG File format used for single-image digital files.

Kilopascal (kPa) The unit of measurement commonly used for elastic modulus in soft tissues such as muscle and liver, equivalent to 10^3 pascal. *See* 'elastic modulus'.

Laminar flow Relating to blood flow; where at low velocities fluid elements follow well-defined paths with little mixing between adjacent layers. *See* 'disturbed flow', 'turbulence'.

Lateral Relating to an ultrasound beam; the direction within the 2D scan plane which is perpendicular to the beam axis. *See* 'axial (1)', 'elevation'.

Lateral resolution *See* 'spatial resolution'.

LCD *See* 'liquid crystal display'.

LED *See* 'light-emitting diode'.

Light-emitting diode A device for producing light or infra-red radiation; used in optical tracking systems for freehand 3D ultrasound. *See* 'optical tracking system'.

Line Portion of the ultrasound image along a single direction radiating from the transducer, corresponding to ultrasound information acquired from a single beam, also known as 'scan line'.

Line density Number of scan lines per unit distance of the transducer for a 2D ultrasound system, or number of scan lines per unit area of a 2D array transducer.

Linear array Transducer consisting of many (128–256) elements arranged in a line, enabling electronic beam forming. There is some beam-steering capability (typically 20°–30°), sufficient to enable image compounding and production of steered Doppler beams. *See* '1D array', '1.5D array', 'array'.

Linear format Relating to an ultrasound image, in which the scan lines are parallel and directed vertically down from the transducer (usually a linear array).

Liquid crystal display Widely used image-display device, which uses liquid crystal technology.

Lithotripsy *See* 'extracorporeal lithotripsy'.

Longitudinal wave Wave, such as ultrasound or sound, in which the direction of motion of the local particles is in the direction of travel of the wave. *See* 'transverse wave'.

Low-MI techniques Contrast-specific imaging techniques which rely upon the non-destruction of microbubbles. *See* 'contrast-specific imaging'.

M-mode Ultrasound mode in which a selected line from the B-mode image is swept across the display to show interface movement. Useful for accurate measurement of the change in the dimensions of moving structures, e.g. in the heart.

Mach cone Expanding shear wave produced by a supersonic source. *See* 'supersonic imaging'.

Magnetic tracking system Device for tracking the position and orientation in space of a transducer, e.g. using a small transmit device to generate a 3D magnetic field within which ultrasound scanning takes place, and sensor coil attached to the transducer which sends electrical signals to a receiver;

used in freehand 3D ultrasound. *See* 'freehand 3D ultrasound system'.

Magnitude of velocity *See* 'velocity magnitude'.

Main lobe Relating to the detailed distribution of ultrasound energy from the transducer which has high intensity in several specific directions, called lobes. The main lobe corresponds to the direction intended to have the highest intensity, and is another name for the ultrasound beam. *See* 'side lobe'.

Matched filter In coded excitation; design of the decoding filter which results in compression of the energy of the received echo into a signal with maximum signal-to-noise ratio. *See* 'coded excitation'.

Matching layer A component of most ultrasound transducers, designed to maximize the transmission of acoustic energy into the tissues; achieved by suitable choice of acoustic impedance.

Matrix array *See* '2D array'.

Maximum (Doppler) frequency Peak value of the Doppler frequency at any one time. *See* 'maximum frequency waveform'.

Maximum frequency envelope *See* 'maximum frequency waveform'.

Maximum frequency waveform Maximum Doppler shift versus time waveform derived from the Doppler spectral waveform; used in estimation of some waveform indices; usually acquired in stand-alone Doppler systems where the beam-vessel angle is unknown. *See* 'pulsatility index', 'resistance index'.

Maximum velocity waveform Maximum velocity versus time waveform; identical in shape to the maximum frequency envelope but has units of velocity; used in estimation of some waveform indices; acquired using duplex systems where the beam-vessel angle can be measured. *See* 'flow rate', 'pulsatility index', 'resistance index'.

MDD The European Medical Device Directive, which sets conditions for the manufacture and sale of medical equipment, including ultrasound scanners, for the European Community.

Mean Doppler frequency *See* 'mean frequency (1) and (2)'.

Mean frequency (1) Average value of the detected Doppler frequencies at any one time.

Mean frequency (2) One of the values which is produced from autocorrelator-based frequency estimators of colour flow systems, equal to the mean Doppler frequency. *See* 'power (2)', 'variance'.

Mean frequency waveform Waveform of mean Doppler frequency versus time. May be used to estimate waveform indices, but this is generally not advised as the mean Doppler frequency is prone to error. *See* 'maximum frequency envelope', 'maximum velocity waveform'.

Measurements Values of specific indices or quantities measured using the ultrasound system (*see* 'area', 'distance', 'elastic modulus', 'resistance index', 'velocity', 'volume', 'volumetric flow') or relevant to the performance of the ultrasound system (*see* 'intensity', 'mechanical index', 'power (1)', 'thermal index', 'spatial resolution').

Mechanical index (MI) An indicator calculated from the ratio between the peak rarefactional pressure (MPa) and the square root of acoustic frequency (MHz). MI may be used as a general indicator of the risk of cavitation in human tissues and the behaviour of ultrasound contrast agents when exposed to ultrasound.

Mechanical scanner Ultrasound system in which the beam is swept through the tissues by mechanical movement of the element(s).

Mechanical wave Wave that requires a medium such as tissue for propagation; examples include shear waves, sound and ultrasound. *See* 'shear wave', 'sound wave', 'ultrasound'.

Mechanically steered array Transducer used for 3D imaging which consists of a curvilinear or linear array located within a fluid-filled housing, where the array is swept to and fro mechanically, hence sweeping the scan plane through the tissue to collect a series of 3D volumes. *See* 'Mechanically steered array 3D ultrasound system'.

Mechanically steered array 3D ultrasound system System for 3D ultrasound in which 3D data are collected using a mechanically steered array transducer. *See* 'mechanically steered array'.

Megahertz (MHz) The unit of measurement commonly used for ultrasound frequency, equivalent to 10^6 hertz.

Megapascal (MPa) The unit of measurement commonly used for acoustic pressure and related quantities such as elastic modulus, equivalent to 10^6 pascal. *See* 'elastic modulus', 'pressure'.

Memory General: an area of the computer used to store information. Medical ultrasound: an area of the computer used to store ultrasound images.

MI *See* 'mechanical index'.

Microbubble The most common configuration of contrast agents, in which a small gaseous bubble is encapsulated by a thin shell. Usually between 2 and 10 μm in diameter and used to enhance contrast in ultrasound imaging. *See* 'contrast agent'.

Mirror-image artefact Appearance of a second version of a region of tissue where there is a strong reflector present; scattered echoes in front of the interface are received in the normal manner, and also via the interface, resulting in their display below the interface.

Misregistration Relating to 2D or 3D images; positioning of ultrasound data (B-mode, colour flow, etc.) at an incorrect location within the displayed image. Usually occurring as a result of speed-of-sound artefacts. *See* 'multiple scattering artefact', 'range error', 'refraction artefact', 'speed-of-sound artefacts'.

Mixing Relating to processing of the ultrasound signal; where one signal is combined with a second reference signal, to produce a composite signal with a low-frequency component and a high-frequency component which can be removed by high-pass filtering. *See* 'Doppler demodulation'.

Modulus *See* 'elastic modulus'.

MPEG File format used for digital video files.

Multi-line acquisition Technique to increase the frame rate by the simultaneous acquisition of several (e.g. four) receive beams for each broadened transmit beam. *See* 'high-frame-rate imaging'.

Multi-path artefact *See* 'mirror-image artefact'.

Multiple scattering artefact Related to contrast agents; the ultrasound beam can be scattered from many microbubbles as it travels through a region of interest. This means that it will take longer to be detected and the ultrasound scanner will place the echoes deeper than their source because of the delay in detecting. *See* 'misregistration'.

Multiple-zone focusing Method to improve the transmit beam formation and spatial resolution; by dividing the field of view into several depth regions (zones) each acquired sequentially using a separate ultrasound pulse, with each transmit pulse focused for each zone. A consequence of this method is a reduction in frame rate.

Near-field Term applied to that part of the acoustic field near to an unfocused transducer where destructive interference may occur. *See* 'far-field'.

NEMA The National Electrical Manufacturers Association in the United States.

Noise Generally, any feature of the image which is unwanted which obscures the feature of interest. The dominant noise in ultrasound imaging is speckle and electronic noise, and also clutter breakthrough for Doppler and colour flow. *See* 'colour speckle', 'clutter breakthrough', 'Doppler speckle', 'electronic noise'.

Noise reduction Image-processing method which reduces the noise in an image, usually at the expense of some loss of spatial resolution. *See* 'frame averaging', 'image processing'.

Non-inertial cavitation A class of acoustic cavitation associated with low acoustic pressures, characterized by small-scale stable bubble oscillations. Collapse of the bubble or gas body does not occur.

Non-linear distance Distance along the perimeter of a curved structure, e.g. abdominal circumference.

Non-linear imaging A method of constructing images from non-linear signals detected from non-linear scatterers. Useful for imaging at depth and contrast agents. *See* 'contrast-specific imaging'.

Non-linear propagation Movement of waves through a tissue characterized by changes in the ultrasound pulse shape with time due to different parts of the pulse travelling at different speeds, resulting in a steepening of the pulse with the generation of additional frequencies or harmonics.

Non-uniform insonation Relating to the insonation of blood flow in arteries; where the beam-width is less than the vessel diameter resulting in large parts of the vessel not being insonated.

NPL The National Physical Laboratory, Teddington, United Kingdom.

Nyquist limit Relating to estimation of frequency using regular sampling of the signal, where the upper limit of frequency estimation is half the sampling frequency. For Doppler ultrasound, see 'aliasing'.

O-ring rubber Filament used in a string phantom which gives Doppler signals similar to those from blood, as opposed to spiral-wound filaments such as cotton or silk which give preferential scattering at certain angles. *See* 'filament', 'string phantom'.

Optical tracking system Method for tracking the position and orientation in space of a transducer, e.g. using a pair of infra-red sensors to record the position of infra-red LEDs attached to the transducer; used in freehand 3D ultrasound. *See* 'freehand 3D ultrasound'.

Oscillation The sequence of expansion and contraction of the microbubble. *See* 'resonance'.

Out of phase Relating to two waves of the same frequency, where one wave is displaced by half a wavelength (phase difference = 180°). *See* 'in phase'.

Output Display Standard A document prepared by the AIUM and NEMA in the United States which defines the Safety Indices and their use. Now included in IEC Standard 60601-2-37.

Output power *See* 'power (1)'.

PACS *See* 'picture archiving and communication system'.

Parallel beam forming *See* 'multi-line acquisition'.

Partial volume effect Generally, change in the value or image display when the quantity being calculated does not occupy the entire pixel or voxel. In power Doppler, reduction in the displayed power near the edge of a vessel when the sample volume only partially covers the region of flowing blood.

Peak negative pressure *See* 'rarefaction pressure'.

Pencil probe Simple ultrasound transducer used for Doppler, with no imaging capability, usually consisting of one element (PW Doppler) or two elements (CW Doppler), about the size of a pencil.

Penetration Ability of an ultrasound system to visualize deep tissues. *See* 'penetration depth'.

Penetration depth Maximum depth at which ultrasound information can be obtained; may be different for each modality (B-mode, spectral Doppler, colour flow). *See* 'penetration'.

Perfluorcarbon *See* 'fluorocarbon'.

Perfusion Blood flow to organs and tissues. Perfusion is important as a means to limit temperature rise caused by the absorption of ultrasound.

Persistence *See* 'frame averaging'.

Petechiae Rounded spots of haemorrhage on a surface such as skin, mucous membrane, serous membrane or on a cross-sectional surface of an organ.

Phantom Construction or device for testing the properties of an ultrasound system; may attempt to mimic the acoustic properties of human tissue, and other properties, e.g. the viscous properties of blood in the case of the flow phantom. *See* 'flow phantom', 'string phantom', 'thermal test object', 'tissue-mimicking', 'tissue-mimicking phantom'.

Phase The position in time or distance for a single sinusoidal wavelength; mathematically expressed as an angle so that 0° is the start of the wave, 360° is the end of the wave. *See* 'in phase', 'out of phase'.

Phase aberration Changes in the shape of the wave front from the transducer which occur as a result of non-uniformity of speed-of-sound values within the beam; typically occurring as the beam propagates through the subcutaneous fat layer resulting in defocusing of the beam and loss of spatial resolution.

Phase domain Relating to estimation of blood or tissue velocity, through manipulation of the phase of the detected ultrasound signals. *See* 'time domain'.

Phase-domain systems Relating to PW Doppler systems for measuring blood or tissue velocity, where the estimation of target velocity is undertaken using phase-domain techniques. *See* 'time-domain systems'.

Phase-inversion imaging *See* 'pulse-inversion imaging'.

Phased array Transducer consisting of many (128–256) narrow elements arranged in a line, enabling electronic beam steering to be performed. The array is relatively short overall enabling the formation of sector images, mostly used in cardiology.

Picture archiving and communication system Network-based system which allows the transfer of images from several different types of medical imaging system to: workstations (for reporting, viewing and further processing), printers for hardcopy and storage systems for data archival. *See* 'DICOM'.

Piezoceramic Ceramic material manufactured with piezoelectric properties, for use in the manufacture of the piezoelectric plate of an ultrasound transducer. *See* 'piezoelectric plate'.

Piezoelectric Property of a material, describing its ability to change its dimensions on the application of an electrical voltage, and vice versa, hence able to produce ultrasound waves when stimulated appropriately; the main physical phenomenon behind medical ultrasound imaging.

Piezoelectric plate Component of a transducer which converts the electrical voltage changes into an ultrasound signal, and vice versa; the most important component of the ultrasound system. *See* 'element', 'transducer'.

Pixel 'Picture element'; the smallest component of a digital image; typically rectangular, with typical ultrasound images consisting of 500 by 500 pixels.

Plane disc source Ultrasound transducer consisting of a single disc shaped element.

Plane wave Wave which propagates through space as a flat sheet. *See* 'spherical wave'.

Plane-wave imaging Ultrasound imaging technique in which all array elements are activated in each transmission to produce a plane wave in order to insonate the whole field of view in one pulse, hence allowing very high-frame-rate imaging of several thousand frames per second; used to visualize the movement of shear waves in shear-wave elastography. *See* 'high-frame-rate imaging', 'shear-wave elastography'.

Plate *See* 'piezoelectric plate'.

PMN-PT Abbreviation for lead titanate doped with lead, magnesium and niobium; the material from which the piezoelectric plate of some (wideband) transducers is constructed. *See* 'piezoelectric plate'.

Point target In relation to spatial resolution measurement; target whose dimensions are much less than the spatial resolution.

Portable Able to be carried; usually a low-weight ultrasound system or phantom which can be easily carried by a typical operator.

Power (1) The amount of energy transferred per unit time; unit is watt (W) or Joules/second; e.g. typical B-mode imaging systems deliver an acoustic power of 0.3–285 mW.

Power (2) One of the values which is produced from autocorrelator-based frequency estimators of colour flow systems, related to the amplitude of the received

Doppler signals. *See* 'mean frequency (2)', 'power Doppler', 'variance'.

Power Doppler 2D imaging of blood flow using the power of the Doppler signal. *See* 'colour Doppler', 'directional power Doppler', 'power (2)'.

Power modulation *See* 'amplitude modulation'.

Pre-amplifier Device(s) for increasing the amplitude of the electrical signal detected by the element(s) from the received echoes in order to allow subsequent processing. *See* 'receiver'.

Pressure Force per unit area in a direction at 90° to the surface of an object; unit is pascal (Pa) or $N\ m^{-2}$; e.g. the peak rarefaction pressure for pulsed Doppler systems is in the range 0.6–5.3 Pa.

Pressure-strain elastic modulus Index of arterial stiffness which does not take account of wall thickness.

Pressure wave Wave, such as a sound wave or ultrasound wave, in which local pressure changes travel through a medium such as gas, liquid or solid; where the speed of propagation is controlled by the local density and the local bulk modulus. *See* 'bulk modulus', 'compressional wave', 'longitudinal wave'.

PRF *See* 'pulse repetition frequency'.

PRI *See* 'pulse repetition interval'.

Printer Hard-copy device connected to an ultrasound scanner, or more commonly part of a picture archiving and communication system. *See* 'picture archiving and communication system'.

Priority encoder *See* 'blood-tissue discriminator'.

Propagation Movement of a wave through a medium or tissue.

Propagation artefact Transient, decreased brightness distal to contrast-filled regions due to attenuation of the contrast microbubbles within the scan plane. *See* 'contrast-specific imaging'.

Pulsatility index (PI) An index measured from the Doppler waveform, usually from the maximum frequency envelope, which provides information on the degree of diastolic flow. *See* 'resistance index'.

Pulse Ultrasound wave of short duration which, for the purposes of ultrasound imaging, is transmitted along a beam.

Pulse-echo Technique used in ultrasound imaging to provide information on the depth from which received echoes arise, involving timing the delay between transmission and reception, dividing by 2, and multiplying this by the assumed average velocity of $1540\ m\ s^{-1}$.

Pulse inversion amplitude modulation (PIAM) Relating to contrast agents. A combination of pulse inversion and amplitude modulation for imaging contrast agents. *See* 'amplitude modulation', 'pulse inversion'.

Pulse-inversion imaging Method for improving image quality which uses two consecutive pulse-echo lines, with the second pulse an inverse of the first. When received echoes are combined the harmonic content remains. Used in harmonic imaging and contrast imaging. *See* 'contrast-specific imaging', 'harmonic'.

Pulser Component of an ultrasound system which produces electrical signals at radio frequencies (e.g. 10 MHz), where the signals are then modified by the transmit beam-former to produce the electrical pulses applied to the transducer for generation of the ultrasound beam. Note that there may be several pulsers to cover a range of ultrasound transmit frequencies.

Pulse repetition frequency (PRF) Number of ultrasound pulses transmitted per second; value will be different for each modality (B-mode, spectral Doppler, colour flow).

Pulse repetition interval (PRI) Time between consecutive transmitted pulses; the inverse of pulse repetition frequency. *See* 'pulse repetition frequency'.

Pulsed-wave (PW) Doppler Doppler ultrasound technique in which ultrasound pulses are used, as opposed to a continuous-wave technique. *See* 'colour flow', 'spectral Doppler'.

Pulsed-wave (PW) Doppler signal processor Part of the machine which processes the ultrasound signals when the machine is used in PW Doppler ultrasound mode.

Pulsed-wave (PW) ultrasound An ultrasound system in which ultrasound is generated as pulses. *See* 'colour flow', 'pulse echo', 'pulsed-wave Doppler'.

Pushing beam In acoustic radiation force imaging, a high-output beam which is used to produce deformation of tissue at the focal region. *See* 'acoustic radiation force imaging'.

PW *See* 'pulsed-wave ultrasound'.

PZN-PT Abbreviation for lead titanate doped with lead, zinc and niobium; the material from which the piezoelectric plate of some (wideband) transducers is constructed. *See* 'piezoelectric plate'.

PZT Abbreviation for lead zirconate titanate; the material from which the piezoelectric plate of many transducers is constructed. *See* 'piezoelectric plate'.

Quad processing Multi-line acquisition technique in which four scan lines are interrogated in parallel.

Quality assurance General: set of actions whose aim is to produce items or service of a specified quality. In ultrasound: assessment of equipment performance using test objects to ensure compliance with relevant standards.

Quality control General: set of activities designed to evaluate a developed product. In ultrasound: often used interchangeably with quality assurance. *See* 'quality assurance'.

Radial format Arrangement in which scan lines are produced 360° around a central transducer. See 'endoprobe', 'intravascular ultrasound'.

Radiation force The force exerted on a medium by an ultrasound wave. Radiation force may be exerted on a surface, on an object such as a microbubble or throughout the medium. The radiation force acts in the direction of wave propagation.

Radio frequency (RF) Frequency of received echoes once they have been converted into electrical signals; MHz electrical signals fall within the radio portion of the electromagnetic spectrum.

Random error Relating to measurement error (e.g. of distance, volume); where the error varies over a small range, with an average value of zero for the error when the measurement is repeated many times. *See* 'error (2)', 'measurement', 'systematic error'.

Range ambiguity Relating to spectral Doppler and colour flow systems, detection of received echoes from the previous (not the current) echo(es), resulting in the potential display of blood flow data from deep structures rather than from the intended location.

Range artefact In coded excitation; increase (worsening) of the axial resolution which may occur after decoding. *See* 'coded excitation', 'Golay code'.

Range error Errors in ultrasound imaging occurring as a result of incorrect assumption of the speed of sound. Occurs when the true speed of sound is not equal to 1540 m s^{-1}. Can result in information displayed at the incorrect depth, distortion of interfaces and errors in the measured size. *See* 'speed-of-sound artefact'.

Rarefaction Regions in which there is increased volume (decrease in density) due to decrease in local pressure; in the context of an ultrasound wave, the low-pressure part of the wave.

Rarefaction pressure The magnitude of the negative acoustic pressure in an ultrasound wave. In pulsed beams the value of peak rarefactional pressure, that is, the greatest negative value of the acoustic pressure, is often given. *See* 'rarefaction'.

Read zoom *See* 'digital zoom'.

Real time Occurring now, as in 'real-time ultrasound', which provides information with negligible delay between acquisition and visualization; as opposed to off-line imaging such as computed tomography where the data are acquired then visualized with a delay of a few seconds or minutes.

Receive beam The region of space within the tissue from which ultrasound signals will contribute to the beam.

Receive beam-former Component of an ultrasound system that deals with formation of the receive beam; involving adjustments of amplitude of the signal and imposition of time delays from each element before combination. Details of beam forming are usually different for each modality (B-mode, spectral Doppler, colour flow). *See* 'beam-former'.

Received ultrasound Ultrasound echoes received by the transducer.

Receiver Component of the ultrasound system which amplifies the detected RF signals from each of the individual elements (or groups of elements) of the transducer, to a level where they can be further processed. The output from the receiver is a series of analogue RF signals which (in modern systems) is usually digitized then passed to the beam-former.

Reception The process of detection of ultrasound echoes arriving at the face of the transducer, following transmission.

Rectification Process applied to ultrasound signals after conversion to electrical signals, in which the negative portion of the cycle is inverted, resulting in a signal with only positive components. *See* 'amplitude demodulation'.

Rectified diffusion A process which can cause bubbles to expand over time in an ultrasound field, by allowing greater inward diffusion when the bubble is large (in rarefaction) than outward diffusion when the bubble is small (in compression).

Reflection General: change in direction of a wave after it has encountered an interface (dimensions $\gg$ wavelength) between two materials with different impedance, with a portion of the wave returning to the medium from which it came. In medical ultrasound: change in direction of an ultrasound beam at the interface (dimensions $\gg$ wavelength) between tissues of different acoustic impedance, with a portion of the beam returning to the medium from which it came.

Reflection coefficient For a wave which is partially reflected; ratio of the pressure of the reflected to the incident wave; values vary between 0 and 1, with 0 indicating that no reflections will occur.

Refraction General: change in direction of a wave after it has encountered an interface between two materials with different wave speed, with the portion of the wave which has entered the second medium being shifted in direction. In medical ultrasound: change in direction of an ultrasound beam after it has encountered an interface between two tissues with different sound speed, with the portion of the ultrasound beam which has entered the second medium being shifted in direction. *See* 'Snell's law'.

Refraction artefact Artefact occurring in ultrasound imaging as a result of refraction, where image features may be displaced from their correct relative position. *See* 'refraction'.

Registration Relating to 2D or 3D images, positioning of ultrasound data (B-mode, colour flow etc.) at the correct location within the displayed image. *See* 'misregistration', 'range error', 'refraction artefact', 'speed-of-sound artefacts'.

Resistance Relating to blood flow, the ratio of pressure drop to flow rate; a measure of the force needed to pump blood through arteries.

Resistance index (RI) An index measured from the Doppler waveform, usually from the maximum frequency envelope, which provides information on the degree of diastolic flow. *See* 'maximum frequency envelope', 'pulsatility index'.

Resolution *See* 'spatial resolution', 'temporal resolution', 'velocity resolution'.

Resonance The vibration of a system when subject to an oscillating force at a specific frequency which is at the resonance frequency of the object. *See* 'resonance frequency'.

Resonance frequency The frequency at which a material or structure naturally vibrates; e.g. when struck. *See* 'resonance'.

Retrospective transmit beam forming Method for producing a transmit beam with improved spatial resolution.

Reverberation Persistence of ultrasound in a spatial region due to ultrasound waves repeatedly travelling back and forth between two interfaces where there are large changes in acoustic impedance between the materials or tissues; examples include reverberation within the PZT plate of a transducer which has no backing layer, and between the front and back walls of a cyst. The effect of ultrasound persisting may be referred to as 'ringing'.

Reverberation artefact Generation of one or more additional copies of a structure at deeper depths, occurring as a result of reverberation between two almost parallel surfaces (e.g. anterior surface of the bladder and the surface of the transducer).

Reverse flow Relating to blood flow in arteries; flow in the direction towards the heart, a normal occurrence in healthy peripheral arteries for a part of the cardiac cycle. *See* 'forward flow'.

RF *See* 'radio frequency'.

Ringing *See* 'reverberation'.

Risk An assessment of the importance of a hazard, which takes account of its nature, the severity of any effect, and the probability of its occurrence. *See* 'hazard'.

Sample volume The region within the Doppler beam from which the Doppler ultrasound signals will be detected. *See* 'Doppler beam', 'gate'.

Scalar A quantity which has magnitude but not direction (e.g. mass, time, volume).

Scale Relating to Doppler spectral display or colour Doppler display; machine control enabling the operator to adjust the maximum Doppler frequency shift which is displayed; note the scale control affects the PRF and so affects aliasing.

Scan plane The region in the tissue from which the image is produced in 2D ultrasound.

Scatterer Small region of tissue which gives rise to scattering of ultrasound.

Scattering General: generation of a wave which travels in all directions, after an incident wave has encountered a small object (dimensions ≪ wavelength), where the object has an impedance different to the surrounding material. For medical ultrasound: generation of a wave which travels in all directions, after an incident beam has encountered a small object (dimensions ≪ wavelength), where the object has an acoustic impedance different to the surrounding tissue. *See* 'backscatter', 'spherical wave'.

Second-harmonic imaging An imaging mode designed to receive and construct an image from signals scattered at the second harmonic frequency. *See* 'contrast-specific imaging', 'harmonic', 'harmonic imaging'.

Sector format Arrangement of scan lines from a small transducer, such as a phased array, in which the scan lines diverge strongly with depth, producing a field of view which is approximately triangular in shape. *See* 'phased array'.

Segmentation Division of a 2D or 3D image into separate regions, demarcated by a contour or surface. In ultrasound this may be performed to identify key structures from which measurements are made, such as the fetal abdomen for circumference measurement, or the left ventricular chamber for volume measurement; ideally segmentation is performed automatically using image-processing techniques. *See* 'image processing'.

Shaded surface display Display method used in 3D ultrasound, in which the boundaries of structures are displayed, with the grey level dependent on the orientation of the surface; this gives a solid appearance to the object and provides an intuitive way for the operator to visualize the data. *See* 'surface shading'.

Shadowing Loss of ultrasound information below a structure which either completely absorbs or reflects the ultrasound beam, such as a gallstone, calcified

arterial plaque or bowel gas, leaving no transmit beam to insonate deeper tissues.

Shear force A force applied parallel to a surface which causes the surface and the material underlying the surface to be dragged in the direction of the force. *See* 'shear modulus', 'shear strain'.

Shear modulus A measure of the ability of a material to withstand a shear force; defined as the shear stress divided by the shear strain; unit is pascal (Pa); e.g. the shear modulus in healthy liver is 800–1200 Pa. *See* 'shear force', 'shear strain'.

Shear strain A measure of the degree of distortion to a material caused by a shear force, defined as the horizontal distance a sheared face moves divided by the vertical distance. *See* 'shear force', 'shear modulus'.

Shear stress A sideways or tearing force; defined as the shear force divided by the area over which the force is applied. *See* 'shear force'.

Shear wave Wave, occurring in an elastic medium, characterized by motion of the particles with no change in local density (as opposed to pressure waves, where there is change in local density). *See* 'shear-wave elastography', 'transverse wave'.

Shear-wave elastography Techniques that provide information related to the elastic modulus of tissue *in vivo*, based on the measurement and display of shear-wave velocity. *See* 'strain elastography', 'transient elastography'.

Shell The surface of a microbubble which surrounds the gas. The shell is usually made from a biocompatible material. *See* 'gas', 'microbubble'.

Side lobe Relating to the detailed distribution in space of ultrasound energy from the transducer, where there is high intensity in several specific directions, called lobes. The side lobes are all lobes other than the main lobe. The side lobes are not useful in the formation of the ultrasound image, giving rise to loss of image contrast and other artefacts. *See* 'main lobe'.

Sine wave Wave with a single frequency whose amplitude as a function of time (or distance) is expressed by the mathematical sine function. *See* 'cosine wave'.

Single-element transducer Transducer consisting of one element which is used for both beam transmission and then reception; used in pulsed-wave ultrasound. Is now seen mainly in stand-alone pulsed-wave Doppler systems. *See* 'pencil probe'.

Slice thickness Width of the beam in the elevation direction. *See* 'spatial resolution'.

Small vessel imaging Methods to visualize small vessels; typically down to around 50 microns diameter.

Snell's law For a wave which is refracted; a law relating the change in direction of the wave to the speed of sound in the two tissues through which the wave passes.

Soft-tissue thermal index (TIS) The thermal index calculated using a uniform soft-tissue model.

Soft tissues Tissues of the body which are not hard or fluid (e.g. not bone or blood); includes muscle, liver, kidney, brain, pancreas etc.; this is the general group for which elastographic techniques are designed. *See* 'elastography'.

Sonoporation The transient creation of gaps or pores in a cell membrane when exposed to ultrasound. Sonoporation is associated with stable cavitation.

Sound Vibrations in the form of pressure waves which travel within a medium (solid, liquid or gas); divided into infrasound (frequency < 20 Hz), audible sound (frequency 20 Hz–20 kHz), ultrasound (frequency > 20 kHz).

Sound wave Pressure waves within a medium such as a solid, liquid or gas; divided into infrasound (frequency < 20 Hz), audible sound (frequency 20 Hz–20 kHz), ultrasound (frequency > 20 kHz).

Spatial resolution Minimum separation in space for which two separate point or line targets can be identified, or size on the image of a point object. There are three values, one for each of the principal beam directions, required for a full reporting of spatial resolution; 'axial resolution', 'lateral resolution' and 'elevation resolution'; the latter may be called 'slice width', 'slice thickness' or 'azimuthal resolution' for 2D ultrasound systems.

Spatio-temporal image correlation Relating to 3D imaging of the fetal heart; where the 2D image plane is swept slowly over several seconds through the fetal heart, and the acquired 2D data are reorganized to provide a series of 3D volumes through the cardiac cycle.

Speckle Noise appearing on ultrasound images, arising from variations in the position and scattering strength of the various scatterers within the beam. *See* 'colour speckle', 'Doppler speckle'.

Spectral analysis *See* 'spectrum analysis'.

Spectral broadening Increase in the range of Doppler frequencies observed on the Doppler waveform in spectral Doppler. Has a variety of origins, some associated with disease (e.g. presence of disturbed flow and turbulence), some associated with the ultrasound system (e.g. geometric spectral broadening). *See* 'disturbed flow', 'geometric spectral broadening', 'intrinsic spectral broadening', 'spectral broadening index', 'turbulence'.

Spectral broadening index An index which describes the range of frequencies present in the Doppler ultrasound waveform; has been used in attempts to quantify the degree of stenosis. *See* 'spectral broadening'.

Spectral display 2D image of Doppler frequency shift against time showing spectral Doppler waveforms in real time, usually in a scrolling format moving from right to left with the most recent waveform data displayed on the right side of the display. *See* 'spectral Doppler'.

Spectral Doppler Doppler ultrasound technique which produces Doppler frequency shift versus time waveforms with full Doppler frequency shift data estimated and presented at each time point. *See* 'spectral display'.

Spectrum Graph of amplitude versus frequency; showing the frequency components of a waveform, obtained using spectrum analysis.

Spectrum analyser Device for performing spectrum analysis. *See* 'spectrum analysis'.

Spectrum analysis Process for estimating the frequency components of a wave, usually using a variant of the Fourier transform called the fast Fourier transform (FFT). In Doppler ultrasound; the method used for estimation of the Doppler shift frequency components at each time point in the cardiac cycle. *See* 'fast Fourier transform', 'spectral Doppler'.

Specular reflection Enhanced brightness occurring when the interface between two regions of different acoustic impedance is perpendicular to the beam direction. *See* 'diffuse reflection', 'reflection'.

Speed of a wave Distance that the wave crest of a wave (or other similar point) travels per second.

Speed of sound Distance that the crest of the sound wave (or other similar point) travels per second. Values in soft tissue are 1400–1600 m s^{-1}, with an average value of about 1540 m s^{-1}.

Speed-of-sound artefacts Artefacts in ultrasound imaging occurring as a result of differences between the actual speed of sound in the tissues and that assumed by the machine (1540 m s^{-1}). *See* 'misregistration', 'range error', 'refraction artefact'.

Spherical wave Wave, generally from a small source, which propagates through space as an expanding sphere. *See* 'plane wave'.

Stable cavitation *See* 'non-inertial cavitation'.

Stand-alone continuous-wave Doppler system Ultrasound system used only for continuous-wave Doppler, with no imaging capability. *See* 'pencil probe'.

Stand-alone pulsed-wave Doppler system Ultrasound system used only for pulsed-wave Doppler, with no imaging capability. *See* 'pencil probe'.

Standards Widely accepted devices or methods which when adopted allow: (i) the production of equipment which is able to be connected to or integrated with equipment from a wide range of other producers; (ii) the measurements made by equipment to be defined consistently and quantified. For ultrasound the IEC is the most important source of standards, which cover all aspects of the design and testing of ultrasound systems. Other national bodies may also set their own standards, which are sometimes adopted internationally. *See* 'AIUM', 'BMUS', 'IEC', 'IPEM'.

Static elastography *See* 'strain elastography'.

Steering Process by which a beam is transmitted and received at angles other than 90° with respect to the transducer face.

Steering angle (1) The angle of the beam with respect to the transducer face. *See* 'steering'.

Steering angle (2) Machine control enabling the operator to adjust the steering angle; e.g. in colour flow. *See* 'steering angle (1)'.

Stenosis A narrowing within an artery, usually caused by atherosclerosis.

Stereoscopic viewing Relating to 3D ultrasound, where the operator visualizes 3D by the use of special glasses or other means, requiring modifications to the displayed 3D data on the screen. Similar technology to that used in 3D films in cinemas.

STIC *See* 'spatio-temporal image correlation'.

Stiffness Description of a material concerning its ability to withstand deformation by a force. *See* 'elastic', 'elastic modulus'.

Stimulated acoustic emission (SAE) Relating to contrast imaging. When imaging at high acoustic pressures (high MI) with flash pulses, the contrast agent is forced to collapse, releasing the gas contained within. These gas bubbles with no shell give a much enhanced backscattered signal. *See* 'flash imaging'.

Strain Related to the distension of a tissue along a particular direction by a force or stress applied along that direction, where strain is the change in length divided by the original length of the material.

Strain elastography Techniques that provide information related to the elastic modulus of tissue *in vivo*, based on the measurement and display of strain. *See* 'shear wave elastography', 'strain'.

Strain ratio Measurement made in 2D strain elastography, where the strain in the lesion is divided by the strain in a reference region ideally placed at the same depth in order to avoid depth artefacts.

Stress For a force which is applied to a tissue; the local force divided by the area; units are pascals (Pa).

String *See* 'filament'.

String phantom Device for testing Doppler ultrasound systems in which the movement of blood is simulated by the movement of a filament, usually O-ring rubber. *See* 'filament', 'O-ring rubber'.

Sub-dicing Dividing of an element into smaller parts, for the purpose of reducing unwanted lateral modes of vibration.

Subharmonic A frequency which is a fraction of and less than the fundamental frequency F_o, e.g. 1/2 F_o. *See* 'harmonic'.

Supersonic Related to the production of waves by a moving source, where the velocity of the source is greater than the velocity of the waves; producing a wave with a characteristic shape called a mach cone. *See* 'mach cone', 'supersonic imaging'.

Supersonic imaging Ultrasound elastography technique which generates shear waves by several sequentially placed sources; equivalent to a high-output source moving through the tissues at a speed greater than that of the shear waves which are generated. The use of several sources enables shear waves of higher amplitude to be produced without increasing tissue heating. *See* 'supersonic'.

Surface shading Method used to allocate image grey level to each pixel of a surface of an organ or structure in 3D ultrasound, in order to make the structure look realistic. *See* 'shaded surface display'.

Swept array 3D transducer *See* 'mechanically steered array'.

Swept array 3D ultrasound system *See* 'mechanically steered array 3D ultrasound system'.

Swept gain *See* 'time-gain compensation'.

Synthetic-aperture imaging Ultrasound imaging technique in which a single array element, or a very small number of array elements, produces a spherical wave in transmission in order to insonate the whole field of view in one pulse, hence allowing very high-frame-rate imaging. *See* 'high-frame-rate imaging'.

Systematic error Relating to the error when a measurement is made (e.g. of distance, volume). The measurement is consistently larger or smaller than the true value by a specific amount; the average value is not zero for the error when the measurement is repeated many times. *See* 'error', 'random error'.

Target In Doppler ultrasound, the moving blood or tissue within the sample volume.

TDI *See* 'Doppler tissue imaging'.

Temporal resolution Minimum separation in time for which two separate events can be identified.

Teratogen Any agent with the potential to cause developmental defects in the fetus or embryo. For example, heat.

Test object *See* 'phantom'.

TGC *See* 'time-gain control'.

Thermal index (TI) A number, closely associated with temperature rise, used to indicate thermal hazard, and shown on the screen of ultrasound scanners.

Thermal test object (TTO) A device designed to measure directly the temperature rise associated with ultrasound exposure in either a soft-tissue or bone mimic material.

TI *See* 'thermal index'.

TIFF (tagged image file format) File format used for single-image digital files.

Time-average velocity (TAV) Relating to spectral Doppler; average value of mean blood velocity over the cardiac cycle. May be obtained by averaging the mean frequency waveform, or more accurately by averaging the maximum frequency waveform and dividing by 2; used in the estimation of flow rate. *See* 'flow rate', 'maximum frequency waveform', 'mean frequency waveform'.

Time domain Relating to estimation of blood or tissue velocity, through manipulation of the time characteristics of the detected ultrasound signals. *See* 'phase domain'.

Time-domain systems Relating to PW systems for measuring blood or tissue velocity, where the estimation of velocity is undertaken using time-domain techniques. *See* 'phase-domain systems'.

Time-gain compensation Process by which received ultrasound signals from deeper tissues are amplified by a greater amount in order to compensate for the decrease in received echo size with increasing depth which occurs as a result of attenuation. *See* 'attenuation'.

Time-gain control Ultrasound system control, whereby the operator is able to adjust the amplification of the ultrasound signal from different depths, to account for the reduction in received ultrasound signal amplitude with signals from deeper tissues, usually by the use of a set of sliders.

Time-intensity curve A curve showing how the backscattered intensity from microbubbles in a region of interest changes with time, usually as the blood flows through.

Tissue Doppler imaging *See* 'Doppler tissue imaging'.

Tissue-mimicking Relating to materials used in the construction of phantoms, where they have similar acoustic properties to human tissue. Typically the speed of sound, attenuation and backscatter are replicated. *See* 'blood mimic', 'phantom', 'tissue-mimicking phantom'.

Tissue-mimicking material (TMM) Material which has the same acoustic properties as soft tissue, generally with a speed of sound of 1540 m s^{-1} and attenuation in the range $0.3–0.7 \text{ dB cm}^{-1} \text{ MHz}^{-1}$.

Tissue-mimicking phantom Phantom in which the components which are insonated are tissue-mimicking. *See* 'tissue-mimicking'.

TO *See* 'transverse oscillation'.

Transcutaneous Meaning 'through the skin or other outer body part', relating to the acquisition of ultrasound by application of the transducer to an external body part. *See* 'endoprobe'.

Transducer Component of the ultrasound system which generates and receives ultrasound pulses and echoes, and sweeps the beam through the tissues to produce an image.

Transducer bandwidth Frequency range (MHz) over which the transducer can transmit and receive ultrasound.

Transducer self-heating Heating, especially associated with the front face of an ultrasound transducer arising from the inefficient conversion of electrical to acoustic energy.

Transient cavitation *See* 'inertial cavitation'.

Transient elastography Shear-wave elastography in which an external vibrator induces a shear-wave pulse. *See* 'shear-wave elastography', 'vibrator'.

Transmission The process of production of ultrasound from the transducer.

Transmit beam The region of space within the tissue which the ultrasound produced by the transducer passes through; this is to some extent an idealized region in which the main energy of the ultrasound pulse is not refracted or reflected by tissues.

Transmit beam-former Component of an ultrasound system that deals with formation of the transmitted beam; involving adjustments to the amplitude and delay in time of the electrical signals applied to the transducer for the purposes of production of the ultrasound pulse. Details of beam forming are usually different for each modality (B-mode, spectral Doppler, colour flow). *See* 'beam-former'.

Transmit frequency Nominal frequency of the transmitted ultrasound pulse; e.g. 6 MHz.

Transmit power Machine control enabling the operator to increase or decrease the output power; generally there is independent control for each ultrasound modality.

Transmitted ultrasound Ultrasound which has been produced by the transducer.

Trans-oesophageal probe Endoprobe designed for insertion down the oesophagus to allow imaging of the oesophagus and other organs such as the heart.

Trans-rectal probe Endoprobe designed for insertion into the rectum to allow imaging of the rectum, prostate and other organs.

Trans-vaginal probe Endoprobe designed for insertion into the vagina to allow imaging of the vagina, uterus, ovaries and other pelvic structures in women.

Transverse oscillation Method for estimation of blood velocity magnitude and direction.

Transverse wave Wave, such as shear waves, in which the local direction of motion of the particles is at 90° to the direction of motion of the wave. *See* 'shear wave'.

Trapezoidal format Arrangement of scan lines from a linear array, in which the scan lines diverge with distance from the transducer, producing a wider field of view with depth. *See* 'linear array'.

Turbulence Relating to blood flow; where at high velocities fluid elements follow erratic paths and where there is mixing between adjacent layers. May lead to increased spectral broadening. *See* 'disturbed flow', 'laminar flow', 'spectral broadening'.

UCD *See* 'ultrafast compound Doppler'.

Ultrafast compound Doppler Doppler ultrasound method involving use of plane waves.

Ultraharmonic A frequency which is a rational (ratio of two integers) multiple of the fundamental frequency F_o, e.g. 3/2 F_o, 5/2 F_o. *See* 'harmonic'.

Ultrasonic Relating to ultrasound. *See* 'ultrasound'.

1D ultrasound Ultrasound system or component of an ultrasound system in which information only along a single line is used, usually in the form of an amplitude-depth plot; rarely used now.

2D ultrasound Ultrasound system or component of an ultrasound system in which information in a 2D plane is used or displayed; the most common method of operation of modern ultrasound systems.

3D ultrasound Ultrasound system or component of an ultrasound system in which information in a 3D volume is used or displayed.

4D ultrasound Ultrasound system capable of producing real-time 3D ultrasound; (i.e. three dimensions in space, one dimension in time).

Ultrasound Generally: sound waves with frequencies too high to be heard by the human ear. For medical ultrasound: sound with a frequency used for diagnosis or therapy in humans and animals (currently 0.5–60 MHz).

Ultrasound contrast agent (UCA) *See* 'contrast agent'.

Ultrasound system Machine using ultrasound for the purposes of diagnosis or therapy.

Variance One of the values which is produced from autocorrelator-based frequency estimators of colour flow systems, related to the variation in the amplitude of the received Doppler signal amplitude. *See* 'autocorrelator', 'mean frequency (2)'.

Vector A quantity which has magnitude and direction (e.g. displacement, velocity, force).

Vector Doppler Doppler ultrasound technique in which both velocity magnitude and direction are estimated.

Vector flow imaging Blood flow imaging technique in which both velocity magnitude and direction are estimated and displayed.

Velocity Distance an object or wave travels per second; units are m s^{-1}; e.g. the velocity of blood in arteries is 0–6 m s^{-1}, of ultrasound in soft tissue is 1400–1600 m s^{-1}, of shear waves in soft tissue is 1–10 m s^{-1}.

Velocity component Projection of the velocity on the *x*, *y* or *z* axis.

Velocity magnitude Overall value of velocity; the value of velocity in the direction of motion.

Velocity profile Relating to blood flow in an artery, graph of velocity against distance along the diameter of the vessel.

Velocity ratio Ratio of the blood velocity estimated within the stenosis to that in an undiseased region, such as proximal to the stenosis; used in the estimation of the degree of stenosis.

Velocity resolution The minimum difference in blood or tissue velocity which can be visualized using an ultrasound system.

Velocity vector Velocity (e.g. relating to blood motion) in which the velocity has an overall magnitude and direction.

Vessel axis Line passing through the centre of each cross section of a vessel.

Vibrator Device for producing low-frequency (10–500 Hz) vibrations in tissue, used in transient elastography. *See* 'transient elastography'.

Video clip *See* 'cine loop'.

Viscoelastic Behaviour of a material which stretches or deforms in response to an applied force, but when the force is applied or removed the strain is not achieved immediately, but after a brief time delay. *See* 'elastic'.

Volume Quantity representing the 3D size of an object or image; may be measured using advanced measurement facilities available on some 3D ultrasound systems.

Volume flow *See* 'flow rate'.

Volumetric flow *See* 'flow rate'.

Vortex Region of recirculating flow, which may be stable (contained at one location), or shed (travels downstream for a few diameters, after which it dies out). *See* 'disturbed flow'.

Vortex shedding Relating to blood flow; where a local region of recirculating flow, e.g. immediately distal to a stenosis, may be shed and travel downstream for a few diameters. *See* 'disturbed flow'.

Vortices Plural of vortex. *See* 'vortex'.

Voxel Volume element; i.e. building block of a 3D data set which is composed (usually) of several thousand voxels.

Wall filter (1) Relating to Doppler ultrasound detection of blood flow; signal-processing step which attempts to remove the clutter signal, leaving behind the Doppler ultrasound signal from the moving blood, by removing low-frequency signals; this term usually reserved for 'spectral Doppler'. *See* 'clutter filter'.

Wall filter (2) Relating to spectral Doppler systems; machine control enabling the operator to adjust the velocity or Doppler frequency below which the signal is removed. *See* 'wall filter (1)'.

Wall filter value Value of the wall filter, typically ranging from 50 Hz (obstetrics) to 300+ Hz (cardiology). *See* 'wall filter (2)'.

Wall motion Cyclic motion of arterial walls associated with change in diameter occurring during the cardiac cycle.

Wall motion measurement Measurement of the motion of arterial walls using ultrasound. *See* 'wall motion'.

Wall motion tracking *See* 'wall motion measurement'.

Wall thump filter *See* 'wall filter (1)'.

Wash-in The time required for the contrast agent to enter a region of interest.

Wash-out The time required for contrast to clear, either by dissolution or by blood flow, from the region of interest.

Wave Generally, a periodic disturbance which travels through space, with transfer of energy but not mass. *See* 'pressure wave', 'shear wave', 'sound', 'ultrasound'.

Waveform In general: for a quantity which varies periodically in time, the display of the value of the quantity with time for one period. In blood flow: the velocity-time or flow rate versus time data for a single cardiac cycle. In Doppler ultrasound: the Doppler frequency shift versus time spectral display for a single cardiac cycle.

Waveform indices Relating to Doppler ultrasound; quantities which are derived from the spectral Doppler waveform which may be useful in diagnosis. *See* 'pulsatility index', 'resistance index', 'spectral Doppler'.

Wavelength Property of a wave; the distance between two consecutive crests or other similar points on the wave.

WFUMB World Federation for Ultrasound in Medicine and Biology.

Write zoom Magnification of a part of the displayed image, performed in real time, enabling the ultrasound field to be restricted to only that part displayed, hence allowing increase in frame rate and/or scan line density. *See* 'digital zoom'.

Young's modulus Property of an elastic material; defined in terms of the stretch of a thin sample of the material, resulting from a force which is applied to one end of the material (when the other end is tethered); a quantity describing the change in dimensions (strain) that occurs due to an applied force (stress). *See* 'bulk modulus', 'elastic modulus', 'shear modulus'.

Zoom Magnification of the displayed image. See 'digital zoom', 'write zoom'.

Answers

CHAPTER 2

Multiple Choice Questions

Q1. c	Q2. a, c	Q3. b	Q4. e	Q5. a
Q6. a, b, c	Q7. b	Q8. d	Q9. b	Q10. d, e
Q11. a	Q12. a, d	Q13. c	Q14. a	Q15. b
Q16. a, b	Q17. c	Q18. d	Q19. b, d	Q20. b, d

Short-Answer Questions

Q1. The acoustic impedance of a medium is a measure of its response to a pressure wave in terms of velocity of movement of the particles of the medium. The characteristic acoustic impedance of a medium is given by the product of its density and speed of sound or by the square root of the product of density and stiffness.

Q2. Acoustic impedance determines the way in which particles of a medium move in response to a pressure wave and the strength of reflections at an interface. Acoustic absorption is a process by which energy in the ultrasound wave is lost to the medium and converted into heat.

Q3. In reaching the target at 5-cm depth, the transmitted ultrasound pulse will be attenuated by 5 cm $\times$ 3 MHz $\times$ 0.7 dB cm^{-1} MHz^{-1} = 10.5 dB. The returning echo will also be attenuated by 10.5 dB on its way back to the transducer, giving a round-trip attenuation of 21 dB.

By a similar calculation, the round-trip attenuation in reaching the target at 1-cm depth will be 4.2 dB. Hence the target at 5 cm will be attenuated by 16.8 dB with respect to the target at 1 cm.

Q4. The amplitude reflection coefficient at an interface between tissue and bone is approximately 60%. The attenuation coefficient in bone is given as 22 dB cm^{-1} MHz^{-1}. Most of the ultrasound energy reaching the interface is reflected back into the tissue. The energy transmitted into the bone is rapidly attenuated. At an interface between tissue and gas, the reflection coefficient is 99.9%, so virtually no ultrasound energy is transmitted into the air. Hence, it is not possible to image targets beyond interfaces between tissue and bone or gas.

Q5. Acoustic impedance is given by the product of speed of sound and density. For tissue (a) this is 1.659×10^6 kg m^{-2} s^{-1} and for tissue (b) 1.33×10^6 kg m^{-2} s^{-1}. The amplitude reflection coefficient for a large interface between them is given by their difference divided by their sum, which is $0.329/2.989 = 0.11$.

Q6. Refraction refers to the change in the direction of propagation of a wave as it crosses a boundary between two media with different speeds of sound. This occurs for non-normal angles of incidence at the boundary. As the ultrasound imaging system assumes that ultrasound waves travel in straight lines, echoes received via refracted waves will be misplaced in the image.

Q7. The near field length is given by a^2/λ, where a is the disc radius and λ is the wavelength. From above, $a = 7.5$ mm and $\lambda = 0.5$ mm (c/f). The near field length $= (7.5)^2/0.5 = 112.5$ mm. The angle of divergence θ in the far field is given by $\theta = \sin^{-1}(0.61\lambda/a) = \sin^{-1}(0.04) = 2.3°$.

Q8. Focusing of an ultrasound beam can be achieved by transmitting curved wave fronts that converge to a focal point. Curved wave fronts can be produced from a curved transducer or by adding a converging acoustic lens to the face of a flat transducer.

Q9. The beam width W at the focus for a focal length F is given approximately by $W = F\lambda/a$ where λ is the wavelength and a is the source radius. In this case, $W = 60$ mm $\times$ 0.5 mm/7.5 mm $= 4$ mm.

Q10. A harmonic image is formed using the second harmonic frequency component in the returned echoes. Harmonic frequencies are formed due to non-linear propagation, which occurs only in high-amplitude parts of the transmitted beam. Side lobes and reverberations may produce acoustic noise in the fundamental image. However, as they are low in amplitude, they do not generate harmonic frequencies and so do not appear in the harmonic image.

CHAPTER 3

Multiple Choice Questions

Q1. c	Q2. d, e	Q3. a, b, c	Q4. a
Q5. d, e	Q6. a, d	Q7. a, c, e	Q8. d
Q9. a, e	Q10. c, e	Q11. a, d	Q12. c
Q13. a, b, e	Q14. b, c, d	Q15. a, c	Q16. c, d
Q17. a, b, c, e	Q18. a, d, e	Q19. c, d, e	Q20. d

Short-Answer Questions

Q1. The single element transducer contains a piezoelectric plate which changes thickness when an electrical signal is applied across it, generating an ultrasonic wave. Conversely, it produces electrical signals when ultrasound waves (echoes) cause it to alternately contract and expand. A backing layer is attached to the rear of the piezoelectric layer to reduce ringing and thus make possible a short ultrasonic pulse. A matching layer is bonded to the front face to improve transmission of ultrasound into, and out of, the lens and patient. A lens is usually attached between the matching layer and the patient to provide focusing in both the scan plane and elevation plane.

Q2. Resonance occurs because sound waves reverberating back and forth across the plate will reinforce each other if the round-trip distance is precisely one wavelength (twice the half wavelength plate thickness).

Q3. To form an ultrasonic beam, electrical transmit pulses are sent to a group of adjacent array elements (e.g. elements 1–20). This active group transmits an ultrasound pulse along the first scan line. When the transmit-receive sequence for this first line is complete, the active group is stepped along the array to the next line position by dropping off an element from one side of the group adding another to the other side (i.e. using elements 2–21). This stepping and transmission process is repeated along the array to form a set of B-mode scan lines.

Q4. To form a transmit focus, pulses from the different elements in the active transmit group must all arrive at the chosen focus at the same time. Because the outer elements in the group have further to travel to this point than do those from the more central elements, it is necessary to fire the elements at different times, starting with the outermost pair of the active group, then the next pair in towards the centre of the group, and so on.

Q5. The transmit focal depth is set by the user at a single fixed depth at the centre of the depth range of interest. In order to have a reasonably large range of depths over which the transmit beam is narrow, this beam is only weakly focused. This is achieved by using a moderately narrow transmit group of elements. The receive focus, however, is continuously and automatically advanced as echoes arrive from deeper and deeper. At every moment, the receive beam can be strongly focused, by using a wide receive aperture, to give a beam that is as narrow as possible at the source depth of echoes arriving at that moment. The effective receive beam is, therefore, a narrow cylinder consisting of a sequence of closely spaced narrow focal zones.

Q6. Multiple-transmit-zone focusing uses several transmit pulses to interrogate each scan line, with each transmit pulse focused to a different depth. The image is then constructed by stitching together the sections of each scan line corresponding to the several transmit focal zones. The effective transmit beam is thus narrow over a greater range of depths than could be achieved with a single transmit focus, leading to good resolution over a larger range of depths. However, the use of several transmit-receive cycles for each scan line results in a reduced frame rate. Also, the stitching process may affect image uniformity.

Q7. A beam in the straight ahead direction can be produced by firing all elements in the array at the same time. The beam can be steered away from the straight ahead direction by firing the elements in sequence from one side of the array to the other with a small delay between each. The deviation of the beam from the straight ahead direction is increased by increasing the delay between elements.

Q8. The narrow array elements of an array transducer are able to transmit and receive over a wide range of angles. Echoes may arrive from certain directions such that they arrive at one array element exactly one wavelength ahead of, or behind, their arrival at an adjacent element. Echo signals from adjacent elements will then be in phase with each other and add constructively to give a strong response to echoes from that direction. Similar constructive reinforcement occurs if an echo arrives two, three or any integer number of wavelengths ahead of, or behind, its arrival at an adjacent element.

The directions of strong response are those of the grating lobes.

Q9. The width of a focused ultrasound beam in the focal region is determined by the ratio of focal depth to aperture width. A wide aperture leads to a narrower beam (for a given focal length). In the centre of the phased-array image, the apparent aperture (the width of the transducer), as seen from the target, is at its widest. When the beam is steered to the edge of the sector image, the apparent width of the aperture is less, leading to weaker focusing. Also, the array elements are less sensitive at such angles and grating lobes stronger leading to noisier images.

Q10. In line-by-line scanning, each transmit-receive cycle leads to one image line. The time required to interrogate each line is determined by the imaged depth and the speed of sound in the tissue, which is, of course fixed. Hence, large image depths lead to low frame rates. A wide field of view requires the interrogation of more scan lines, which also leads to a lower frame rate. The number of transmissions used to interrogate each scan line (for example when using multiple zone focusing in transmission) affects the image quality, as does the density of scan lines. However, increasing either the number of transmissions per line or the line density in order to improve image quality reduces the frame rate.

Q11. A transmission focus can be achieved at each point in the field of view by retrospective transmit focusing. A sequence of wide, overlapping transmit beams is used to form a set of images using parallel receive beam forming with as many as 16 receive beams served by one wide transmit beam. Where the curved pulses within the overlapping transmit beams intersect, a focus can be synthesised by coherent addition of the beam-formed RF echoes for this point, even though these echoes were obtained from different transmit pulses, fired at different times. This process can be applied to every point in the field of view.

CHAPTER 4

Multiple Choice Questions

Q1. d	Q2. e	Q3. d, e	Q4. b
Q5. a	Q6. e	Q7. b, d	Q8. a, c, d
Q9. b, d	Q10. b, c	Q11. c	Q12. b
Q13. a	Q14. d	Q15. a, d, e	Q16. e
Q17. b, c, e	Q18. a	Q19. b, c, d	Q20. a, d

Short-Answer Questions

Q1. The B-mode image brightness increases with overall gain and transmit power. The operator should use the minimum power that gives the required penetration and increase overall gain to compensate for a reduction in transmit power.

Q2. Echoes will be attenuated by $5 \times 0.7 \times 2$ dB for each centimetre of depth. The scanning system compensates by increasing the gain by 7 dB for each centimetre of depth. The operator sets the TGC to give uniform brightness with depth for a given tissue.

Q3. The dynamic range of echoes is the ratio of the largest echo amplitude that can be processed to smallest that can be detected above noise. A wide dynamic range of echoes must be displayed to include echoes from organ boundaries at the same time as scattering from tissues.

Q4. The dynamic range of echoes received at the transducer must be compressed to fit into the limited dynamic range of the display monitor. This is achieved using logarithmic amplification so that small echoes are amplified more than large ones. Increasing the dynamic range setting allows a wider range of echo amplitudes to be displayed at the same time.

Q5. The operator must trade off the improved resolution of higher transmit frequencies against reduced penetration due to their increased attenuation. Coded excitation can help by enabling the detection of weaker echoes, so that a higher frequency can be used for a given depth of penetration.

Q6. A high-amplitude pulse is transmitted at a relatively low frequency, the fundamental frequency. Harmonic frequencies are generated due to non-linear propagation of the pulse. The fundamental component of the received echoes is removed and the image is formed from the harmonic frequencies.

Q7. Tissue harmonic imaging suppresses clutter in the image that is caused by side lobes and grating lobes and low-amplitude multiple reflections and reverberations of the pulse. The method can also help to reduce the distorting effects of superficial fat layers on the transmitted beam.

Q8. Interpolation is a technique used during scan conversion to create brightness values for pixels in the image matrix which fall between B-mode image lines. The interpolated pixel value is calculated from the B-mode image line samples which lie on each side of the pixel.

Q9. Write zoom images a selected region of interest on an expanded scale before the image is written to the image memory. Read zoom is used to view a selected region of interest of the stored image as it is read out from the image memory. Write zoom does not record any information outside the region of interest.

Q10. Frame averaging reduces random electronic noise in the image which changes from frame to frame. Echo patterns from static scattering features are reinforced. Persistence in the image causes smearing of echoes from rapidly moving targets such as heart valve leaflets.

CHAPTER 5

Multiple Choice Questions

Q1. c	Q2. a, d	Q3. b, d	Q4. d, e
Q5. b, d	Q6. c, e	Q7. b, c, d	Q8. a, c
Q9. a, b, c, d	Q10. a	Q11. a, b, d	Q12. d
Q13. c, e	Q14. b, e	Q15. a, c, d	

Short-Answer Questions

Q1. Lateral resolution can be defined as the minimum separation at which two point reflectors at the same depth in the image can be resolved. The width of the lateral brightness profile through the image of a point reflector is determined by the beam width. If the beam width is small, then the point reflectors can be brought closer together before their brightness profiles merge.

Q2. In the slice thickness direction, the aperture is limited to the width of the transducer, resulting in weak focusing. Larger apertures can be used in the scan plane allowing stronger focusing. Hence the slice thickness is greater than the beam width in the scan plane.

Q3. Axial resolution is defined as the minimum separation at which two point reflectors on the beam axis can be resolved. If two point reflectors are separated by half a pulse length, then the returned echoes will be separated by one pulse length and will be just resolved separately in the image.

Q4. Speckle is the result of coherent addition of echoes from small-scale structures within the sample volume formed by the beam width/pulse length. As the scatterers are random in position and scattering strength, a random pattern of brightness variations is produced. The typical speckle dimensions are related to the sample volume size, a narrow beam and short pulse leading to finer speckle.

Q5. In classic beam forming, each pulse-echo cycle produces one image line. The time to form a single image frame is determined by the imaged depth, the number of image lines and the speed of sound in the tissue. Increasing the imaged depth increases the time to form a frame hence decreasing the frame rate.

Q6. The beam in the centre of the image will pass through the fat layer at normal incidence (90°) and be undeviated. The sector scanner beams transmitted to each side of the central beam will be refracted away from the normal at the lower surface of the fat. The ends of the femur will be imaged via such refracted beams and the images of the femur ends will be displaced towards the central beam. Hence the measured length of the femur will be less than the true length.

Q7. In forming a B-mode image, time gain compensation is normally set for the whole image to compensate for attenuation in a typical soft tissue, e.g. liver. The reflection at the surface of a calcified gallstone is very strong and the attenuation through it is very high compared to tissue. Hence, any echoes that are received from beyond the gallstone are very weak compared to those from a similar depth in tissue and show as an acoustic shadow.

Q8. The large smooth interface will give rise to specular reflection. The beams in the centre of the image will strike the interface at near normal incidence and be reflected back to the active part of the array, giving rise to a bright image. The angled beams incident on the outer parts of the interface will be reflected away from their original aperture and not show in the image.

Q9. Reverberations may occur in the presence of a strongly reflecting interface parallel to the transducer face. The initial echo received from the interface may be partially reflected by the transducer face back to the interface producing a second echo which is displayed at twice the depth. The operator can identify the second echo as a reverberation by pressing the transducer down gently into the skin surface. The reverberation will rise in the image at twice the rate of genuine echoes.

Q10. In the presence of a large, strongly reflecting interface, a lesion within the tissue may be imaged via a path that involves reflection of the beam by the interface. As the imaging system assumes that all echoes are received from the original beam direction, the tissue echoes from the reflected beam are displayed in the image beyond the reflecting interface. Mirror image artefact is most commonly seen beyond the diaphragm.

CHAPTER 6

Multiple Choice Questions

Q1. c	Q2. a, e	Q3. d	Q4. b, d
Q5. b, e	Q6. c, e	Q7. d	Q8. a, b, c, d
Q9. b	Q10. a, b, c, e		

Short-Answer Questions

Q1. Time of flight (go and return time) and speed of sound in the axial direction. Distance = SoS × time of flight/2. Transducer geometry in the lateral direction. For linear arrays the element pitch or line separation. For curved or sector arrays the beam angle relative to the central axis is used to calculate lateral distance at any depth.

Q2. Circumference is the distance around the perimeter of a structure. Options for measurement are direct tracing; ellipse fitting; calculation from orthogonal diameters; and sum of point-to-point measurement.

Q3. Random errors are accidental and distributed above and below the true measurement. Systematic errors are consistent in direction and size. Causes for human error: three of the following: lack of training; lack of experience; absence of standards; failure to follow procedures.

CHAPTER 7

Multiple Choice Questions

Q1. c, e	Q2. a	Q3. b	Q4. a
Q5. b	Q6. d, e	Q7. a, d	Q8. b, e
Q9. d	Q10. c	Q11. a, e	Q12. a, e
Q13. b, c, d	Q14. b, c	Q15. a, d, e	Q16. b, c, e

Short-Answer Questions

Q1. The Doppler effect concerns the change in frequency between a source and an observer. For ultrasound the Doppler effect is the change in frequency of the ultrasound between the transmitted ultrasound beam and the received ultrasound beam. This change in frequency (Doppler shift) arises from scattering of the ultrasound from a moving object such as blood.

Q2. The spectral Doppler display is a real-time display of Doppler frequency shift against time and provides information on blood velocity as a function of time. The colour flow display is a 2D colour display of moving blood overlaid on a greyscale B-mode image.

Q3. The Doppler shift which arises from moving blood is dependent on the cosine of the angle between the beam and the direction of motion of the blood. The Doppler shift is maximum when the angle is zero (beam and direction of motion aligned), and takes a minimum value when the beam is at 90° to the direction of motion.

Q4. The Doppler signal processor takes the received RF data and converts this into Doppler data which may be displayed on the screen. The three component parts are the demodulator, the high-pass filter and the frequency estimator.

Q5. Any two of:
   i. CW Doppler involves continuous transmission and reception of ultrasound, whereas PW Doppler involves ultrasound in the form of pulses.
   ii. CW Doppler does not suffer from aliasing whereas PW Doppler does suffer from aliasing.
   iii. In PW Doppler blood-flow signals can be acquired from a known depth, whereas for CW Doppler the range of depths is fixed and broad.
   iv. PW Doppler needs a minimum of one element in the transducer (used for both transmission and reception), whereas CW Doppler needs a minimum of two elements (one for transmission and one for reception).
   v. CW Doppler works by the Doppler effect, whereas this is an artefact in PW Doppler.

Q6. The received ultrasound signal in Doppler ultrasound is an RF (radiofrequency) signal containing both the clutter signal from tissue and the Doppler signal from blood. Demodulation is the process whereby the RF signal is removed leaving a signal in the audio range (kHz) still containing the clutter signal from tissue and the Doppler signal from blood.

Q7. The receive Doppler signal consists of high-amplitude clutter signal from stationary or slowly moving tissue and Doppler shifted signal from the moving blood. The clutter filter is a high-pass filter which suppresses frequencies below a certain value, hence removing the clutter leaving only the signal from moving blood.

Q8. In PW Doppler (both spectral Doppler and colour Doppler) there is an upper limit to the Doppler frequency shift which may be estimated which is half the pulse repetition frequency. This in turn places an upper limit on the blood velocity which may be estimated; velocities higher than the upper limit are estimated incorrectly.

CHAPTER 8

Multiple Choice Questions

Q1. c, e   Q2. b   Q3. a, b, c, d   Q4. d   Q5. d

Short-Answer Questions

Q1. Laminar flow is associated with movement of blood in layers with one layer sliding over the other. Turbulent flow is associated with random movement of blood in all directions but with an overall forward flow. Disturbed flow concerns the generation of eddies of vortices.

Q2. Three situations at which flow reversal may occur are (i) during normal diastolic flow in a peripheral artery such as the superficial femoral artery, (ii) within a normal carotid artery bulb, and (iii) beyond a stenosis.

Q3. As the degree of stenosis increases from 0% to around 75% (by diameter) the flow rate is unchanged but the maximum velocity increases. From 75% the flow rate decreases

to zero at 100% stenosis. Blood velocity reaches a maximum at around 90% stenosis and decreases after this.

Q4. The return of venous flow to the heart is affected by the cardiac cycle, respiration, posture and calf muscle pump action.

CHAPTER 9

Multiple Choice Questions

Q1. c, d   Q2. a, b, d   Q3. b, c, d, e   Q4. d
Q5. b, c   Q6. a, c, d

Short-Answer Questions

Q1. PW Doppler systems sample the blood flow, rather than measure continuously, so suffer from aliasing and therefore there is an upper limit to the velocity that can be measured. However PW Doppler is able to select or identify the source (i.e. depth) from which the signal has arisen.

Q2. Aliasing is due to under-sampling of the signal when the pulse repetition frequency (scale) used is set too low. It can be overcome by increasing the PRF. It can be difficult to overcome when measuring high velocities at depth.

Q3. Sample volume length (also known as gate size) is used by the operator to select the region within the imaging plane from which the blood velocity is recoded. The size of the sample volume may affect the degree of spectral broadening seen due to the velocities detected.

Q4. Intrinsic spectral broadening is a broadening of the range of velocities detected due to properties of the scanner, rather than the blood flow, generally relating to the size of the aperture (group of elements) used to form the Doppler beam. It can lead to an over-estimation in measured velocities.

Q5. The angle of insonation is measured using the angle correction cursor, by lining up the angle correction cursor with the path of the blood flow. The effect of an error in measuring the angle of insonation is greater at greater angles of insonation therefore reducing the angle of insonation will reduce possible errors in velocity measurements due to incorrect estimation of the angle of insonation.

CHAPTER 10

Multiple Choice Questions

Q1. b	Q2. d	Q3. c, e	Q4. b, d, e
Q5. a, b, e	Q6. a, b, d	Q7. a, e	Q8. b, e
Q9. a	Q10. b, c		

Short-Answer Questions

Q1. The main three modes are colour Doppler, power Doppler and directional power Doppler. Colour Doppler displays mean Doppler frequency, power Doppler displays the power of the Doppler signal, directional power Doppler shows the power of the Doppler signal but with direction also displayed (e.g. as red/blue).

Q2. The colour flow processor takes the received radio-frequency (RF) data and converts this into Doppler data which may be displayed on the screen in colour at relevant pixels of the image. The five component parts are the demodulator, the clutter filter, the Doppler statistic estimator, the post-processor and the blood-tissue discriminator.

Q3. The number of ultrasound pulses to produce each colour flow line is in the range 2–20, with around 10 pulses being typical, so the frame rate would be around 1/10 of the equivalent B-mode frame rate if colour was displayed throughout the whole FOV. Restricting the width and depth of the colour box allows production of clinically usable frame rates.

Q4. In the colour flow box two complete images are produced; a B-mode image and a colour flow image. The blood-tissue discriminator decides for each pixel within the 2D box whether B-mode data or colour data are displayed.

Q5. Frame rate is determined by the time taken to collect ultrasound data within the colour box. An increase in colour box width results in an increased number of lines (and hence greater time to collect data). Similarly an increase in depth results in increased time for each line. Increase in colour box depth and width therefore leads to decreased frame rate.

Q6. Colour Doppler displays mean frequency and therefore demonstrates a cosine angle dependence and suffers from aliasing. Power Doppler displays Doppler power and does not suffer from aliasing, and also does not display angle dependence. However power Doppler may display no data for angles near 90° due to low Doppler frequency shifts which fall below the clutter filter.

Q7. The clutter filter removes high-amplitude low-frequency signals which arise from stationary or slowly moving tissues. Tissues moving at high velocities can produce Doppler frequencies which exceed the clutter filter cut-off and which are then displayed on the image as colour. This commonly occurs in cardiac motion, vessel wall motion and may also be produced by movement of the transducer.

Q8. Any three of:

i. Electronic noise. Produced from electronics within the Doppler system; resulting in random noise if the colour gain is set too high.

ii. Clutter breakthrough arising from motion of tissues. This is noise overlaying tissues due to motion of tissues when tissue velocity is high and not suppressed by the clutter filter.

iii. Clutter breakthrough arising from echoes from calcified regions (e.g. kidney stones) which appears as a twinkling colour region.

iv. Audio sound. Sound produced within the body (e.g. coughing) which is detected by the colour flow system resulting in colour in tissues.

v. Flash artefacts. Areas of colour produced by movement of the transducer.

vi. Speckle. Random variations in the colour image associated with variations in the autocorrelator output.

CHAPTER 11

Multiple Choice Questions

Q1. b, c, e	Q2. b, e	Q3. a	Q4. c	Q5. b, c
Q6. a, d	Q7. a, c	Q8. b	Q9. e	

Short-Answer Questions

Q1. Vector Doppler involves measurement of the Doppler shift in two directions from which the velocities in these two directions can be estimated. The two velocity components are compounded in a vector manner which produces the true velocity value in the scan plane and also the direction of motion.

Q2. Each B-flow image arises from the sum of two separate B-mode images. Each B-mode image is acquired using a coded pulse, with the second pulse the inverse of the first. The summed echoes from stationary tissue cancel while the summed image from moving blood results in a signal. The overall effect is an enhancement of the echoes from flowing blood.

Q3. A series of angled plane wave images is used to produce a single compound image. A series of compound images are then produced. These form the Doppler data which passes to the colour flow processor. For equivalent colour flow image quality the frame rate is approximately 10 times higher than for conventional colour flow.

Q4. Simultaneously a large number of Doppler samples over the 2D image and at a high frame rate are produced. This allows the design of improved clutter filters based on temporal filtering and spatial filtering. This reduces the minimum velocity which can be detected from around 5 mm.s^{-1} (conventional colour flow) to around 0.5 mm. s^{-1}, hence allowing visualisation of smaller vessels down to around 50 micron.

Q5. Backscatter from tissues is much higher than from blood so sensitivity is reduced. Velocities of moving tissue are much lower than from blood so the pulse repetition frequency and velocity scale are reduced. The

clutter filter is deactivated as it is the clutter which is of interest. The ensemble length is also reduced.

Q6. The equation for elastic modulus requires information on diameter and wall thickness (obtained from the B-mode image), on the change in diameter from systole to diastole (obtained using a wall motion measurement system), and on systolic and diastolic pressure (measured using an arm cuff assuming that the pressure is the same as in the artery of interest).

CHAPTER 12

Multiple Choice Questions

Q1. a, b, c, d, e	Q2. b, d	Q3. a, c	Q4. a, e
Q5. e	Q6. b, c	Q7. a, e	Q8. b, d, e

Short-Answer Questions

Q1. Optical tracking involves the attachment to the transducer of a device on which are mounted several LED sources. The positions of these LED sources are tracked using an infrared camera, usually containing at least two heads. From the acquired data the position and orientation of the transducer are estimated.

Q2. The 2D array is capable of electronically steering the beam within a 3D volume allowing a 3D dataset to be acquired. Advantages are reduced size and weight of the transducer which is easier for the operator, no mechanical moving parts which leads to improved reliability, and improved image quality.

Q3. An endoprobe transducer commonly collects 2D data by rotation of the beam in a circle round the axis of the transducer. Slow retraction of the transducer by hand or mechanically allows for collection of a series of 2D images which form a 3D volume. Examples include IVUS, trans-rectal scanning and trans-oesophageal scanning.

Q4. In surface shading the surface of a 3D object is displayed on a 2D screen. This method gives a life-like appearance and allows easy interpretation of image features. The surface shading is performed using image processing, and works best when there is high contrast between the organ and the surroundings. Examples of clinical use include visualisation of the fetus and of cardiac chambers.

Q5. Either of:

    i. In a sequential series of 2D slices the operator manually delineates the boundary of the organ. Each area is converted into a volume accounting for the inter-slice separation and the organ volume is calculated as the sum of the slice volumes.

    ii. The organ volume is estimated using an automated image analysis method. This may involve the operator identifying points on the image such as an internal point within the organ and several points on the boundary of the organ.

CHAPTER 13

Multiple Choice Questions

Q1. a, b, c, e   Q2. a, b, c, d   Q3. c   Q4. a, b, c, d
Q5. a, b, d   Q6. a, c, d

Short-Answer Questions

Q1. As the radius of the microbubbles decreases the frequency at which resonance occurs increases.

Q2. As acoustic pressure increases the microbubble response becomes more non-linear and asymmetrical. At very high acoustic pressures the shell of the bubble can rupture.

Q3. Harmonic imaging makes use of the second harmonic. Echoes received at the fundamental imaging frequency are suppressed and echoes received at the second harmonic, which is twice as high as the fundamental frequency, are displayed. This removes most of the echoes generated from soft tissue from the image.

Q4. Harmonic imaging works well with ultrasound contrast agents as their non-linear oscillations contain the second harmonic frequency. This can be detected and hence the echoes from the microbubbles can be displayed in preference to echoes from the tissue.

Q5. The main risks and hazards associated with contrast agents are embolic risk, allergic reaction, toxicity and biological effects due to acoustic cavitation.

Q6. Ultrasound contrast agents are licenced principally for use in the liver, breast or for cardiac endocardial border definition, guidelines are also provided for assisting in studies which are not currently licenced (off-licence) including renal, spleen and pancreatic, transcranial, urological and cardiac perfusion studies.

CHAPTER 14

Multiple Choice Questions

Q1. b, c   Q2. d, e   Q3. a, e   Q4. c, d   Q5. b, e
Q6. e   Q7. d   Q8. a   Q9. a, c, e

Short-Answer Questions

Q1. When a material is stretched a pulling force is applied to the material resulting in an extension of the material in the direction of the force. The greater the force, the larger is the extension. Generally the material will contract in the direction perpendicular to the force. Stress is the ratio of applied force over cross-sectional area of the material. Strain is the ratio of extension over original length.

Q2. For an elastic material Young's modulus is defined as the ratio of stress (force/area) divided by strain (extension/original length). Young's modulus may be measured in a tensile testing system where narrow strips of the material are subject to a known load (force) and the extension is measured. The load is gradually increased to breakage of the material. The Young's modulus is measured from the early part of the graph as the ratio of stress divided by strain.

Q3. (i) Strain elastography. Tissues are deformed by an applied force and the resulting deformation is measured. The measured deformation is used to calculate local strain (strain = deformation divided by original length) which is displayed on-screen. (ii) Shear-wave elastography. This relies on induction of shear waves and the measurement of local shear wave speed. The elastic modulus can be estimated from shear wave speed. The elastography display is of shear wave speed or calculated elastic modulus.

Q4. In strain elastography the strain ratio is the ratio of strain in a reference region divided by strain in the lesion. The strain ratio is used as a measure of elastic modulus enabling comparison between different patients. As strain generally decreases with depth from the transducer the two regions should be positioned at a similar depth.

Q5. (i) Induction of shear waves which may be undertaken using a high-power ultrasound beam; the deformations in the focal region produce shear waves. (ii) High-frame-rate ultrasound is used to measure the deformations produced by the shear waves, (iii) the local shear wave velocity is measured, (iv) estimation of local elastic modulus from an equation (elastic modulus = three times assumed density times shear wave velocity squared).

Q6. A-line echo data are recorded before and after compression of the tissue. The compression may be a deliberate pushing of the transducer or for more sensitive systems it may be the natural changes in position of the transducer or movements of tissue caused by the patient during breathing, swallowing or arising from beating of the heart. Signal processing is used to estimate the amount by which each echo has moved. The local strain is calculated from the estimated deformation divided by the original length (along the A-line).

Q7. Strain elastography produced using ARFI requires a minimum of three ultrasound beams; a pushing beam in which the tissue is deformed in the region of the focal zone, and two imaging beams to record echo positions, for example immediately before and immediately after the pushing beam.

Q8. In strain elastography a high-power beam may be used to locally deform the tissue in the region of the focus. Imaging beams are used to record the echo position pre- and post-deformation, from which local strain is calculated. In shear-wave elastography a high-power beam is used to deform the tissue in the region of the focus which results in shear waves which pass through the tissue. High-frame-rate imaging is then used to visualise shear-wave movements from which local shear-wave velocity is estimated which is displayed on-screen.

CHAPTER 15

Multiple Choice Questions

Q1. a, d, e	Q2. b, c	Q3. a, c, e	Q4. a, c, d, e

Short-Answer Questions

Q1. Measure the reverberation depth from the probe surface to the deepest visible reverberation line in the middle third of the image. The loss or addition of a reverberation line will be reportable. Turn the overall gain down until the deepest reverberation line disappears across the whole width of the image and note the gain. This may be checked by turning the gain up by one increment and the echo should reappear.

Q2. A series of bright lines parallel to the probe surface and each of uniform brightness. Small lateral variations in brightness may be seen as a result of minor irregularities in the lens; with experience it should be possible to differentiate between this minor non-uniformity and unusual patterns of non-uniformity. Three fault conditions: element failure (dropout); delamination; variation in lens thickness (lens wear).

Q3. The distance from the TMTO surface to the depth where speckle disappears. Use a clinical preset appropriate to the probe, with maximum output, TGC and focus in the default positions, speed of sound correction, automatic gain and automatic image optimization disabled. Image scale and overall gain should be adjusted to show the depth where the unvarying speckle pattern disappears and the image becomes dominated by randomly varying noise (or in a system with very low noise, where the image becomes dark).

Q4. Acoustic safety tests should include a visual assessment of displayed thermal and mechanical indices (TIs and MIs), referring to Chapter 16 for further guidance. User manuals will show the conditions for maximum TI and MI values and the visual assessment should include attempting to reproduce these, as well as checking that TI and MI vary as expected with settings. Unexpected results should be discussed with the supplier.

Q5. A water tank, filled with water or speed of sound corrected fluid, contains a string (rubber O-ring) mounted on pulleys and driven by a motor, usually mounted on the side of the tank. The motor is electronically driven and string velocity may be varied; some systems include physiological drive waveforms as well as constant flow. A probe mounting system is required to hold the probe and allow accurate positioning.

Q6. Compare the stated constant velocity of the string with the mean velocity reported by the scanner. Use low output and gain to minimise saturation of the large signal obtained. Sample volume is positioned at the level of the moving string, using the B-mode image for guidance. Automatic trace of maximum and mean velocities is preferred; otherwise the measured mean string velocity may be taken as the mean of the maximum and minimum velocities from manual calliper measurements. A possible tolerance for Doppler mean velocity accuracy is $\pm5\%$.

CHAPTER 16

Multiple Choice Questions

Q1. a	Q2. e	Q3. c	Q4. d
Q5. c	Q6. b	Q7. a, b, d, e	Q8. a, b, c, d, e
Q9. d	Q10. a, d		

Short-Answer Questions

Q1. Hazard describes the nature of the threat, while the associated risk takes into account the potential consequences of the hazard and the probability of occurrence.

Q2. It is the calculated prediction of an acoustic quantity such as intensity or acoustic pressure at a position within the body. It is commonly calculated from a measurement of intensity or pressure in water, reduced by a factor to allow for the attenuation due to the tissue path. The factor is typically 0.3 dB/cm/MHz.

Q3. Soft-tissue thermal index (TIS) relates to all exposures in the absence of bone. Bone at focus (or just bone) thermal index (TIB), relates to exposure of bone at the focus of the beam. Cranial thermal index (TIC) relates to conditions where there is bone adjacent to the skin surface, such as transcranial Doppler.

Q4. Most of the electrical energy driving the transducer is converted to heat its face. This heat energy then conducts into the tissue, adding to the temperature rise caused by ultrasound absorption, particularly close to the transducer. Conditions of concern include transvaginal scanning, neonatal cranial scanning through the fontanelle and ophthalmic scanning.

Q5. The neonatal lung is at risk from exposure to ultrasound due to the potential for lung capillary rupture and bleeding. Neonatal lung can be exposed during neonatal cardiac examination. The likelihood of capillary rupture increases as the acoustic pressure gets larger. MI is the only index that depends on acoustic pressure.

Index

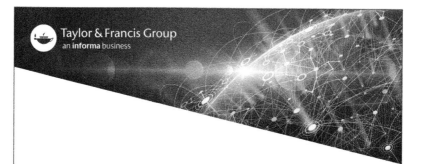

Printed and bound by CPI Group (UK) Ltd, Croydon, CR0 4YY

22/10/2024

01777614-0006